Encyclopedia of
the History of Arabic Science

Encyclopedia of the History of Arabic Science

Volume 1

Edited by
ROSHDI RASHED
in collaboration with
RÉGIS MORELON

ROUTLEDGE

LONDON AND NEW YORK

First published in 1996
by Routledge
11 New Fetter Lane, London EC4P 4EE
29 West 35th Street, New York, NY 10001

Typeset in $10/12\frac{1}{2}$ Times Compugraphic by
Mathematical Composition Setters Ltd, Salisbury, UK

Printed in Great Britain by
Clays Ltd, St Ives plc

British Library Cataloguing in Publication Data

A catalogue record for this book is available from the British Library.

Library of Congress Cataloguing-in-Publication Data

A catalogue record for this book is available on request.

ISBN 0–415–12410–7
3 volume set ISBN 0–415–02063–8

Contents

VOLUME 1

VOLUME 2

VOLUME 3

Preface

Ever since the history of science emerged as a discipline at the heart of the Age of Enlightenment in the eighteenth century, Arabic science[1] – or at least certain sectors of it – have constantly been cited by the philosophers and historians of science. For the former, such as Condorcet, it was a guarantee of the continued progress of enlightenment during a period dominated by 'superstitions and darkness'; for the latter, notably Montucla, Arabic science was necessary not for the sketching of a historical picture, but in order to establish the facts of the history of the mathematical disciplines. But philosophers and historians alike had received only the echoes of Arabic science, which had reached them through ancient Latin translations. We must, of course, beware of over-generalization or errors of perspective, and bear in mind that the sciences do not all maintain the same connection with their history; thus, of the mathematical sciences at least, astronomy is the one most firmly linked with its history, if only on account of the values of the observations that were recorded in books over the course of time and consulted by successors. Consequently Arabic astronomy assumed a privileged position, fairly rapidly attracting the attention of historians such as Caussin de Perceval, Delambre and, above all, J.-J. Sédillot – to name but French scholars – at the beginning of the nineteenth century.

Later in the course of the same century, the image of Arabic science began to change and to become shrouded with nuance. German Romantic philosophy, and the German school of philology which it engendered, had given considerable impetus to the philological and historical disciplines. The history of Arabic science gained from this rapid expansion, before becoming its victim: the study of Greek or Latin scientific texts could no longer eschew the Arabic works;[2] but the snare of history through languages – which we have stressed elsewhere[3] – enmeshed the history of Arabic science and bore it into retreat. *De jure*, therefore, it lost its right to exist, while *de facto* it was indispensable to historians, who referred to it increasingly.

This paradox, which is apparent not only in second-order studies but permeates a major work like *Le Système du Monde* by Pierre Duhem, is in fact merely the expression of a profound necessity: the historian of classical science, whatever his doctrinal views, cannot avoid Arabic science when he reviews the facts of the discipline whose history he is retracing. Following in the wake of the Western doctrine of classical science, he can view Arabic science as a repository of Hellenic science, a belated Hellenic science as it were: science as theory is Greek and as experimental method it was born in the seventeenth century. According to this doctrine Arabic science constitutes an excavation site, in which the historian is the archaeologist on the track of Hellenism. This approach has frequently ended up misrepresenting the results of Greek science as well as those of the seventeenth century, a necessary distortion if one wishes to link the two ends of the chain in a continuous history; on the other hand, and not without coincidence, it has led to some famous blunders affecting not only interpretation but comprehension too. These doctrinal views prevented Carra de Vaux (who translated the astronomical treatise of Naṣīr al-Dīn al-Ṭūsī) and the eminent historian P. Tannery (who quotes it) from grasping the innovation that it entailed and which Neugebauer was to emphasize much later. But the historian of classical science has also managed to break away from this doctrine: the other historical practice, contemporaneous with the former, came into being with the work of Alexander von Humboldt, under whose influence certain scholars became involved in the direct and innovative study of the history of Arabic sciences: F. Woepcke and L. A. Sédillot, for example, whose work was later followed up by Nallino, Wiedemann, Suter, Ruska, Karpanski, Hirschberg, Kraus, Luckey, Nazif, etc., resulting in an unprecedented acceleration of this line of research from the 1950s onwards.

Built up over the decades, this work opened the way to a better knowledge of Arabic science and of its contribution to classical science; it also enabled the understanding of one of its essential features, which had hitherto been obscured. In Arabic science a potentiality of Hellenic science was realized: the tendency beginning to germinate in the Greek scholars, to go beyond the frontiers of an area, to break the bounds of a culture and its traditions, to take on world-wide dimensions, was fulfilled in 'a science developed around the Mediterranean not as such but as a forum of exchange of all the civilizations at the centre and at the periphery of the ancient world' (Rashed 1984).

Arabic science was 'international', one can say today, as much on account of its sources as through its developments and extensions. Even if the majority of those sources were Hellenistic, they also comprised writings in Syriac, Sanskrit and Persian. The weights of these different contributions were, of course, unequal, but this does not detract from the fact that their multiplicity was essential to the evolution of Arabic science; and even in the case of mathematics, which no one would deny to be the 'heir' of Greek science, it is essential to go back to other sources for a true understanding. We can see, for example, in the chapter devoted to astronomy the importance of Indian and Persian roots, not only in the development of an astronomy of observation and of astronomical calculation, but also for the new configuration of Ptolemaic astronomy.

Within this new framework, the transmission of findings mattered less than the opportunity which occurred to bring together different scientific traditions, henceforth united within the scope of Islamic civilization. The novelty of this phenomenon was that it was not the fruit of chance meetings, of the regular or unexpected passage of caravans or seafarers; it was the deliberate result of a massive movement of scientific and philosophical translation, undertaken by professionals – sometimes rivals – supported by power and stimulated by the research itself. From this movement was born a library on the scale of the world of its time. Thus traditions from different origins and languages became elements of one civilization whose scientific language was Arabic, and found ways of reacting together to bring about new methods, and sometimes even initially unforeseen new disciplines – see, for example, the chapter on algebra (volume II, chapter 11). The social study of Arabic science will one day enlighten us about the role of Islamic society and of Islamic cities in this historic movement; we may then understand how previously independent scientific currents were able to meet and combine.

This characteristic of Arabic science, which was already marked in its earliest phase, became even more pronounced later. The scholars of the eleventh and twelfth centuries continued to discuss results obtained elsewhere, extending them and integrating them into theoretical structures often foreign to their area of origin. Seen in medicine, in pharmacology or in alchemy, this phenomenon also affected the mathematical sciences, as shown later in the works of al-Bīrūnī or of al-Samaw'al on the Indian methods of quadratic interpolation, or in the formulation by Ibn al-Haytham of the theorem of the Chinese remainder.

With Arabic science it became possible to read in one language the translations and the scientific work of the ancients, as well as the advanced research of the moderns. The latter was produced in Arabic at Samarkand as in Granada, by way of Baghdad, Damascus, Cairo or Palermo. Even when a scholar wrote in his mother tongue, notably Persian – like al-Nasawī or Naṣīr al-Dīn al-Ṭūsī – he undertook to translate his own work into Arabic. In short, from the ninth century onwards, the language of science was Arabic, and that language had in turn acquired a universal dimension: it was no longer the language of one people but of several; it was no longer the language of a single culture but of all learning. Thus previously inexistent channels opened up to facilitate immediate communication between scientific centres from central Asia to Andalusia and exchanges between scholars. Two practices then underwent unprecedented expansion. First, scientific journeys as a means of learning and teaching – ample evidence of which can be found in the biographies of scholars bequeathed to us by the ancient bio-bibliographers – such as those of Ibn al-Haytham between Basra and Cairo; of Maimonides from Córdoba to Cairo; and of Sharaf al-Dīn al-Ṭūsī going from Ṭūs to Damascus, through Ḥamadhān, Mosul and Aleppo. Second, scientific correspondence, a new literary genre, with its usages and its standards, became an instrument for collaboration and the diffusion of research. Arabic science, then, commensurate with the world of its time was, as we see, accompanied by a succession of changes: relations between the old traditions were modified, the composition of the scientific library altered, and the mobility of scholars and ideas was on a different scale.

It is surprising that such a fundamental and obvious feature of Arabic science should have remained obscured and escaped the attention of historians. One can, of course, relate this to the oblique viewpoint of an historical ideology which views classical science as the achievement of European humanity alone. But two considerations need to be added to that: one pertaining to the history, and the other to the historiography, of science. It is a question, first, of the privileged links that unite Arabic science with its Latin extensions and, more generally, with the science developed in western Europe up to the seventeenth century. In fact from the twelfth century onward, Latin science could not be understood without Latin translations from Arabic; nor could the most advanced research in Latin – such as that of Fibonacci and of Jordan of Nemours in mathematics, that of Witelo or of Theodoric of Freiberg in optics – be appreciated without reference to al-Khwārizmī, Abū Kāmil and Ibn al-Haytham. These close links captured the attention of historians and overshadowed the connections which unite the Arabic sciences with other parts of the world, notably India and China. The historiographical fact is the pre-eminence of the science of the seventeenth century. The latter, which is considered – wrongly moreover – to be all of a piece and revolutionary throughout, was invested in the writings of historians with an a-historic transcendence, becoming an absolute reference for situating all previous science. Presented as a postulate, and in the absence of authentic knowledge of the works of the school of Marāgha and of its predecessors in astronomy – of al-Khayyām and of Sharaf al-Dīn al-Ṭūsī in algebra and algebraic geometry, of the Arabic infinitesimalists from Ibn Qurra to Ibn al-Haytham – this absolute preeminence has naturally created a vacuum prior to the works of the seventeenth century, and has resulted in a model of Arabic science that flattens its most remarkable peaks of achievement.

It is not that a good knowledge of Arabic science will detract from the innovations of Kepler in astronomy, of Galileo in kinematics or of Fermat in number theory; on the contrary it will enable us to situate them more exactly, by seeking them where they are and not, as is often the case, where they are not. The progress of this knowledge will lead us to a more profound and more rigorous perception of the scientific activities of this great century and of the preceding century. It will encourage us to revise certain representations and certain historiographical methods, and will guard us against ideas of doubtful validity, notably that of the scientific Renaissance, whilst engaging us in the examination of others, like that of the scientific Revolution. But Arabic science must, in turn, recover its cosmopolitan character; this means following its Latin and Italian extensions, also those in Hebrew, Sanskrit and Chinese, not to mention achievements in the languages of Islamic civilization, notably Persian. Finally, for a satisfactory knowledge of Arabic science, it is necessary to restore it to its context, to the society which witnessed its birth, with its hospitals, its observatories, its mosques, its schools ... . How indeed can one understand certain of its developments if one forgets the Islamic city and its institutions, the function that science fulfilled there and the importance of the role that it could play? This necessary reflection will not be slow to dispel the erroneous but still flourishing views engendered by ignorance which confine science to an alleged marginality at the outermost limits of the city, or detect an illusory scientific decadence from the twelfth century onward as the effect of an imaginary theological counter-revolution.

Only at this price will the history of Arabic science accomplish its two principal tasks: to open the way to a genuine understanding of the history of classical science from the ninth to the seventeenth century; and to contribute to the knowledge of Islamic culture itself by according it a dimension which has never ceased to be its own: that of scientific culture.

This book has been conceived and realized to make its contribution towards a history of Arabic science that meets the demands outlined above. It is in fact the *first synthesis* ever carried out in this area and in this spirit, and if such a synthesis is possible today, it is thanks to the research accumulated since the last century and stimulated from the 1950s onwards. The specialists whom we have invited to contribute the different chapters of this edition are writing for the knowledgeable layperson and not merely an inner circle of colleagues, without however over-popularizing their subject; our aim has been to produce a genuine work of reference. We have tried to restore to Arabic science its true aspect and place by emphasizing the analysis of ancient sources and by devoting some chapters to its extensions in Latin and Hebrew. Because of a lack of specialists, other areas of extension have been less favoured. The book as a whole covers the history of Arabic science over about seven centuries.

But a synthesis, and particularly an initial synthesis, cannot precede effective research. This is far from having achieved the same level in the different sectors of science – whence the absence of certain areas of Arabic science, notably the earth and life sciences. Faced with the constraints imposed by the number of pages at our disposal, we have opted for work in depth at the expense of some gaps, rather than producing a so-called comprehensive, but necessarily superficial and insubstantial text. Throughout the work, we have assured ourselves of every humanly possible guarantee: each chapter has been submitted to two other specialists, members or not of the group of co-authors. Among these I should like to thank, in addition to the co-authors themselves, J. Vuillemin, G. Simon, H. Rouquette, E. Poulle, S. Matton, C. Houzel and K. Chemla. My thanks go also to A. Auger.

Roshdi Rashed
Bourg-la-Reine, February 1993

NOTES

Between the time the manuscript was ready and the printing of this work, five authors died: G. C. Anawati, H. Grosset-Grange, D. Hill, A. S. Saidan and A. Youschkevitch. I would like to pay homage to these highly talented authors.

1 By this expression we mean science written in Arabic, in the sense that one speaks of Greek science or Latin science.
2 See, for example, the work of G. Libri, B. Boncompagni, M. Curtze and J. L. Heiberg later.
3 Rashed (1984).

The Editor expresses his gratitude to the Publishers, who generously ordered and supervised the translation from French for a number of chapters.

1

General survey of Arabic astronomy

RÉGIS MORELON

Interest in astronomy has been a constant feature of Arabic culture since the end of the second century AH (eighth century AD), and it is the quantity of study which strikes us first when we begin exploring this subject: the number of scientists who have worked on theoretical astronomy, the number of treatises which have been written in this field, the number of private or public observatories which have been successively active and the number of precise observations recorded there between the ninth and the fifteenth centuries.

This chapter is exclusively concerned with astronomy as an exact science, without considering the question of astrology. In fact, although the same authors sometimes wrote treatises in both disciplines, they never mixed purely astronomical reasoning and purely astrological reasoning in the same book and in most cases the titles of the works indicate unambiguously whether their contents relate to one discipline or the other.

The science of astronomy is chiefly defined by two terms: *ʿilm al-falak*, or 'science of the celestial orb', and *ʿilm al-hayʾa*, or 'science of the structure (of the universe)'; the second term can be translated in many cases as 'cosmography'. In addition, many astronomical works are identified by the word *zīj*, a term of Persian origin corresponding to the Greek *kanôn*; in its proper sense it denotes collections of tables of motion for the stars, introduced by explanatory diagrams which enable their compilation; but it is also often used as a generic term for major astronomical treatises which include tables.[1]

The astronomical term which is generally used to refer to the stars is *kawkab*, *kawākib*, while a word of similar meaning, *najm*, *nujūm*, has a more astrological connotation, and astrology is described with the aid of expressions based on the latter term: *ʿilm aḥkām al-nujūm*, *ṣināʿat al-nujūm*, *tanjīm* . . . ;[2] however, *ʿilm al-nujūm*, 'the science of the stars', can include both astronomy and astrology, as two different approaches to the same reality.[3]

In the Arabian peninsula, as in all of the ancient Near East, traditions of observing the heavens went back a very long way; one of these traditions is of particular note, having become well-known through its revival in what Arab astronomers called the *Treatises on the Anwā'*.

The term *anwā'* is the plural of *naw'*; it describes a system of computation associated with observation of the heliacal risings and acronycal settings of certain groups of stars, permitting the division of the solar year into precise periods. The appearance of stars on the horizon at a given time of year was considered to be a sign of meteorological phenomena signalling a change of weather, so much so that the term *naw'* acquired the meaning of rain or storm. A brief reminder of the heliacal risings and acronycal settings of the fixed stars is contained in Figure 1.1, which shows a rough projection on the prime vertical of the apparent trajectory of the sun.

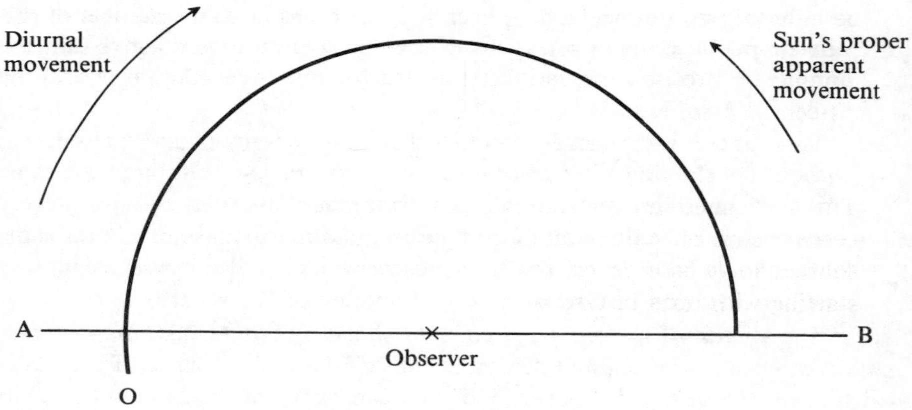

Figure 1.1

AB is the line of the horizon and O is the position of the sun under the horizon before sunrise, so that a star at A, next to the ecliptic, is at the limit of visibility when it rises, and a star at B is at the limit of visibility when it sets, according to the luminosity of the sky on the horizon just before sunrise. This situation shows the heliacal rising of star A and the acronycal setting of star B. The next day, because of the 'apparent movement of the sun' (approximately one degree per day), the sun will be further away from the horizon when A and B are in the same situation, and these two stars will be more visible since the horizon will be less luminous. About six months later, A and B will have exchanged their positions and B will be rising with A setting.

Originally the observation of these phenomena for definite groups of stars allowed the solar year to be divided into fixed periods, probably twenty-eight in number. After the eighth century, under the influence of Indian tradition, this system of calculation became combined with that of the twenty-eight 'lunar mansions' (*manāzil al-qamar*), groups of fixed stars close to the ecliptic, delineating the zones of the sky in which the moon is found night by night during the lunar month. The *Treatises on the Anwā'* which have been handed down – in written form from the ninth century – are like a series of almanacs giving the solar calendar dates for the heliacal risings and acronycal settings of stars which correspond to the lunar mansions, together with the meteorological phenomena that are traditionally associated with them. Under this system the year was divided into twenty-eight periods of thirteen or fourteen days.[4]

This ancient tradition, empirical in origin, was revived as a scientific procedure by Arab astronomers within the framework of their studies concerning the appearance and disappearance of stars on the horizon at the moment of the rising or setting of the sun, which were based in part on the *Phaseis* by Ptolemy, discussed below.[5]

SOURCES OF ARABIC ASTRONOMY

The first scientific astronomical texts translated into Arabic in the eighth century were of Indian and Persian origin, and in the ninth century, Greek sources took precedence. We shall discuss them in chronological order, starting with texts in Greek.

Greek sources

Greek texts were of two types: 'physical' astronomy, in the old sense of the word, and 'mathematical' astronomy.

The aim of 'physical' astronomy was to arrive at a global physical representation of the universe by means of purely qualitative thought; this astronomy was dominated by the influence of Aristotle, with his coherent organization of the world into concentric moving spheres, ranging from a common centre, the earth, and stable at that point. The first celestial sphere was that of the moon – the sub-lunar world being one of generation and corruption, the supra-lunar world one of permanence and uniform circular motion, the only motion that could befit the perfection of the celestial bodies – while each star had its own sphere to move it, and so on out to the sphere of the fixed stars which enclosed the universe.

'Mathematical' astronomy sought a purely theoretical, geometrical representation of the universe, based on precise numerical observations, disregarding if necessary its compatibility with a coherent world of the 'physical' type: to find the geometrical parametric models capable of accounting for measured celestial phenomena, enabling the calculation of the position of the stars at a given moment and the compilation of tables of their movements.

The history of ancient scientific astronomy is built in part on the tension between these two approaches to the same science.

'Mathematical' astronomy developed within the framework of Hellenistic astronomy – especially from the time of Hipparchus (*fl.* 160–126 BC), adapting the work of Apollonius from the previous century – but it was the work of Ptolemy in the second century AD which represented its crowning achievement in the Greek language.

Ptolemy is the scientist whose works have been the most studied, revised, commented on and criticized by later astronomers, until the seventeenth century. His four works on astronomy, in the order of their composition, are the *Almagest*, the *Planetary Hypotheses*, the *Phaseis* and the *Handy Tables*. The first two are the most important.

The *Almagest*, or *Great Mathematical Compendium*, handed down in the original Greek and in several Arabic translations, is regarded as the standard manual, which has served astronomy in the same way as Euclid's *Elements* served mathematics. Suffice it to say that within this monumental work of thirteen volumes Ptolemy synthesized the research of his predecessors, modifying it according to his own observations, and refining the old geometrical models or creating others. It was no accident that the word 'mathematical' was included in the title of the work, because Ptolemy made little reference therein to the 'physical' situation of the universe, even though he took this implicitly into account; he established and detailed the geometrical procedures capable of accounting for observed phenomena, on the basis of two postulates of ancient astronomy: the earth is stable at the centre of the world, and all celestial motion must be explained by a combination of uniform circular movements. He defined his method thus:

1 To collect the greatest possible number of precise observations
2 To identify anomalies in the movements thus observed in relation to uniform circular motion
3 To determine experimentally the laws governing the periods and the magnitudes of the anomalies
4 To combine uniform circular motions with the aid of concentric or eccentric circles and epicycles to account for the observed phenomena

5 To calculate the parameters of these movements in order to compose tables for calculating the positions of stars.

Ptolemy's method was therefore defined very precisely, but his desire to 'save the phenomena' led him in practice to infringe certain of his basic principles and to allow empiricism to intrude on some of his demonstrations, as he states himself in the last volume of his work: 'Each of us must endeavour to make the simplest hypotheses agree with the celestial movements as best he can, but if this is not possible he must adopt the hypotheses which fit the facts'.

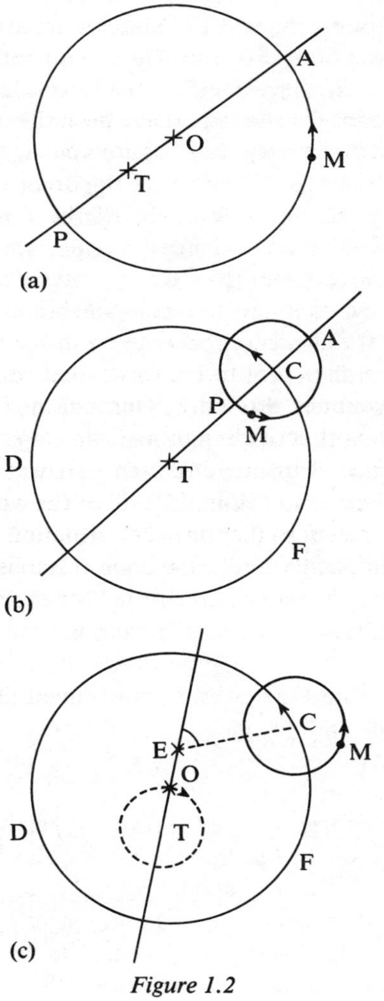

Figure 1.2

Ptolemy based the research for his geometrical models on work carried out by Hipparchus – drawing in turn from Apollonius – when he had developed the system of epicycles and eccentrics. Let the earth be stationary at T, the position of the observer. In the simple eccentric system (Figure 1.2(a)) a star at M travels on the circle MAP in uniform circular motion about the centre O, but the observer notices a different apparent speed when the star is at the apogee A or the perigee P. This geometric model can be applied to account for the apparent movement of the sun. In the simple epicycle system (Figure 1.2(b)), we imagine the observer at T, the centre of a circle CDF (the deferent), on which there travels a small circle with centre C (the epicycle), on the circumference of which moves a star M, the two circular motions being uniform and the angular speed of the centre C corresponding to the mean motion of the planet. This epicycle system, like that of the eccentric, can explain the difference in distance to the earth, but, above all, it can account for the apparent retrograde motion of the planets in a much more convincing way than a pure system of concentric physical spheres: when the planet is at P and its apparent angular speed on the epicycle is greater than that of C, it has an apparent retrograde motion; on the other hand, when it is at A, the two speeds sum and, to the observer at T, it appears to move faster than C.

This system of epicycles is very versatile and lends itself to a more complex combination of the elements concerned: the deferent CDF can be considered as eccentric with respect to the earth (Figure 1.2(c)), and makes in its turn a circular movement around T. One can thus arrive at highly complicated models, such as that of the moon or Mercury. For the larger planets (Mars, Jupiter, Saturn), Ptolemy takes an eccentric deferent CDF, with centre O, with the observer still situated at T, but he asserts that the uniform motion of the centre C of the epicycle is not around O but around the point E such that O is in the middle of TE; the point E is called the 'equant point'. This expedient leads to a better agreement between the theoretical model and the observations but contradicts the basic principle of uniform circular motion.[6]

It is thus possible to find the position of different planets in the heavens; it only requires calculation, based on observations, of the different parameters in each case: eccentricities, relative size of the radii, and angular velocities on the different circles.

The *Planetary Hypotheses* has been preserved partly in Greek (a little less than a quarter of the work) but there is a complete Arabic version.[7] It is much shorter than the *Almagest*, and its general tone is very different. First, Ptolemy calculates the maximum and minimum distances of the stars in terms of the data in the *Almagest* and thus divides the universe into concentric zones, each corresponding to the area in which a given star could move, placing the spheres of fire, air, water and earth under the sphere of the moon, in accordance with Aristotle. Thereafter, his point of view becomes 'physical' in the Aristotelian sense of the term rather than 'mathematical'. He seeks to describe the form of the physical bodies within which the circles which account for the various movements can be conceived, as an expression of the constitution of the real physical universe. He divides the 'ether' into thick globes tangential to one another, recalling the Aristotelian system of homocentric spheres; but Ptolemy also uses eccentric spheres and adds a further arrangement of tori and discs. The result is a kind of highly complex compromise between a purely geometrical system and a coherent physical system such as that defined by Aristotle. Ptolemy had thus attempted to embody his theory in a concrete 'physical' system, but the *Planetary Hypotheses* was to have less influence than the *Almagest*, apart from his calculations of the distances and sizes of stars which would be largely accepted by later astronomers.

The *Phaseis* treats of the appearance and disappearance of fixed stars just before sunrise or just after sunset (heliacal rising and acronycal setting). This work is in two parts, only the second of which is preserved in Greek and which contains a calendar of appearances and disappearances of stars on the horizon in the course of the year. The contents of the first part, a purely theoretical analysis of this particular phenomenon, is only known through an Arabic text.[8]

The *Handy Tables* has been handed down in Greek in Theon of Alexandria's fourth-century *Commentary on the Handy Tables*. It represents a rethinking in practical form of the theoretical results of the *Almagest* through the creation of detailed tables, with modification of certain parameters in accordance with the results in the *Planetary Hypotheses* and in the *Phaseis*.

All these works are cited by Arab astronomers as far back as the ninth century, together with the commentaries on the *Almagest* composed by Pappus and by Theon of Alexandria, and also a series of Greek treatises known as the 'Small astronomy collection' because it was regarded as an introduction to the reading of the *Almagest*: the *Data*, the *Optics*, the *Catoptrica* and the *Phenomena* of Euclid;[9] the *Spherics*, *On Habitations* and *On Days and Nights* of Theodosius;[10] *On the Moving Sphere* and *On Risings and Settings* by Autolycus;[11] *On the Sizes and Distances of the Sun and Moon* by Aristarchus of Samos;[12] *On the Ascensions of Stars* of Hypsicles;[13] and the *Spherica* by Menelaus.[14]

Indian and Persian sources

Three Indian astronomical texts are cited by the first generation of Arab scientists: *Aryabhatiya*, written by Aryabhata in 499 and referred to by Arab authors under the title *al-arjabhar*; *Khandakhadyaka* by Brahmagupta (d. after 665), known in Arabic under the title *zīj al-arkand*; and *Mahassidhanta*, written towards the end of the seventh or at the beginning of the eighth century, which passed into Arabic under the title *Zīj al-Sindhind*.[15]

These texts are based on the yearly cycles corresponding to Indian cosmology, and their scientific tradition is linked with an earlier period of Hellenistic astronomy than that of Ptolemy; they thus preserve a certain number of elements that can be traced back to the time of Hipparchus. They contain few theoretical developments but methods of calculation for creating tables and numerous parameters of the movement of stars. The major scientific innovation of the Indian scientists in this field is the introduction of the *sine* (half-chord of the double arc) in trigonometric calculations, which makes these much less cumbersome than the chords of arcs used in Greek astronomy since Hipparchus (see vol. II, chapter 15).

In Persia, under the Sasanids (AD 226–651), some activity in scientific astronomy developed in the Pahlavi language, under both Indian and Greek influence (Ptolemy's *Almagest* was translated into Pahlavi in the third century). This work seems to have been primarily oriented toward astrology, and the only traces which remain are found in Arabic texts from the end of the eighth century onward; these refer in particular to the 'Royal tables' (*zīj al-Shāh*), several successive versions of which are reported: from 450, 556 and 630 or 640 (under Yazdegerd III). These tables depended principally on Indian parameters.[16]

The chapters which follow detail how the Arab astronomers worked with these different sources.

OBSERVATIONS AND OBSERVATORIES

Small portable instruments and sundials are described in Chapters 4 and 5. Here we shall confine ourselves to a brief presentation of observatories and their large-scale instruments.[17]

Ibn Yūnus reports that astronomical observations were carried out at Gundīshāpūr at the end of the eighth century by al-Nihāwandī (d. AH 174 (AD 790)), whose work has been lost.[18] But the earliest precise observational results to have come down to us were recorded first in the al-Shammāsiyya quarter in Baghdad, and then on Mount Qāsiyūn at Damascus, in the final years of the reign of Caliph al-Ma'mūn (813–33) and through his impetus. They involved a precise programme dealing particularly with the sun and the moon, and at Damascus there was a complete year of continuous observation of the sun in AH 216–17 (AD 831–2). The work does not appear to have continued at these two sites after the death of al-Ma'mūn.

Apart from the numerical results found in later texts, we know little about these two observatories – their functioning, their size, etc. – except that Yaḥyā b. Abī Manṣūr, who was in charge of the observation work at Baghdad, belonged to the famous 'house of wisdom' (*bayt al-ḥikma*), and that the caliph himself had demanded that the instruments used should be the most precise possible. There is no explicit mention of the type of instruments used, but the form of the results and the kind of observations carried out are the same as Ptolemy's, which indicates that the instruments were similar to those described in the *Almagest*, i.e. the equatorial or equinoctial armilla, the meridian armilla, the equatorial quadrant (the plinth), the parallactic rods, the large gnomons, the dioptra of Hipparchus for measuring apparent diameters, and the armillary sphere (Singer *et al.* 1957: III, 586–601); these were the classic instruments of ancient astronomy and were gradually improved by Arab scientists, who sought in particular to construct larger and larger circles to achieve greater precision.[19]

In the wake of the first series at Baghdad and Damascus, a number of other observations were recorded during the course of the ninth century by Ḥabash al-Ḥāsib, the Banū Mūsā, al-Māhānī, Sinān b. Thābit, etc. In the majority of cases only the place is mentioned (Baghdad, Damascus, Sāmarrā or Nishāpūr, for example) with no indication of the setting in which these observations were made, which indicates that they were carried out from private observatories, outside any collective structure.

All these accumulated observations had not yet been organized systematically, but, by way of comparison, it should be noted that Ptolemy based all the work in his *Almagest* on ninety-four observations made between 720 BC and AD 141, the oldest having been recorded in Babylon and the latest (thirty-five in all) being due to Ptolemy himself (Pedersen 1974: 408–22). It is therefore evident that, from the ninth century, the Arabic astronomers had at their disposal the results of a far greater number of recent observations than those available to Ptolemy when creating his work.

At the turn of the ninth and tenth centuries, al-Battānī emerged as one of the major observers of the first period of the history of Arabic astronomy. For a period of about thirty years he followed a systematic programme of observations at Raqqa in the north of present-day Syria, and in the context of locating the first crescent moon on the horizon, he made what appears to be the first reference to 'observation tubes' in an astronomical treatise in the Greco–Arabic tradition.[20] These tubes, without lenses, enabled the observer to focus on a part of the sky by eliminating light interference.[21] Al-Battānī only mentions them, but the work of al-Bīrūnī includes an exact description of this type of apparatus, in a section that is also dedicated to verifying the presence of the new crescent on the horizon:[22]

> This tube is fixed on a column and is capable of two movements: the first is the movement of the column itself, enabling one to turn the tube in all directions; the other is around an axis so that the tube moves in the plane of the circle of elevation in which it lies. The tube must be not less than five cubits in length and one cubit in section. The view is concentrated and strengthened because of the shadow of the tube and its darkness, augmented by its internal blackness. When the column is placed at the centre of the Indian circle, it can be turned round until the plumbline fixed at the end of the tube is in line with the azimuth of the crescent; then the other movement is used until the tube makes an angle with the surface of the earth equal to the height of the crescent; this is simple with a quadrant divided into 90 degrees attached to the column and turning with it parallel to the tube.

This observation tube, whose use is thus attested in the Arabic world from at least the end of the ninth or the beginning of the tenth century, passed into the medieval Latin West where it became a standard astronomical instrument.[23]

Numerous other observations were recorded in the East in the course of the tenth century. Let us briefly mention in particular the work carried out at the end of that century by al-Qūhī and Abū al-Wafā' al-Būzjānī from the large observatory built in the gardens of the royal palace at Baghdad under Sharaf al-Dawla (AH 372–9 (AD 982–9)); that of ʿAbd al-Raḥmān al-Ṣūfī (d. AH 376 (AD 986)), who systematically observed the fixed stars at Isfahan, measured their position, and published as a result his famous catalogue of stars, which was a complete revision of Ptolemy's;[24] and that of Ibn Yūnus at Cairo, at the turn of the tenth and eleventh centuries.[25] But let us look more closely at the observatory of Rayy.

It was at Rayy (12 km south of Teheran), in the reign of Fakhr al-Dawla (AH 366–87 (AD 977–97)) who subsidized it, that al-Khujandī (d. *c.* AH 390 (AD 1000)) devised and built a very large sextant for solar observations, based on the principle of the black box: a dark room with a small opening in the roof (Bruin 1969).

The building was oriented north–south along the meridian; it was composed of two parallel walls, 3.5 m apart, about 20 m in length and 10 m high (see Figure 1.3); it was devoid of light, but a small opening was made in the southern corner of the roof of the building. The ground was partially excavated between the two walls so that a sextant of 20 m radius could be drawn with the opening in the roof as its centre. The interior of the arc of the sextant was covered in copper plate where the image of the sun formed when it was at the meridian, and the markings permitted measurement of its height above the horizon or its distance at the zenith. Each degree measured approximately 35 cm; it was divided into 360 parts of 10 seconds each, and the image of the sun passing at the meridian formed a circle about 18 cm in diameter; by finding the centre of the circle, a precise angle could be read off the copper surface. In 994, al-Khujandī measured the obliquity of the ecliptic as 23; 32, 19 and the latitude of Rayy as 35; 34, 39, but we have no other point of reference to indicate for how long a period this sextant was used.

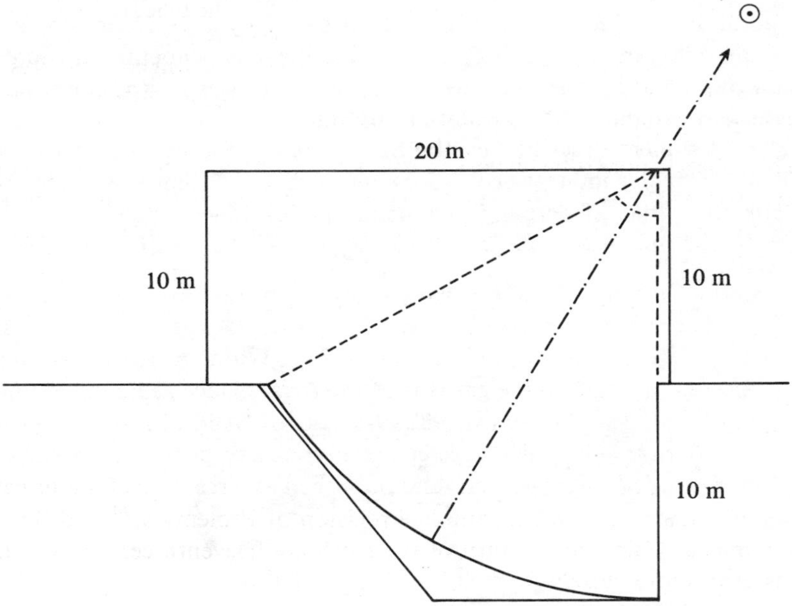

Figure 1.3

There are several allusions to large-scale instruments in various earlier observatories – for example, a construction of spherical shape, 12.5 m in diameter, in the observatory of Sharaf al-Dawla at Baghdad, for following the path of the sun – but the description of the great sextant at Rayy is the first to be given in such precise detail about a large-scale structure within the environment of a permanent observatory; most instruments of Hellenistic design were portable or could be made in one place and transported to another for ongoing use there, including large-sized copper circles or tubes like those of al-Battānī.

One other instrument of great size, cut into a permanent base of masonry, is described by Ibn Sīnā (AH 370–428 (AD 980–1037)) in his treatise *Maqāla fī al-ālāt al-raṣadiyya*.[26] On the top of a circular wall about 7 m in diameter lay a completely horizontal graduated circle. At the centre of the circle was a pillar bearing a double, vertically jointed rule, which could pivot horizontally around the centre. The lower rule lay on the graduated circle and allowed measurement of the azimuth; the upper rule carried a sighting system, and the angle between the two rules gave the height of the object observed. This construction was therefore based on a similar principle to that of the 'observation tube' described by al-Bīrūnī. About two centuries later, at Marāgha, Ibn Sīnā's instrument was further developed by the addition of a second set of jointed rules – or by an analogous arrangement of two vertical sighting devices pivoting independently around the centre of the large stone circle – enabling simultaneous measurement of the height and azimuth of two celestial objects.

The instrument described by Ibn Sīnā – and probably invented by him – is of particular interest because its new sighting system was much more precise than that of earlier instruments, giving independent readings of degrees and minutes. The upper rule was equipped with two identical

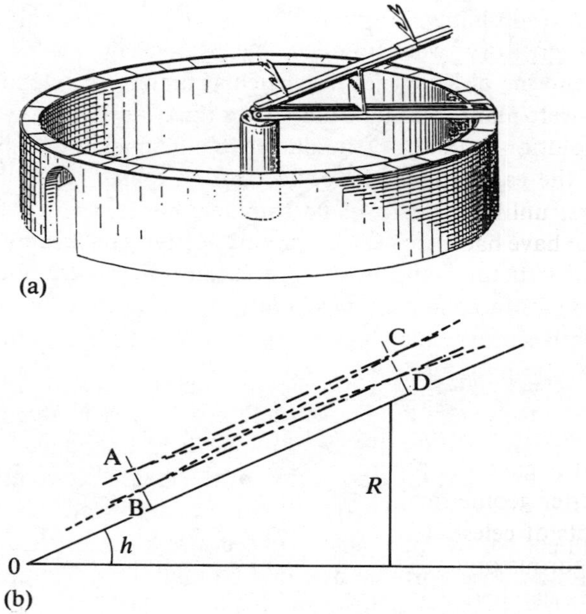

Figure 1.4

movable sights, each comprising two superposed aligning grooves (Figure 1.4(b)), A and B on the first sight, and C and D on the second, so that AB = CD. Calling the angle CAD *a* and the angle CBD *b*, we know these two angles by the respective positions of the two sights, read from the upper rule. If we focus on a star through the two grooves A and C – or B and D – the required height of the celestial object being observed will be the angle *h*, determined by the position of the smaller rule R on the lower rule. If we observe the same object through the two grooves A and D, the position of R will need to be altered to give an angle at O of value h_1 such that $h = h_1 - a$; if we sight through grooves B and C, we must again modify the angle at O to a value h_2 such that $h = h_2 + b$. It is therefore possible in this way to bring the small rule R to a position corresponding to the whole number of degrees that is closest – h_1 greater or h_2 less – to the true height of the observed object, and then to manipulate the position of the two sights to observe the star through A and D or B and C, so that one only has to subtract an angle *a* or add an angle *b*, according to the particular case, these angles being less than a degree and being accurately determined on the upper rule. The position of the scale small rule R thus gives the number of degrees, and the position of the sights AB and CD the number of minutes. This procedure represented a major advance in the precision of recorded measurements.

Around 1074, probably in the region of Isfahan, a large and highly organized observatory was founded by Malikshāh (AH 465–85 (AD 1072–92)), counting al-Khayyām in particular among its scientists. Observations there were planned to take place over thirty years, the period of one complete revolution of Saturn, the planet then considered to be the most distant from the earth (Sayili 1960: 160–6). In fact it only operated for eighteen years, until the death of its founder, but it was the first official observatory to have had such long continuous activity backed by such a precisely planned structure, and it was specifically in this tradition that the well-documented Marāgha observatory was constructed in the second half of the thirteenth century, marking an important turning point in the history of Arabic astronomy (Sayili 1960: 188–223; Vardjavand 1980).

The observatory at Marāgha (in northwest modern Iran) enabled the creation of a new set of astronomical tables, known as the 'Ilkhanian tables' but above all it gave the scientists who worked there the opportunity of producing better geometrical models than those of Ptolemy to account for the movements of celestial bodies, thanks to the high quality of its instruments, the rigorous organization of the work and the number of extremely high-calibre researchers who were able to work there simultaneously. Naṣīr al-Dīn al-Ṭūsī (AH 597–672 (AD 1201–74)) had chief responsibility for the work, and al-ʿUrḍī (d. AH 664 (AD 1266)) undertook the design of the instruments. The building was financed by Hūlāgū Khān (d. AH 663 (AD 1265)), who assigned the observatory large sums of revenue from a protected legacy (*waqf*) for its maintenance. This is the first time, to our knowledge, that an observatory was accorded this privilege, and it explains how work was able to continue there following the death of its founder Hūlāgū, finances not having been abruptly terminated by the disappearance of the princely patron, as had happened with the observatory of Malikshāh, for example.

Building began at Marāgha in AH 657 (AD 1259) and seems to have been completed in AH 661 (AD 1263). The group of buildings was situated over an area of 280 m × 220 m; in addition to the various instruments, it included a very important scientific library and a foundry for the construction of the copper apparatus. The instruments designed by al-ʿUrḍī were those that were already known, improved in size and precision, except for one which seems to have been created for Marāgha: the azimuthal circle equipped with two quadrants, permitting the simultaneous measurement of the height of two stars above the horizon.

A programme of continuous observations was intended by Naṣīr al-Dīn al-Ṭūsī to last for thirty years, as at the observatory of Malikshāh and for the same reason, but was reduced to twelve years, the period of rotation of Jupiter, and the 'Ilkhanian tables' were in fact published after this period. A great many scientists worked at Marāgha – the most famous being Naṣīr al-Dīn al-Ṭūsī and Mu'ayyid al-Dīn al-ʿUrḍī themselves, and Muḥyī al-Dīn al-Maghribī and Quṭb al-Dīn al-Shīrāzī, who will be covered in the following chapters – all of whom participated in the task of extending the astronomy of Ptolemy. Thus a veritable 'school' grew up around Marāgha which would have an important influence on all later developments in astronomy in the East.

Traces of activity at the observatory last until AH 715 (AD 1316), the date of the death of its last known director, Aṣīl al-Dīn, who was in charge from AH 704 (AD 1304), but the buildings were in ruins by about 1350. We are therefore sure that Marāgha functioned for more than fifty years, although it is not possible at present to date the ending of work at the site precisely.

This observatory had a marked influence, not only due to the importance of the scientific work that it nurtured, which will be explained below, but also because it acted as a model for the large observatories built later, of which the most celebrated, because of the quality of their instruments, were those at Samarkand and Istanbul. The observatory at Samarkand was founded in AH 823 (AD 1420) by the sovereign Ulugh Beg, who was also a scientist of high standing, and it remained active until nearly 1500 (Sédillot 1853). The one at Istanbul was built by the astronomer Taqī al-Dīn from AH 982 (AD 1575) and only functioned for a few years (Sayili 1960:

259–305). The last great observatories in the Marāgha tradition were founded in India in the eighteenth century by Jaï Singh, notably the one at Jaipur (1740), most of whose instruments are still in place.

This brief survey has offered us some idea of the evolution of observatories in the East. In the Muslin West, Andalusia and the Maghreb, astronomical observation was far less developed; it did not form part of an ongoing tradition and there is no trace of organized public observatories. The only precise observations that have survived were carried out from private observatories, at the end of the fourth century AH (tenth century AD) by Maslama al-Majrīṭī and in the fifth century AH (eleventh century AD) by al-Zarqāllu, whose 'Toledan tables' had a marked influence in the medieval Latin West.[27]

PROBLEMS OF PRACTICAL ASTRONOMY

From the end of the eighth century, with the development of the exact sciences in the particular context of an organized Muslim society, scientists from various disciplines were called upon to resolve a number of practical questions relating to social or religious matters. It therefore fell to astronomers, for example, to respond technically to the demands of the astrologers, whose official social role was important; the astronomical tables for calculating the position of the heavenly bodies were set up in part for this purpose. But above all the astronomers were required to help solve practical problems of calendars, time, or bearings on land or sea. This is illustrated by Ibn Yūnus at the start of his 'Hakemite tables' written at the beginning of the eleventh century:

> The observation of heavenly bodies is connected with religious law, since it permits knowledge of the time of prayer, of the time of sunrise which marks the prohibition of drinking and eating for him who fasts, of the moment when daybreak finishes, of the time of sunset whose ending marks the start of the evening meal and cessation of religious obligations, and moreover knowledge of the moment of eclipses so that the corresponding prayers can be made, and also knowledge of the direction of the Kaʿba (towards Mecca) for all those who pray, and equally knowledge of the beginning of the months and of days involving doubt, and knowledge of the time of sowing, of the pollination of trees and the harvesting of fruit, and knowledge of the direction of one place from another, and of how to find one's way without going astray.[28]

All these subjects gave rise to important theoretical developments which went far beyond the bounds of the practical problems involved. They will be discussed in detail in the following chapters on gnomonics and the science of time, the question of the 'qibla' for determining the direction of Mecca from a given place, calculation of the visibility of the crescent, mathematical geography and the computation of the latitude and longitude of a place, nautical science for navigating at sea, etc. Let us give some attention here to the question of calendars.

In the Arab world, the official calendar is lunar. Year one of the Muslim era began on Friday 16 July AD 622, date of the Hijra (hence the European custom of referring to Muslim years as AH), and the lunar year is made up of twelve months of twenty-nine or thirty days; the change in date takes place at sunset, and the passage to the following month occurs when the first crescent moon is sighted on the horizon just after sunset. Ptolemy had passed on a very accurate value for the average length of the lunar month at a little over twenty-nine and a half days (by about forty-four minutes); a lunar year of twelve months is therefore equal on average to 354.367 days. This value was verified and re-adopted from the ninth century by Arab astronomers who then introduced a cycle of thirty years to create an official calendar with alternating months of twenty-nine and thirty days, eleven of the years in this cycle having an additional day in the last month (which normally consisted of twenty-nine days); these were the years 2, 5, 7, 10, 13, 16, 18, 21, 24, 26 and 29 of the cycle. The astronomic correspondence is thus closely respected in the long term, but the visibility of the first crescent on the horizon on the evening of the twenty-ninth day always brought in a change of month for the place where this observation was made, so that there could be a difference of one unit in the day of the month from one end of the Muslim world to the other. Although actual visibility of the crescent was required in principle by religious law, the question facing astronomers was how to calculate the visibility of the lunar crescent in advance at a given place on the evening of the twenty-ninth day of the month, whatever the reading on the official calendar (which is what Ibn Yūnus meant by 'days involving doubt' in the earlier quotation). This is a difficult problem in view of the number of parameters involved – celestial co-ordinates of the sun and the moon, apparent relative speed of these 'two luminaries', latitude of the place, brightness of the sky on the horizon, etc. – and numerous astronomers studied the question, thereby producing important theoretical developments concerning the visibility of heavenly bodies on the horizon just after sunset.

In Persia the solar calendar was always used in parallel with the lunar calendar and corresponded at first to 'the era of Yazdegerd' which began on 16 June AD 632. As in the 'Egyptian calendar' used by Ptolemy in the *Almagest*, the year was divided into twelve equal months of thirty days, and five extra days – six every four years for leap years – were added at the end of the year; these were called the 'epagomenes days' and allowed the legal year to coincide with the astronomical solar year. This is the calendar which was adopted from the beginning by the astronomers of Baghdad, because the solar cycle is at the basis of astronomical measurements, and it was easier to create tables of the movements of heavenly bodies for months that always equalled thirty days. But the length of the solar year is a little less than 365.25 days, and at the end of the eleventh century Jalāl al-Dawla Malikshāh – founder of the great observatory described above – asked the astronomers whom he had appointed to review the composition of this calendar and make the necessary corrections to avoid accumulating the slight discrepancy with the apparent movement of the sun. Thus began 'the era of Jalālī', instituted in AH 467 (AD 1075) and comprising eight leap years in thirty-three years – instead of the thirty-two years in the earlier computation – which corresponded well with the astronomical calculations. This correction was of the same order as the one which waited until 1582 in the West, when the Julian calendar changed to the Gregorian calendar.[29]

But, apart from what we have called practical astronomy, the most important contribution of Arab astronomers is found in the arena of pure theoretical astronomy, which is not unrelated to the above.

GREAT PERIODS IN THE HISTORY OF ARABIC ASTRONOMY

The history of Arabic astronomy can be broadly divided into two great periods, the eleventh century being at the turning point between the two.

From the ninth to the eleventh century, the work was almost exclusively in the area of geometrical models inherited from Ptolemy, reworked and criticized on the basis of new observations, and in the eleventh century Ibn al-Haytham (AH *c.* 354–430 (AD *c.* 965–1039)) made an evaluation of the scientific papers accumulated for two centuries in his work *al-Shukūk ʿalā Baṭlamyūs* ('Doubts concerning Ptolemy').[30] He drew up a catalogue of all the still unresolved inconsistencies to be found in three of Ptolemy's works, the *Almagest*, the *Planetary Hypotheses* and the *Optics* – but without proposing solutions.

This critical assessment led to a temporary impasse, since solutions could only be found outside the framework in which astronomy had confined itself. Solutions of two very different kinds were therefore sought, one in the Muslim West and the other in the East.

In Andalusia there was a proposal to re-adopt Aristotelian principles by abandoning epicycles and eccentrics and returning to homocentric spheres, which would be much more consistent from a 'physical' astronomy point of view. The most characteristic representative of this school was al-Biṭrūjī (end of the twelfth century), but his bases were almost entirely philosophical, and it was impossible to make any calculations from his conclusions or to verify them by numerical observations. This approach was therefore unproductive, even though the underlying philosophical processes remain interesting.

In the East the response was scientific and gave rise to what we have called the second great period of Arabic astronomy when the search took place to account for the movement of heavenly bodies by means of new geometrical models of epicycles and eccentrics that were geocentric but non-Ptolemaic. The essential part of that work was carried out by the team connected with the Marāgha observatory, described above.

The history of the development of theoretical astronomy in the Arab world is therefore divided by the two following chapters in accordance with the two great Eastern periods, and the work of the astronomers in the Muslim West is described in the chapter on Arab science in Andalusia (chapter 7).

NOTES

1 For example, al-Battānī's important work, *al-Zīj al-Ṣābī*, or al-Bīrūnī's *Al-Qānūn al-Masʿūdī* – where a transcription of the Greek term is retained – cited in the bibliography; see also the following chapter.
2 See Rashed's note on the term *munajjim* in Diophante (1984: vol. III, pp. 99–102).
3 See, for example, Abū ʿAbd Allāh al-Khwārizmī, pp. 210ff.
4 For the *Anwā'*, cf. C. A. Nallino (1911: 117–40, conferences 18 and 19) and *The Encyclopaedia of Islam*, I, pp. 523–5. For the lunar mansions, cf. 'Manāzil' in *The Encyclopaedia of Islam*, VI, pp. 374–6.
5 In particular Sinān b. Thabit b. Qurra (d. 331 AH (943 AD)) reproduced part of the second book of *Phaseis* in his *Kitāb al-Anwā'*; see Neugebauer (1971).
6 For a short and precise description of the geometrical planetary models proposed by Ptolemy in the *Almagest*, see Neugebauer (1957: appendix I, French translation, pp. 239–55).
7 See Ptolemy, *Planetary Hypotheses*. I have personally undertaken the edition of the Arabic version of this text (Morelon 1993).
8 The contents of this book were found described in a passage of the work by al-Bīrūnī, *al-Qānūn al-Masʿūdī*; see Morelon (1981).
9 Euclid lived around 300 BC; his *Data* contains diverse definitions of the elements involved in geometry; his *Optics* develops a theory of vision and of perspective; his *Catoptrica* is a study on mirrors; his *Phenomena* contains a geometrical study of the celestial sphere.
10 Theodosius lived in the second century BC; his *Spherics* concerns the geometry of the spheres; in *On Habitations* he shows which portions of the celestial sphere are visible according to the regions of the earth; in *On Days and Nights* he determines the portions of the ecliptic traversed by the sun each day over the whole year.
11 Autolycus lived in the third century BC; in *On the Moving Sphere* he describes the different circles of the celestial sphere and the modification of their respective

positions caused by the movement of the sphere; in *On Risings and Settings* he describes the phenomena of the visibility of the stars on the horizon at their rising or setting.

12 Aristarchus lived in the third century BC and is famous for having proposed a short-lived heliocentric hypothesis; in his treatise *On the Sizes and Distances of the Sun and Moon* he calculates their distance from the earth and their respective size based on their position in quadrature and on eclipses.

13 Hypsicles lived around 150 BC; in his *Ascensions* he determines the rising of the different signs of the zodiac for a given place in terms of the relation between the longest and shortest day at that place.

14 Menelaus lived in the first century AD; his book on the *Spherica* contains the fundamental formulae of spherical trigonometry used by Ptolemy in the *Almagest*, introducing equal proportions between the chords of arcs on a complete spherical quadrilateral (see the chapter on trigonometry in vol. II).

15 See al-Hāshimī, *Book of the Reasons*, pp. 201–11.

16 See 'Astrology and Astronomy in Iran' in *Encyclopedia Iranica* (1987: vol. II, pp. 858–71) and Kennedy (1958).

17 On the question of observatories, see Sayili (1960).

18 See Ibn Yūnus, *Le Livre*, pp. 140–1.

19 In particular at Baghdad and Damascus, from the time of the first observations.

20 See al-Battānī, *Al-Battānī*, vol. 3, pp. 137–8; vol. 1, pp. 91 and 272.

21 See Eisler (1949), 'The polar sighting tube'. These 'observation tubes' are not mentioned explicitly in any of the texts of Hellenistic astronomy that have come down to us, but they have been known in China since the sixth century; see Needham and Wang Ling (1959: 332–4).

22 Al-Bīrūnī, *Al-Qānūn*, p. 964, treatise 8, chapter 14, 2nd section.

23 See Eisler (1949), 'The polar sighting tube'.

24 See al-Ṣūfī, *Kitāb ṣuwar al-kawākib*.

25 See Ibn Yūnus, *Le Livre*.

26 Arabic text edited and translated into German with notes by Wiedemann-Juynboll. The following two figures are taken from this publication; the drawing of the instrument was made by J. Frank from data in the text and from the author's knowledge of the instruments of the observatory of Marāgha.

27 See the entries for these two scientists in the *Dictionary of Scientific Biography*.

28 Ibn Yūnus, *Le Livre*, pp. 60–1.

29 See 'Djalālī' in *The Encyclopaedia of Islam*, II, pp. 397–9.

30 Ibn al-Haytham, *Shukūk*.

2

Eastern Arabic astronomy between the eighth and the eleventh centuries

RÉGIS MORELON

Al-Qifṭī notes that the first Arab scientist to be interested in astronomy was Muḥammad b. Ibrāhīm al-Fazārī in the second half of the eighth century, at the beginning of the reign of the Abbasids.[1] His name is connected with a famous tradition according to which an Indian delegation with an astronomer in its ranks was received in Baghdad by Caliph al-Manṣūr around the year 770; the name of this astronomer is not known but the tradition reports that he had with him at least one astronomy text, written in Sanskrit, which was translated into Arabic under the title *Indian Astronomical Table* (*Zīj al-Sindhind*)[2] by al-Fazārī and Yaʿqūb b. Ṭāriq[3] under the supervision and direction of this Indian astronomer. Whatever the historic value of this tradition as far as its details are concerned, the two mentioned authors have been presented by all their successors as the men who introduced scientific astronomy into the Arab world from its origins in India.

The works of al-Fazārī and Yaʿqūb b. Ṭāriq are lost but a certain number of fragments survive in the work of later authors.[4] It is known that al-Fazārī wrote *The Great Indian Table* (*Zīj al-Sindhind al-kabīr*), and later quotations from this text show that he mixed Indian parameters with elements of Persian origin from *The Royal Table* (*Zīj al-Shāh*). We have traces of three works by Yaʿqūb b. Ṭāriq: *Table Solved in India Degree by Degree* (*Zīj maḥlūl fī-l-Sindhind li-daraja daraja*), *The Composition of Orbs* (*Tarkīb al-aflāk*) and *The Book of Causes* (*Kitāb al-ʿilal*); the basis of his reasoning in these is clearly the same as that of his contemporary. These two authors had the great merit of introducing scientific astronomy into the Arab world but their works, to judge from what remains of them, appear to be a compilation of elements which they had at their disposal, unverified by observation and without any attempt at proper internal coherence.

The first work of Arabic astronomy to have reached us in its entirety is that of Muḥammad b. Mūsā al-Khwārizmī (*c.* 800–*c.* 850) and is also called *Indian Astronomical Table* (*Zīj al-Sindhind*); it is in keeping with the preceding tradition but with the addition of elements from Ptolemaic astronomy. The Arabic text is lost and the work has been transmitted through a Latin translation made in the twelfth century by Adelard of Bath from a revision made in Andalusia by al-Majrīṭī (d. AH 398 (AD 1007–8)).[5]

Al-Khwārizmī is equally renowned as a mathematician for his work in algebra. His treatise on astronomy was written under al-Ma'mūn (813–33) and does not include any theoretical elements: it is a set of tables concerning the movements of the sun, the moon and the five known planets, introduced by an explanation of its practical use. Most of the parameters adopted are of Indian origin, and so are the methods of calculation described, including in particular the use of the sine. But some elements are taken from Ptolemy's *Handy Tables* (Neugebauer 1962a: 101–8), without any attempt by the author to achieve coherence between the differing results drawn from the Indians and from Ptolemy. Here we have the same problem as with al-Fazārī and Ya'qūb b. Ṭāriq in their simultaneous use of Indian and Persian sources.

During the ninth century, the Arab astronomers of Baghdad fairly rapidly assigned the Indian tradition, which comprised only methods of calculation and sets of parameters for the composition of tables, to second place in favour of Ptolemaic astronomy, which was well endowed with theoretical reasoning and therefore enabled the development of astronomy as an exact science. But the Indian tradition continued to have an influence of some significance in the compilation of astronomy tables in the Muslim West (Andalusia and Maghreb) (Kennedy and King 1982).

The introduction of Greek astronomy

The eleven short treatises in Greek listed in the preceding chapter, considered to be a preparation for the reading of Ptolemy and grouped under the title 'Small astronomy collection', were all translated during the ninth century by confirmed scientists with a sound knowledge of both Greek and Arabic: Ḥunayn b. Isḥāq (d. 877), his son Isḥāq b. Ḥunayn (d. 911), Thābit b. Qurra (d. 901) and Qusṭā b. Lūqā (d. *c.* 900).[6]

The four works of astronomy by Ptolemy mentioned in Chapter 1 were also translated into Arabic in the ninth century. The *Almagest* is the most important because of the influence it exerted.[7] Several successive translations were made, as noted by twelfth-century author Ibn al-Ṣalāḥ:

> There were five versions of *Almagest* in different languages and translations: a Syriac version which had been translated from the Greek, a second version translated from Greek to Arabic by al-Ḥasan b. Quraysh, for al-Ma'mūn, a third version translated from Greek to Arabic by al-Ḥajjāj b. Yūsuf b. Maṭar and Halyā b. Sarjūn, also for al-Ma'mūn, a fourth version translated from Greek to Arabic by Isḥāq b. Ḥunayn for Abū al-Ṣaqr b. Bulbul – we have Isḥāq's original in his own hand – and a fifth version revised by Thābit b. Qurra from the translation of Isḥāq b. Ḥunayn.[8]

Three of these versions have been lost: the first, an anonymous Syriac version; the second, the Arabic version of al-Ḥasan b. Quraysh, of which traces remain, particularly in the work of al-Battānī in the tenth century (Kunitzsch 1974: 60–4); and the fourth, the version of Isḥāq b. Ḥunayn before its revision by Thābit. At present we do have the third and the fifth versions, in manuscript form:[9] the translation made by al-Ḥajjāj at the behest of al-Ma'mūn around 827–8 and that of Isḥāq b. Ḥunayn revised by Thābit b. Qurra around 892, both translated from Greek to Arabic. Another revision, or in fact a re-writing, of the *Almagest* should be added to the list drawn up by Ibn al Ṣalāḥ: it was produced after that author's time, in the middle of the thirteenth century, using the Isḥāq–Thābit version, by Naṣīr al-Dīn al-Ṭūsī; it became widely known among Arabic-speaking astronomers from this period onwards.

Let us compare the two versions that we have from the ninth century. That of al-Ḥajjāj is very close to the Greek text, and the sentence structure of the original Greek is largely preserved; the scientific vocabulary used in the Arabic is sometimes vague, making it necessary in certain cases to return to the original text in order to understand the reasoning properly even though it is expressed in Arabic. These deficiencies in the translation of such a fundamental text led to the Isḥāq–Thābit version, made towards the end of the ninth century, after more than fifty years of work in scientific astronomy in the Hellenistic tradition. This version removes all need to refer to the Greek text: its language and vocabulary are perfectly clear and devoid of ambiguity. We thus have two precise points of reference from which to conclude that an Arabic scientific language was developed for astronomy between 827 and 892.

We do not have such precise information about the translation of the three other works of Ptolemy. The *Planetary Hypotheses* is cited in Arabic from at least the middle of the ninth century, under the title *Kitāb al-iqtiṣāṣ*, or *Kitāb al-manshūrāt* (especially by al-Bīrūnī). A complete but hitherto unpublished translation preserves the last three-quarters of the work which have been lost in the original Greek (see pp. 6–7). The name of the translator has not been passed on but one of the two complete manuscripts containing the work states that the text was corrected by Thābit b. Qurra.[10]

The *Book concerning the Appearance of the Fixed Stars*, or *Phaseis*, is quoted by Thābit b. Qurra under the title *Kitāb fī ẓuhūr al-kawākib al-thābita*. Thābit knew Greek and this reference is not enough to confirm that the text had been translated. But the Arabic translation is quoted by al-Masʿūdī (d. *c*. AH 345 (AD 956),[11] and was used by Sinān b. Thābit (d. AH 332) (AD 943)) in his *Kitāb al-Anwāʾ* (see chapter 1). The Arabic translation of the *Phaseis* must therefore have been made at the beginning of the tenth century at the latest, and although we do not have the original translation, we have numerous references to it by Arab astronomers.

Ptolemy's *Handy Tables* was used by al-Khwārizmī, as we have seen, and then by Qusṭā b. Lūqā (in the mid-ninth century),[12] and we find traces of it subsequently in the work of many other authors, but we do not possess the Arabic translation and we have no knowledge of the circumstances in which it was produced.

We should add in the context of Ptolemaic astronomy that the comments on the *Almagest* by Theon of Alexandria were available in Arabic during the ninth century, since we find lengthy literal quotations from them in the work of astronomy *Kitāb fī-l-ṣināʿat al-ʿuẓmā* by Yaʿqūb b. Isḥāq al-Kindī (d. *c*. 873).[13] The Arabic translation of Theon's work has not survived.

As we have said, the evolution of astronomy as an exact science beginning in Baghdad in the third century AH (ninth century AD) was primarily based on the works of Ptolemy. Only very few of the earliest extant Arabic studies of astronomy have yet been edited or undergone detailed scientific comment, and in most cases reference must be to the manuscript sources. All attempts at synthesis must therefore be provisional at this stage and will be continually subject to review in the light of any seriously edited and commented texts that may emerge. Here we shall take some examples of significant works and demonstrations in order to outline the first phase of the evolution of scientific astronomy in Arabic, concentrating more on the progressive transformation of methods of reasoning than on calculations of the different parameters for the movement of the stars despite their specific interest.

ARABIC ASTRONOMY IN THE EAST DURING THE NINTH CENTURY

As an introduction to the early development of Arabic astronomy in the East, we shall group the contributions of the different scientists who began work in this area into subjects of study, from the simplest to the most complex: the dissemination of Ptolemy's astronomy, the critical analysis of his results, and finally the rigorous mathematization of astronomical reasoning; an additional section is devoted to the celebrated astronomer al-Battānī, who worked at the turn of the ninth to tenth century in Raqqa.

The dissemination of Ptolemy's astronomy

From the first half of the ninth century, several treatises were written to present the findings of the *Almagest* in a simple manner, or to summarize it so that this fundamental work could reach the largest possible audience beyond the restricted circle of specialized astronomers. The best-known example of this type of literature is the book by Aḥmad b. Muḥammad b. Kathīr al-Farghānī. His was also the most widely distributed work – first in Arabic (as attested by the large number of manuscripts listed from all periods and all regions), then in Latin (there were two successive Latin translations in the twelfth century) – and it was passed on under several titles, the most common being *Compendium of the Science of the Stars* (*Kitāb fī jawāmiʿ ʿilm al-nujūm*).[14] We know little of the author, except that he was a member of the team of scientists formed by al-Ma'mūn (813–33) and that he died sometime after 861. His book was probably written after 833 and before 857. It is a sort of manual of cosmography, comprising about a hundred pages in its published version and containing thirty chapters in which al-Farghānī explains the state of the universe according to the findings of Ptolemy. It is purely descriptive without any mathematical demonstration. It contains, in the following order: a description of the computations for the months and years of the different eras (Arab, Syrian, Byzantine, Persian and Egyptian); the justification of the sphericity of the sky and of the earth, the latter being stationary at the centre of the universe whereas the sky has two circular movements; the inclination of the ecliptic over the equator; a description of the inhabited part of the earth with seven climates and the different regions and towns; the size of the earth; the movement of the seven 'wandering stars' both in longitude and latitude, explained by the model of eccentrics and epicycles; the precessional movement of the fixed stars; the distances of all the heavenly bodies from the earth and their sizes; the heliacal risings and acronycal settings; the phases of the moon, its parallax and the eclipses of the moon and sun.

This work thus sets out the principal problems of ancient scientific astronomy, which is why it became a subject of repeated commentary by scientists of the highest calibre, including in particular al-Bīrūnī.[15] Al-Farghānī uses Ptolemy as his virtually exclusive source, but corrects him on several points according to the results obtained by the astronomers of al-Ma'mūn, such as the rectification of the obliquity of the ecliptic from 23; 51 to 23; 33, the assertion that the apogees of both the sun and the moon follow the precessional movement of the fixed stars, and the use of the measurement of the earth's circumference determined under al-Ma'mūn. In addition, after claiming that Ptolemy had only calculated the distance and the size of the sun and the moon, which demonstrates that al-Farghānī knew the *Almagest* but not the *Planetary Hypotheses*, he gives numerical values identical to those in the latter book, without indicating the source from which he drew his data.

Several other works of a similar nature have survived, including in particular a treatise by Qusṭā b. Lūqā, as yet unpublished, and two more scientific treatments by Thābit b. Qurra, focusing especially on the movements of the heavenly bodies and taking their reasoning from the first part of the *Planetary Hypotheses*.[16]

These texts made scientific astronomy accessible, bringing together its findings in an understandable form that would nowadays be dubbed a 'popularization of the highest standard', produced by professional astronomers and widely diffused among the educated circles of the time. This tradition continued in all the accounts of the *Almagest* written by the authors of encyclopedias such as Ibn Sīnā, who included a summary in his great survey of philosophy, the *Shifā*'.

Critical analysis of Ptolemy's results

Once the *Almagest* became available in Arabic under al-Ma'mūn, the work of verifying its findings began, and this was the reason for the setting up of the first programme of astronomical observations in Baghdad and Damascus referred to in chapter 1. About 700 years lay between Ptolemy and the al-Ma'mūn astronomers, who found in the *Almagest* schemes of computation and tables permitting the theoretical calculation of the position of the celestial bodies for a given date. The results of these calculations, made for a period of 700 years, were set against the data from the observations recorded at Baghdad and Damascus, and a discrepancy was noted between the two sets of figures obtained.

This discrepancy, inevitable over such a long period of time, led the Baghdad astronomers not just to 're-set their clocks' – i.e. simply to add a correction to all the lines of a table so that it could be used again – but to re-examine the theoretical base of Ptolemy's results, in order to revise the mechanisms he had proposed and recalculate the parameters of the different movements. Three examples will serve to illustrate this work, which was undertaken from the beginning of the ninth century: the 'Verified table' (*al-Zīj al-mumtaḥan*), the 'Book on the solar year' (*Kitāḇ fī sanat al-shams*) and the work of Ḥabash al-Ḥāsib.

The 'Verified table'

The Arabic term *al-zīj al-mumtaḥan* is in itself a generic one, denoting a set of tables compiled on the basis of observations and thereby offering all possible guarantees of scientific accuracy. But the term 'Verified table', used alone, refers specifically to the first set of astronomical tables in Arabic based on the results of observations carried out at the observatories of Baghdad and Damascus; it was Yaḥyā b. Abī Manṣūr (d. AH 217 (AD 832)) who was nominated by al-Ma'mūn to co-ordinate this research. The tables had a great influence in so far as they continued the first series of precise scientific observations recorded since Ptolemy, within the same tradition of Hellenistic astronomy, and they were widely quoted by later Arab astronomers, such as Ibn Yūnus and al-Bīrūnī.

The complete original text has not come down to us.[17] However, the recorded results from it cited by later authors show that the different parameters of the motion of the heavenly bodies had been recalculated.[18] But the most important conclusion from the observations in these tables concerned the movement of the sun: they showed that the apogee of the solar orb was connected with the precession of the fixed stars, contrary to the view of Ptolemy, who considered this apogee to be subject to diurnal movement only.[19]

Although we cannot establish a clear relationship between this conclusion from the 'Verified table' and the 'Book on the solar year', it is the latter text that contains the demonstration of the link between the movement of the sun and that of the fixed stars.

The 'Book on the solar year'[20]

Manuscript tradition attributes this text to Thābit b. Qurra, but close critical analysis reveals that it pre-dates this author and that it probably came from within the team of researchers that formed around the Banū Mūsā prior to the arrival of Thābit, i.e. before the middle of the ninth century.

The author of this treatise defines his position in relation to Ptolemy with regard to the movement of the sun and the calculation of the length of the solar year. Figure 2.1 is a reminder of what the *Almagest* has to say on this subject.

E is the position of the observer on a stationary earth at the centre of the world. The sun moves in a uniform circular orb on an eccentric circle in relation to the earth: circle I, with centre D, of which the most important points are the apogee A and the perigee P. E is also the centre of the ecliptic circle II, the apparent trajectory of the sun in the sky during the year; the reference points of the ecliptic are the equinoxes B and C, and the solstices

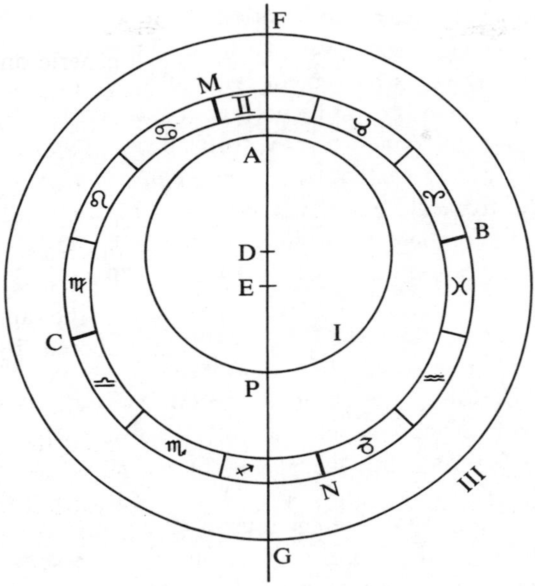

Figure 2.1

M and N. The common plane of these two circles intersects the sphere of the fixed stars according to circle III, also centred on E.

In one year, the sun completes one revolution on its eccentric orb I, with a uniform regular motion. The time of this complete revolution is constant, whatever its starting point. This is the value of the *anomalous year*, the time taken for the sun to return to the same point in its orb. This value is the only one that can be considered as a constant reference; but it is not directly measurable from E, as the eccentric in itself does not contain any sufficiently precise element of reference. The observer must first position circle I clearly in relation to circle II or circle III.

When we observe from E the motion of the sun on circle II, and we measure the time interval between two successive passages of the sun at the same point, for instance B, the spring equinox, we obtain the value of the *tropical year*.

When we observe from E the sun's movement on circle III, and measure the time interval between two successive conjunctions of the sun with the same star, we obtain the value of the *sidereal year*.

If the circles I, II and III were fixed relative to each other, the three values of the solar year defined above would be absolutely identical – but this is not the case. The problem for the ancient astronomers was to try to find, from observations of the irregular motion of the sun on circles II or III, the value of the anomalous year over orb I, the only absolute constant.

The study of the sun's movement is detailed in the third book of the *Almagest*. Here Ptolemy begins by noting, in line with Hipparchus, that the sidereal year is slightly longer than the tropical year, but he concentrates on the latter, in order to prove that this is the sought-for absolute constant. He thus makes the tropical year and the anomalous year coincide, causing circles I and II to combine while circle III moves in relation to them by the motion of the precession of the equinoxes, evaluated by Ptolemy at 1° per century.

To calculate the parameters of the eccentric solar orb, Ptolemy uses the model illustrated in Figure 2.2. ABCD is the circle of the ecliptic, with centre E, the position of the observer; the circle MNPO, with centre G, is the eccentric orb around which the sun moves, A and C are the two equinoxes and B is the summer solstice. The figure is completed by the straight lines MQGP and NGXO parallel to DEB and AEC respectively, and the straight line EGH which intersects the eccentric orb at H, its apogee. By observing the moment at which the sun passes over A, B and C, one can obtain, by a simple calculation based on the mean movement of the sun, the value of the arcs IL, IK, KL, IN, PK, LO on the eccentric,

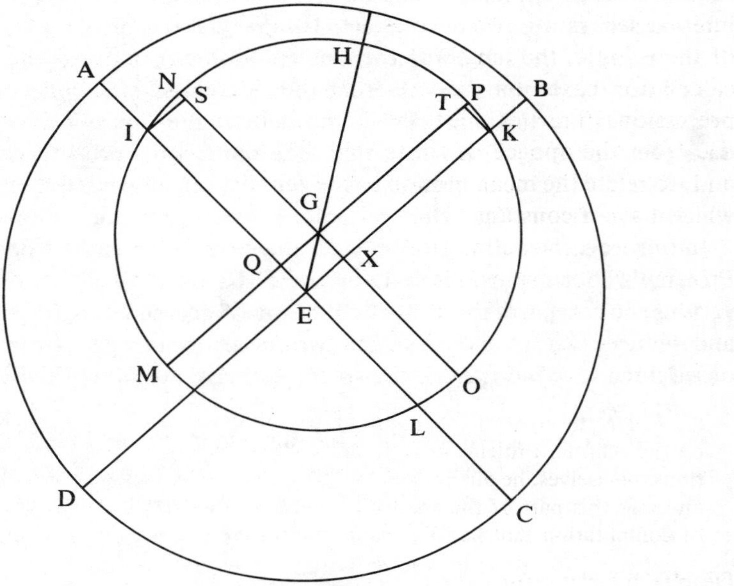

Figure 2.2

and consequently calculate all the parameters. Assuming by convention that the radius of the eccentric orb is 60 'parts', Ptolemy then finds that the eccentricity EG is equal to 2; 30 of these parts, that the apogee H is situated at 5; 30° of Gemini, that this apogee is fixed on the ecliptic, and that the length of the tropical solar year (passage of the sun over the same point of the ecliptic) is constant and equal to 365; 14, 48 days.

Following the observations made in Baghdad between 830 and 832 – some 700 years after the *Almagest* and 950 years after Hipparchus – the author of the 'Book on the solar year' notes that the sun's apogee in his era is at 20; 45° of Gemini and that this shift of 15; 15° since Hipparchus' observations is analogous to that due to the precession of the fixed stars, which had been measured on Regulus as 13; 10°, excepting observational errors of which the author was well aware. This therefore lends him to link circles I and III of Figure 2.1 and to conclude that the apogee of the eccentric orb of the sun is subject to precessional motion. The anomalous year is thus found not in the tropical year but in the sidereal year, the only absolute constant. But since the sidereal year is only a theoretical reference, it is necessary to derive the value of the tropical year from it, as the only practical reference which will permit the marking of terrestrial time throughout the year.

Since the eccentric orb shifts in relation to the circle of the ecliptic, we cannot measure the length of the tropical year directly by observing the time interval separating two successive passages of the sun over the same point of the ecliptic; the length of the tropical year can only be derived from a calculation based on the values of the sidereal year and the constant of precession. In fact, if one takes the mean motion of the sun along the eccentric from the apogee, it shifts slightly because of precessional movement and, to relate the mean motion to the ecliptic, one must add these two shifts which have a constant value.

In this way, the author of the 'Book on the solar year' radically challenges Ptolemy's conclusions, his computations and even the quality of his observations: he compares these last with his own and with those of Hipparchus, and deduces that all of Ptolemy's observations should be rejected in favour of a return to Hipparchus. He concludes his vigorous critique as follows:

> As well as the error of calculating the duration of the solar year from a point on the ecliptic, Ptolemy has created further error as a result of his observations themselves: he did not conduct them as they should have been conducted and it is this part of the error that has most seriously damaged the method of computation that he has proposed.[21]

Despite his criticisms, the author considers that Ptolemy's is still the best geometrical method for calculating the parameters of the sun's orb. He reworks the third book of the *Almagest*, quoting it at length by adopting its geometrical method, partly re-organizing the plan of the book while retaining its content, and relying solely on the observations of Hipparchus and himself. His calculation of the parameters of the solar orb is based on Ptolemy's model as shown in Figure 2.2, but he alters the orientation of the observations: A, B and C no longer correspond to the two observed equinoxes and one solstice for, as the author says:

> Given that observations of solstices are difficult, we shall not include any solstice observation results in our three measurements. In the three measurements from which he identified the solar anomaly, Ptolemy included one measure of the summer solstice. We do not agree with this; on the contrary, we consider that this gave him little safety from error.[22]

During a solstice, the variation of the sun's declination is effectively very slight, and it was difficult to determine the exact moment of the sun's passage at this point. The author of the treatise therefore changes the three observations by 45° and measures the passage of the sun on the ecliptic in the middle of Aquarius, Taurus and Leo. He reformulates the method of computation in the *Almagest* by 'modernizing' it, i.e. by reasoning with sines of arcs instead of with their chords (see vol. II, chapter 15), and he obtains the following results:[23]

Position of the apogee of the sun 20; 54° of Gemini (22; 53°)
Constant of precession 0; 0, 49, 39° per year (0; 0, 50, 1)
Sidereal year 365; 15, 23, 34, 33 days (365; 15, 22, 53, 59)
Tropical year 365; 14, 33, 12 days (365; 14, 32, 9, 20)
Eccentricity of the solar orb 2; 6, 40

In addition to the good level of accuracy of the preceding results, bearing in mind the means of observation of the period, the 'Book on the solar year' is extremely important for the understanding of how Arabic astronomy first developed from the heritage of Ptolemy. This treatise was written in the first half of the ninth century, thus shortly after the Arabic translation of the *Almagest* by al-Ḥajjāj, which is liberally quoted in over a third of the text. It enables us to see first how some Arab astronomers of the first generation used this fundamental text, and second to identify a certain number of scientific innovations which became established as a result of their work.

To summarize the foregoing, we see that the author of the 'Book on the solar year' concludes, on the one hand, that Ptolemy has made some errors of computation, particularly over the precession constant, and on the other hand, that his observations are less reliable than those of Hipparchus. He therefore dismisses the observations and the results. After having established the displacement of the solar apogee and its relationship with the precession of the fixed stars, he works out a method which will permit him to determine the time it will take the sun to return to the same star, in order to calculate the length of the sidereal year. He retains Ptolemy's geometrical reasoning, and all the subjects treated in *Almagest* III, slightly modifying the order by moving two chapters, and redrafts all these elements. The result suggests that the 'Book on the solar year' was not composed as an isolated treatise, but was part of a vast project aimed at rewriting the *Almagest*, keeping its structure and its theoretical reasoning but eliminating Ptolemy's observations and calculations; its author keeps Hipparchus's observations in order to compare them with recent observations made in Baghdad and Damascus, and he creates new methods of computation based on the theories proposed by Ptolemy.[24] We do not know how far this project of a 'new *Almagest*' may have been pursued, but the content and structure of the book we have been discussing clearly demonstrate that this important work was started in Baghdad in the first half of the ninth century in the school formed around the Banū Mūsā.

The 'Book on the solar year' also contains a number of innovations which were adopted by later astronomers. First, following the composition of this treatise it was accepted that the apogee of the solar orb moved in relation to the ecliptic, and that a relationship needed to be established between the sidereal year, the constant of precession and the tropical year (although it would be the end of the eleventh century before the Andalusian astronomer al-Zarqāllu calculated the real supplementary movement of the solar apogee at 19 minutes per century). Next, contrary to Ptolemy, the author of the treatise connects the movement of the apogee of the sun's orb and that of the moon's orb with the precessional motion of the sphere of fixed stars, in the same manner as the apogee of the orbs of all the other planets; the motion of the sphere of the fixed stars therefore carries with it all the celestial spheres; the sun and the moon are not special cases in the universe. In this way, the circle of the ecliptic becomes a purely theoretical circle beyond the sphere of the fixed stars, located by the passage of terrestrial time and the rhythm of the seasons. Finally, the displacement by 45° of the three solar observations, introduced in order to avoid errors in solstice observation, were used by later astronomers in their calculations of the parameters of solar motion.[25]

The work of Ḥabash al-Ḥāsib

We know little about the life of Ḥabash. He was one of the astronomers of Caliph al-Ma'mūn, and he was alive in AH 254 (AD 859), as a calculation is attributed to his name in that year, but we do not know the date of his death. Only one of his original, but incomplete, works has been published:

his short treatise on the sizes and distances of the heavenly bodies, partially preserved in a single manuscript (Langermann 1985). A lengthy work of his, *al-Zīj al-dimashqī* ('Damascus tables'), has come down in two different versions, one to be found in Istanbul and the other in Berlin. The text of the Berlin manuscript has obviously been much revised by later hands; the Istanbul version appears to be quite close to Ḥabash's original work but has not yet been published.[26]

This work is in the Ptolemy tradition, but is not a reworking of the *Almagest* as the 'Book on the solar year' is in part. Ḥabash merely selects those areas which he considers susceptible to modification in the light of his own studies and of the data acquired from the early work of theoretical astronomy at Baghdad and Damascus. His text is therefore for use alongside the *Almagest* and is not intended to replace it. An important part of the 'Damascus tables' concerns trigonometry: Ḥabash 'modernizes' the theories of the *Almagest*, introducing sines, cosines and tangents in place of chords, and proposes complete formulae to be applied in the different astronomical computations. This will be examined in detail in volume II, chapter 15; here we shall consider certain points raised by Ḥabash concerning pure astronomy.

The first section of the 'Damascus tables' deals with chronology and the passage between the different eras for the calculation of the equivalences of dates – under Persian, Egyptian, Greek, Hegirian, etc. calendars – focusing on tables of concordance. But in addition Ḥabash sets out to draw up tables for the motion of the heavenly bodies based on the lunar year, recalculated with great care because this was the legal year in his society; however, this attempt was not pursued by Arab astronomers because the lunar year was much less well adapted to the computations and theories of astronomy than the solar year, which was used both in the Hellenistic world of Ptolemy and by the Persians with their regular months of thirty days.

Throughout his work, Ḥabash compares the parameters calculated by Ptolemy for the motion of the different heavenly bodies with his own calculations and systematically modifies the composition of his tables accordingly for each one, without returning to the theoretical aspect of the geometrical models. But Ḥabash's most important theoretical innovation occurs in his study of the visibility of the crescent moon.

This problem of the visibility of the crescent is not treated in the Greek tradition of astronomy, but some methods of calculation were elaborated in the Indian tradition. Before looking at Ḥabash's solution, let us consider two previous solutions based on various elements of reference on the celestial sphere.

In the situation of a stationary earth at the centre of the world, the sun and the moon each have their 'proper motion' daily, in the opposite direction to the diurnal motion, being slightly less than 1° with respect to the ecliptic for the sun and about 13° for the moon on either side of the ecliptic (its maximum latitude is 5°). Thus each month, the moon 'catches up' with the sun and overtakes it, the crescent then becoming visible again on the western horizon just after sunset, signalling the beginning of a new lunar month. Figure 2.3 shows the moon setting at D; its latitude is DG, and the sun is at O under the horizon. HDA is the horizon from the point of observation, E is the closest equinoctial point (here the autumn equinox), OGE is the ecliptic and MAE is the celestial equator. OM is the position of the horizon at the moment of sunset, OH represents the distance from the sun to the horizon when the moon is setting, OG is the longitudinal distance between the sun and the moon, and the angle A between the horizon and the equator is equal to the complement of the latitude of the place.

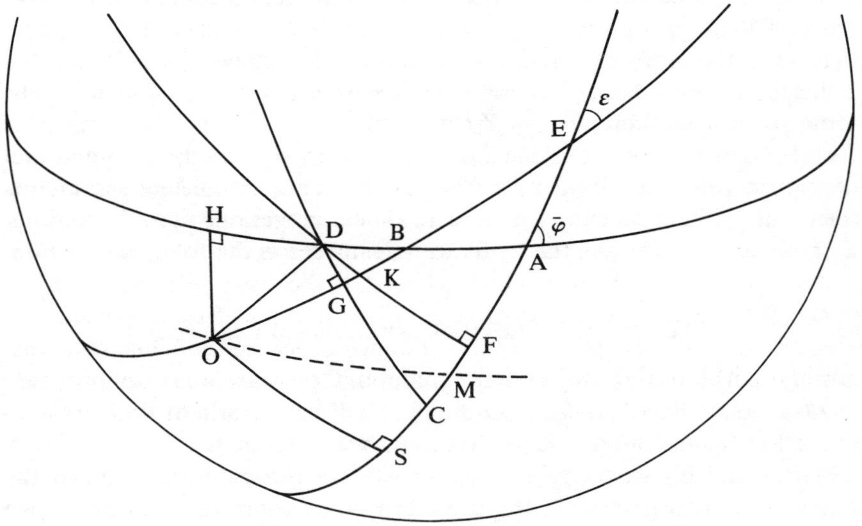

Figure 2.3

Ya'qūb b. Ṭāriq and al-Khwārizmī, whom we mentioned earlier, both adopted an Indian solution based on the time that elapses between sunset and the setting of the moon, i.e. on the arc AM in Figure 2.3.[27] According to them, the crescent will be visible if the calculation for the chosen day shows this arc to equal at least 12°, which corresponds to an interval of 48 minutes between the setting of one and the other body.

Ḥabash follows the tradition created by Ptolemy in his study of the visibility of the fixed stars and of the planets on the horizon.[28] Ptolemy had never been concerned with the visibility of the crescent moon, but he had based his entire study of the visibility of the other heavenly bodies at their rising or setting on the luminosity of the atmosphere on the horizon, and thus on 'the arc of depression of the sun under the horizon' before its rise or after its setting – i.e. OH in the case of Figure 2.3. He had determined the value of this arc necessary for a given body to be visible on the horizon; the arc was later known in the Latin tradition as *arcus visionis*, or 'arc of visibility'. Ḥabash took this idea and applied it to the case of the moon; he determined as a result of observations and calculations that 'the arc of depression of the sun under the horizon', or 'arc of visibility of the crescent', OH, should have a value of at least 10° for the lunar crescent to be visible after sunset, on the twenty-ninth night of the lunar month.

This method of Ḥabash became famous; it was taken up unchanged some two centuries later by al-Bīrūnī, and cited by many subsequent authors as one of the typical means of approaching the difficult problem of the visibility of the crescent.

Ḥabash therefore emerges as an observer who studied the *Almagest* in order to verify its data, pursuing the work begun under al-Ma'mūn by the group which had compiled *al-Zīj al-mumtaḥan*; however, his work goes further than that of his immediate predecessors, as he also adapted and developed certain of Ptolemy's ideas after having completely assimilated them, but without tackling his theoretical demonstrations as such. Another author was to undertake that task, as we shall see in the following section.

The mathematization of astronomical reasoning

A single author is our subject here: Thābit b. Qurra. He was born probably in AH 209 (AD 824) and died in AH 288 (AD 901); he came from Ḥarrān in upper Mesopotamia and his mother tongue was Syriac but he knew Greek perfectly and his working language was Arabic. Joining the team of the Banū Mūsā in Baghdad, Thābit produced original works in all the known sciences of his time. Especially famous as a mathematician, he wrote more than thirty treatises on astronomy, nine of which have been handed down under his name, including the 'Book on the solar year' discussed earlier, which is incorrectly attributed to him; we therefore have only eight of his books from which to judge the work in astronomy of this author.[29] We shall look at just three of these eight treatises: the first concerning the theoretical study of the motion of a heavenly body on an eccentric; the second about the choice of time intervals for determining the different motions of the moon; and the third concerning a method of calculating the visibility of the crescent.

Theoretical study of the motion of a heavenly body on an eccentric[30]

When Ptolemy discusses the motion of the sun on its eccentric orb, he notes the inequality of its apparent motion:

> The greatest difference between the mean motion and the motion which appears irregular, the difference by which we know the passage of the heavenly bodies in their mean distances, occurs when the apparent distance from the apogee is a quarter of a circle and the heavenly body takes longer to go from the apogee to this mean position than from the latter to the perigee.[31]

Ptolemy thus establishes only that the slowest apparent motion occurs on the side of the apogee, the fastest on the side of the perigee, and that between the two there is a point of 'mean passage' at a quarter of a circle from the apogee.

Thābit studies this question and proves Ptolemy right. When any heavenly body, or the centre of an epicycle, moves along an eccentric ABC with centre D with uniform circular motion, this movement is observed from the earth at E on the circle of the ecliptic A′B′C′, and it is an irregular apparent movement. Thābit takes equal arcs on the eccentric, which are thus covered by the moving body in question in equal times: GF on either side of the apogee A, HI on either side of the perigee C, BK on the side of A and LM on the side of C (see Figure 2.4).

Based on the reasoning in Euclid's *Elements* he proves that the arcs of apparent motion observed on the ecliptic are such that $G'F' < B'K' < L'M' < H'I'$, so that he can conclude precisely:

> When the motion of a heavenly body or of the centre of any orb is uniform on an eccentric, its slowest apparent motion, on the ecliptic, will be produced when the moving body is at the apogee of its eccentric, and its fastest apparent movement will occur when it is at its perigee. For the rest, apparent motion is slower when it occurs close to the apogee than when it occurs far away from it.

Note that Thābit refers to the speed of a moving object at the apogee or the perigee. As far as we know, this is the first reference in history to the notion of speed at a point.

This was the first theorem of Thābit's treatise. The second theorem is no less important. Thābit takes an eccentric ABC with centre D, apogee A and perigee C, and he positions points B and F as 'those for which the distance to the apogee, in apparent motion on the ecliptic, is a quarter of a circle' (see Figure 2.5).

He then demonstrates, again with the help of reasoning from the *Elements* of Euclid, that the arc of mean motion IH, the sum of HB

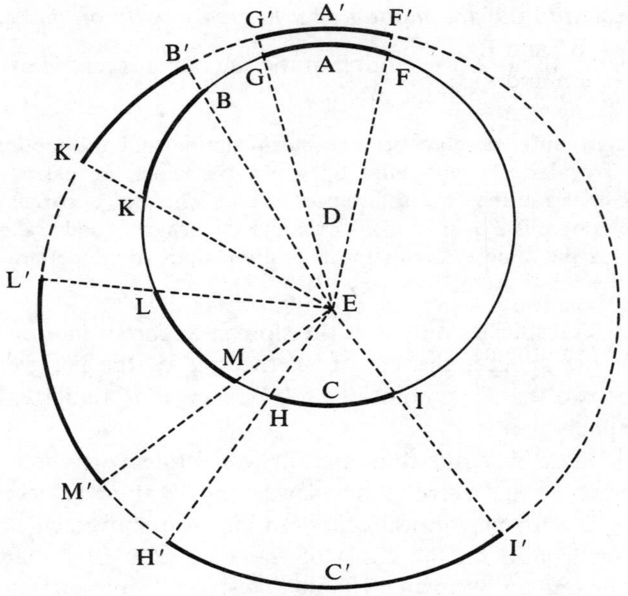

Figure 2.4

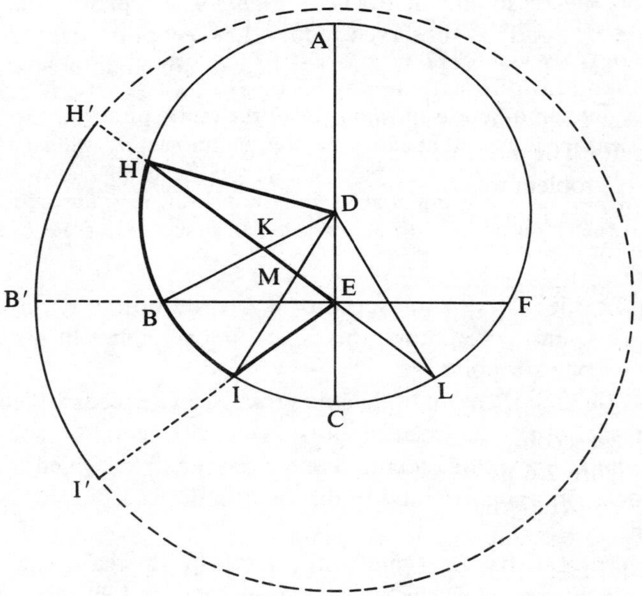

Figure 2.5

and BI, is equal to the arc I′H′, which is the sum of the arcs of apparent movement H′B′ and B′I′, that there is an 'approach to equality between the mean motion and the apparent motion when point B is approached ... and that the same is true when the motion occurs at point F'. He concludes, in conjunction with the preceding theorem:

> The closer the apparent motion to one of these two points, B or F, the closer it is to equality with the mean motion; and each time that two equal arcs of apparent motion on the ecliptic are taken on either side of each of these two points, their sum is in fact equal to the mean motion. These are the two points which resemble two points of mean motion.

This purely mathematical proof permits him to analyse precisely the apparent motion and the mean uniform motion relative to one another and to situate two axes: AC, an axis of symmetry for the mean uniform motion as observed from point E; and BF, an axis of symmetry for the apparent motion on the ecliptic. Thus for Thābit it became possible to analyse theoretically a geometric model as such that had been postulated to account for the movement of a heavenly body, using all the resources offered by the development of mathematics, leading him here to carry out the first mathematical analysis of a movement.

The choice of time intervals for determining the motions of the moon[32]

Here again Thābit takes a problem posed by Ptolemy, this time at the beginning of *Almagest* IV. Ptolemy based his whole study of lunar movements on the observations of lunar eclipses, as these could give the relative positions of the sun and moon without any error of parallax entering in to distort the results. The movement of the sun had been studied in *Almagest* III, and the next problem was to choose the time intervals at the limits of which lunar eclipses are found periodically, so as to ensure that the moon would have accomplished complete revolutions of each of its different orbs; once the number of these revolutions was known, the periodicity of the different movements of the moon could be determined. Before considering how Ptolemy solved this problem, we shall look at the way in which Thābit poses it.

He concentrates first on the sun, taking the two axes of symmetry determined in his preceding treatise for the movement of a body on an eccentric, shown in Figure 2.6 as AP and BC, O being the position of the observer at the centre of the ecliptic, and E the centre of the eccentric. In the first time interval t_1 the sun goes from M_1 to M_2, in a second interval $t_2 = t_1$ it goes from N_1 to N_2, and these two arcs of mean motion on the eccentric are then equal, $M_1M_2 = N_1N_2$, whereas the question remains for the corresponding arcs of apparent motion observed on the ecliptic, $M_1'M_2'$ and $N_1'N_2'$, the relationship between these two arcs being dependent on the respective position of M_1 and M_2 on the eccentric according to the results of the treatise described earlier (see note 30).

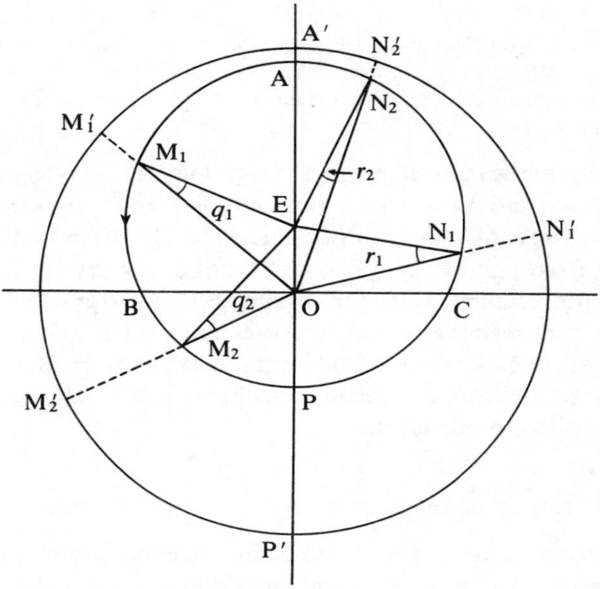

Figure 2.6

Let us call q_1 and q_2, r_1 and r_2 respectively the difference between mean motion and apparent motion for M_1 and M_2, N_1 and N_2; we obtain

$$M_1M_2 - M_1'M_2' = q_2 - q_1$$

and

$$N_1N_2 - N_1'N_2' = r_2 - r_1$$

Keeping $t_1 = t_2$, i.e. for two intervals of equal time, Thābit obtains seven types of combination between the two movements, which can be expressed in a purely theoretical way in terms of relations between $q_2 - q_1$ and $r_2 - r_1$, immediately applicable in the case of the sun.

1 In t_1 the sun leaves M_1 and returns to the same point after a number of complete revolutions; in t_2 it leaves N_1 and returns to it. Obviously we then have $q_1 = q_2$ and $r_1 = r_2$.

2 $q_2 - q_1 = r_2 - r_1 = 0$
3 $q_2 - q_1 = r_2 - r_1 > 0$
4 $q_2 - q_1 = r_2 - r_1 < 0$
5 $|q_2 - q_1| = |r_2 - r_1|$
but each is of opposite sign.
6 $q_2 - q_1 \neq r_2 - r_1$
7 $q_2 - q_1 = 0$ and $r_2 - r_1 \neq 0$

In these two equal time intervals, there is equality between the apparent movements for the cases 1 to 4, and inequality for cases 5 to 7; in cases 1 and 2 there is equality between the mean motion and the apparent motion (case 2 corresponds to the preceding second theorem). Figure 2.6 shows the general case of mode 6.

With the help of the two theorems from the preceding treatise, and with reference to the two axes of symmetry, we can easily situate the points M_1 and M_2, and N_1 and N_2, corresponding to the points of departure and arrival of the sun in the two equal time intervals for each of these seven cases.

The case of the moon is more complex in that it moves on an epicycle, itself mobile on an eccentric. But we are in the situation where eclipses of the moon are at the limits of the two time intervals cited, which allows us to relate the movement of the moon to that of the sun, since they are then in opposition, as shown in Figure 2.7.

The sun being at O and the earth at T, the moon on its epicycle can be found at L or L′ at the moment of opposition. In this situation, Thābit finds seven forms of combination for lunar motion, analogous to those of the sun. If the sun, in each of the two time intervals, has covered equal angular distances in apparent motion, so has the moon. But, in order for all its motions to be restored to its different orbs, it is necessary to eliminate the cases where the moon would pass from L to L′ on its epicycle between the two limits of the time intervals considered. Looking again at the seven cases, cases 5 to 7 are set aside because of the situation of the sun which presents unequal apparent movements at the limits of the time intervals, and cases 2 to 4 are eliminated since the moon would then pass from L to L′ on its epicycle. Only the first case is retained, when the moon and sun leave from the same point of the ecliptic to return there, because only in this situation will they have completed a number of revolutions on their various respective orbs.

Ptolemy had also discussed two analogous time intervals, choosing four cases for the sun.[33]

(a) In t_1 and t_2, it travels complete circles – equivalent to Thābit's case 1.

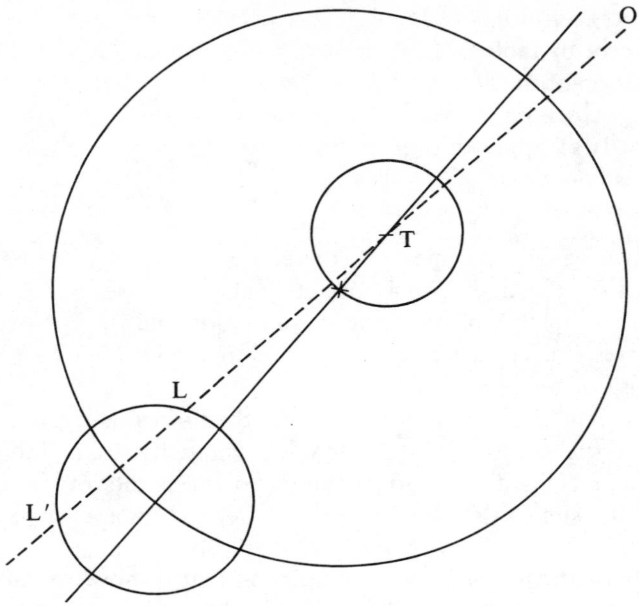

Figure 2.7

(b) In t_1 it goes from perigee to apogee and in t_2 it goes from apogee to perigee – a special case of Thābit's mode 2.
(c) In t_1 and t_2, the sun departs from the same point on the ecliptic – a special case of Thābit's modes 3 or 4.
(d) The point of departure for t_1 is symmetrical, with respect to the apogee or the perigee, to the point of arrival for t_2, and vice versa – equivalent to Thābit's cases 3 or 4.

Ptolemy then considers the question of the moon and eliminates cases (b), (c) and (d), keeping only the first – Thābit's mode 1. The conclusions of both thinkers are similar, but Ptolemy reasons only from particular cases, while Thābit regards the problem in its entirety, his analysis is exhaustive and he reaches a conclusion which becomes irrefutable (in the frame of the geometrical models concerned) due to the impeccable precision of his theoretical analysis.

The visibility of the crescent

Like all the Arab astronomers, Thābit studied the problem of the visibility of the crescent moon, and two of his treatises on this subject have been preserved: 'The visibility of the crescent by calculation' and 'The visibility of the crescent by tables'. The first is purely theoretical, and the second is a simplification of the first for practical application with the aid of tables.[34]

In a general sense, Thābit looks for a quantifiable relationship between the luminosity of the first lunar crescent and that of the horizon just after sunset. As we have seen above, in his study of the visibility of the fixed stars and the planets, Ḥabash took from Ptolemy the notion of the 'arc of visibility' of the crescent, giving it a fixed value of 10°. Thābit followed the same tradition, but his solution is much more complex because, for him, this 'arc of visibility' is no longer a constant and he needs to modify its values by successive calculations based on the four variables that he defines.

The first three variables are the three sides of the fundamental spherical triangle OHD in Figure 2.8, O being the position of the sun under the horizon, H the 'brightest point on the horizon' in a vertical line from the sun, and D the position of the moon at its setting; let us call these three arcs α_1, α_2, α_3 (Figure 2.8).

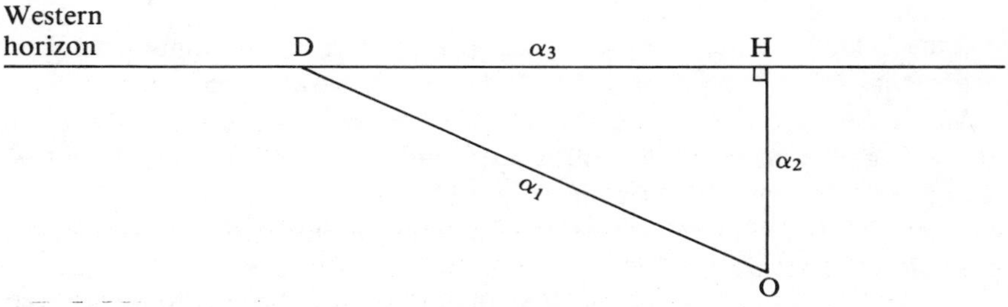

Figure 2.8

The first, α_1, is the angular distance between the moon and the sun; it is this arc which will determine the portion of the crescent, as seen from the earth, that is illuminated by the sun. The luminosity of the sky at point H on the horizon after sunset depends on the second arc, α_2; this is 'the arc of depression of the sun under the horizon'. On the third arc, α_3, depends the luminosity of the sky at the point where the moon sets; it is the distance from D to H, the 'brightest point on the horizon'. This fundamental triangle allows two limit situations (Figures 2.9 and 2.10).

When the moon sets on the vertical line from the sun, at the 'brightest point on the horizon' (Figure 2.9), $\alpha_3 = 0$, the crescent can be visible if α_1 and α_2 have a value at least equal to what we shall call α_0, the precise limit value for these two arcs together; α_0 is the absolute value for the 'arc of visibility' of the crescent, to be determined as a function of the earth–moon

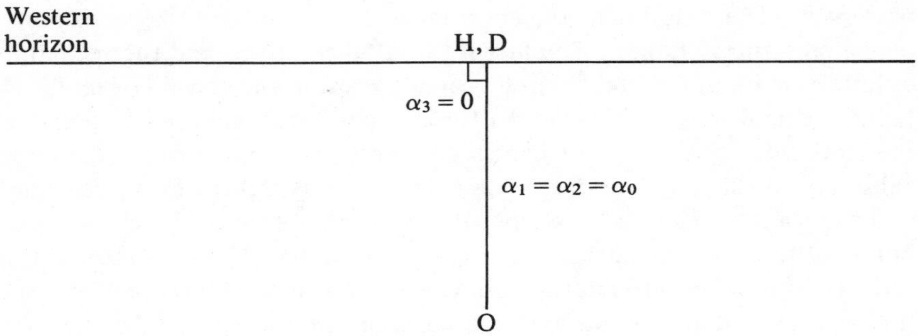

Figure 2.9

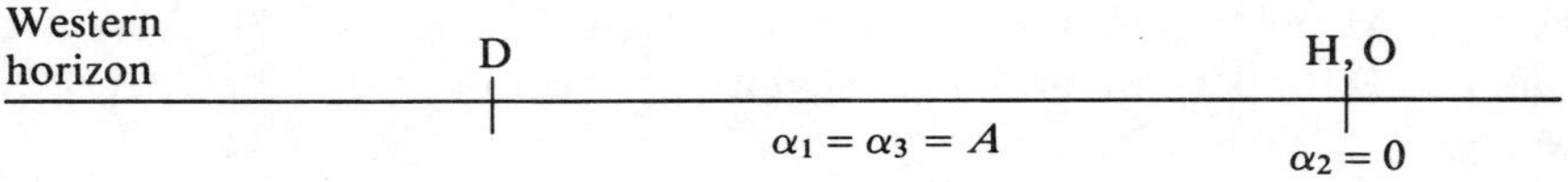

Figure 2.10

distance. Thābit states, without justifying it, that its minimum value in degrees is 10; 52; thus if $\alpha_0 < 10;52$ the crescent will be invisible.

When the crescent is at the limit of visibility while the moon and the sun are setting together, the angular distance between the two must be such that the crescent can be visible by day (Figure 2.10). We then obtain $\alpha_2 = 0$ and $\alpha_1 = \alpha_3 = A$, the limit value beyond which the crescent will be visible in all possible conditions. Thābit states that for $A > 25^\circ$ the crescent will be visible by day whatever the value of the other variables. This upper limit of 25° seems to correspond to an observation; in fact, recent observations show that in the middle of the day the moon is at the limit of visibility when its angular distance to the sun is close to 25°.

The fourth variable refers to the distance between the moon and the earth, on which depends the apparent diameter of the moon and therefore its luminosity for that portion of crescent illuminated. At the moment when the crescent is first visible, the position of the centre of the moon's epicycle could be mistaken for the apogee of its eccentric; the true angle of anomaly a is the only variable which intervenes in the distance from the moon to the earth (Figure 2.11).

The moon is furthest from the earth when $a = 0$ and closest when $a = 180$. If R is the radius of the eccentric orb, e its eccentricity and r the radius of the epicycle, the distance from earth to moon will go from $R + e + r$ to $R + e - r$ as a goes from 0 to 180.

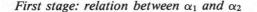

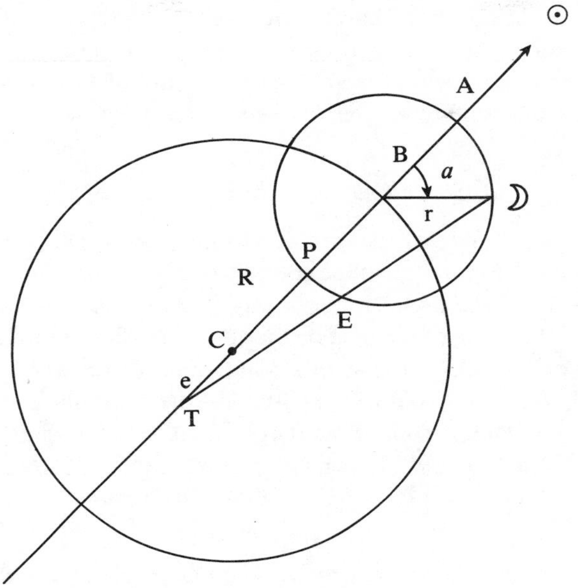

Figure 2.11

First stage: relation between α_1 and α_2

The main point of the discussion concerns arcs α_1 and α_2 of Figure 2.8, which are the two most important variables. If α_1 increases, the crescent becomes brighter, and if α_2 decreases the sky's luminosity will be stronger on the horizon. A balance needs to be found between the variation of these two arcs and will need to be modified in accordance with the other two variables. Let $V(\alpha_1, \alpha_2)$ stand for the relation between a specific pair of these variables when the crescent is at the limit of visibility. Thābit looks for the relation that must exist between the 'increasing' $\Delta\alpha_1$ and the 'decreasing' $\Delta\alpha_2$ so that we can express the following identity:

$$V(\alpha_1, \alpha_2) \Leftrightarrow V(\alpha_1 + \Delta\alpha_1, \alpha_2 - \Delta\alpha_2)$$

The second term of this expression means that the crescent is once again at the limit of visibility for the pair in question. Thābit then states that the relation between $\Delta\alpha_1$ and $\Delta\alpha_2$ is a constant $k = \Delta\alpha_1/\Delta\alpha_2 = (A - \alpha_0)/\alpha_0$, A and α_0 being the values defined above. This constant is found when the crescent passes from one limit to the other (see Figures 2.9 and 2.10), i.e. from $\alpha_1 = \alpha_2 = \alpha_0$ to $\alpha_1 = A$ and $\alpha_2 = 0$. Using the previous expression we can then say that $V(\alpha_0, \alpha_0) \Leftrightarrow V(A, 0)$, with $\Delta\alpha_1 = A - \alpha_0$ and $\Delta\alpha_2 = \alpha_0$ which gives the proposed constant relation k between $\Delta\alpha_1$ and $\Delta\alpha_2$. We then obtain $A = (k+1)\alpha_0$, and Thābit states that this relation k is known. The numerical data in the text give us $k = 1;11,46$, a figure possibly derived from the author's work on the values for the 'arc of visibility' of different planets given by Ptolemy in the *Planetary Hypotheses*.[35]

Second stage: intervention of α_3

α_0 is the absolute value of the 'arc of visibility' of the crescent for, in the limit situation of Figure 2.9, the moon sets in a vertical line from the sun and $\alpha_3 = 0$. When the moon moves away from H, the 'brightest point on the horizon', the value of α_0 must be modified in order to find a new slightly fainter 'arc of visibility', for at this point the horizon will be slightly less luminous than at H. Thābit then applies the formula put forward by Ptolemy in the *Phaseis* for the visibility of the fixed stars at any point on the horizon (Morelon 1981: 3–14), and he gives the first formula for modification of the arc of visibility: $\alpha_0' = \alpha_0(360 - \alpha_3)/360$.

Third stage: intervention of the distance earth–moon (as a function of a*)*

As we have seen, Thābit gives $\alpha_0 = 10;52$ as the absolute minimum of the arc of visibility, and $A = 25$ as the maximum above which the moon is visible by day whatever other conditions might be. Thus for him, $\alpha_0 = 10;52$ corresponds to the best conditions of visibility, and therefore the moon's closest position to the earth ($a = 180$ in Figure 2.11), and $A = 25$ corresponds to the worst conditions of visibility, i.e. when the moon is furthest from the earth ($a = 0$). When the moon moves from one distance to another its apparent diameter changes, and consequently α_0 and A must be calculated again.

To solve this problem, Thābit draws an analogy with the visibility of Venus as explained in the *Planetary Hypotheses*: there Ptolemy determines that the arc of visibility of Venus is 5° when the planet is at its minimum distance from the earth (166 times the earth's radius according to the figures accepted at the time) and 7° at its maximum distance (1,079 times the earth's radius), whilst the figures given for the moon in the same work correspond to $R + e - r = 53$ and $R + e + r = 64$. Without explicitly justifying his calculation, Thābit then declares that the corresponding differences in the arc of visibility in the case of the moon are 0;31 for α_0 and 1;8 for A. He then deduces that $10;52 \leqslant \alpha_0 \leqslant 11;23$ and $23;52 \leqslant A \leqslant 25$ when $0 \leqslant a \leqslant 180$.

The only method of calculation that will enable a close approximation of these figures to be found is to establish, term by term, a corresponding geometrical progression for the distances and an arithmetical progression for the values of the arc of visibility. The result is as follows: for Venus the arithmetical progression is of ratio 1, and the two arcs of visibility are obviously of rank 5 and 7; the geometrical progression is of ratio 2.712 and we find 147 and 1,079 of ranks 5 and 7. For the moon the arithmetical progression is of ratio 0; 31, and we find 10; 51 and 11; 22 of ranks 21 and 22; the geometrical progression is of ratio 64/53, and we find 53 and 64 of ranks 21 and 22. This correspondence is sufficiently close for us to conclude that this was the procedure used. If the relation k is known, as Thābit states, the single value of $A = 25$, corresponding to one observation, is sufficient to find the limit values of α_0 and A.

This term-by-term correspondence of two progressions only yields the extreme values of α_0 and A, those which correspond to $a = 0$ and $a = 180$. To go from one to the other, Thābit uses a simple interpolation formula which Ptolemy calculated by making a table[36] for $I(a)$ such that $0 \leqslant I(a) \leqslant 1$ when $0 \leqslant a \leqslant 180$. He then sets out as a function of a:

$$\alpha_0 = 11;23 - 0;31 I(a)$$

and

$$A = 25 - 1;8 I(a)$$

The discussion then turns to the arc α_2 (arc of depression of the sun under the horizon), in order to compare it with the 'arc of visibility' calculated in stages by giving a fixed value to one or another of the variables.

1 Thābit writes $\alpha_3 = 0$ and $\alpha_1 = 10;52$ (its absolute minimum); he calculates as a function of a the value of $\alpha_0 = 11;23 - 0;31 I(a)$ and concludes that for $\alpha_2 \geqslant \alpha_0$ the crescent will be visible.
2 He takes the real value of α_3 and keeps $\alpha_1 = 10;52$; he calculates the corresponding decrease of the arc of visibility $\alpha_0' = \alpha_0 - \Delta\alpha_0$ using the formula of Ptolemy in the *Phaseis*: $\alpha_0' = \alpha_0(360 - \alpha_3)/360$. He then concludes that for $\alpha_2 \geqslant \alpha_0'$ the crescent will be visible.
3 He substitutes all the variables by their real values and calculates $\alpha_0'' = \alpha_0' - \Delta\alpha_0'$. The corresponding decrease will depend on α_1, the real angular distance between the moon and the sun, giving the true width of the visible crescent, and another factor is added, acting to increase α_1 from its absolute minimum of 10; 52 and introducing A' as the value of A modified like α_0 by the formula from the *Phaseis*. The final expression is as follows:

$$\alpha_0'' = [11;23 - 0;31 I(a)] \, \frac{360 - \alpha_3}{360} \, \frac{A' - \alpha_1}{A' - 10;52}$$

He concludes that for $\alpha_2 \geqslant \alpha_0''$ the crescent will be visible.

This theory of the visibility of the crescent thus involves six elements: an observation, $A = 25$; a constant relation k between the 'increase' of α_1 and the 'decrease' of α_2; a term-by-term correspondence of two progressions, one arithmetical and the other geometrical; the situation of the three main variables with respect to their limit values, $\alpha_0 \leqslant \alpha_1 \leqslant A$, $0 \leqslant \alpha_2 \leqslant \alpha_0$, $0 \leqslant \alpha_3 \leqslant A$; a simple interpolation formula taken from Ptolemy; the formula of the *Phaseis* to modify the result according to the position of the moon over the horizon.

Thābit bases his study on an analogy between the case of the crescent and that of the fixed stars, using the formula from the *Phaseis*, and another analogy with the case of the planets, using the example of Venus. This means that, for him, there is just one problem of visibility for every luminous celestial body on the horizon after sunset or before sunrise: the lunar crescent, fixed stars and planets all take part in this unique phenomenon which Thābit tries to analyse in mathematical terms using a relation between the magnitudes linked to the luminosity of the body in question and to that of the horizon at that moment. He seems, then, to have looked for a general law which he has tried to apply numerically to the case of the crescent.

This author thus attempts to deal with the problems of astronomy in a rigorously mathematical way, looking at them in general terms and studying the models proposed by Ptolemy from a purely geometrical point of view without questioning them as such. He recognizes that the degree of accuracy of pure reasoning cannot always match that of observation, for as he says, 'what is perceived by sense does not lend itself to such precision' (Thābit, p. 108, l. 6). Verification by observation will always be necessary, and the conclusion of his purely theoretical treatment of the visibility of the crescent is entirely devoted to this matter, including conditions of observation and personal factors associated with the quality of the observer.

Al-Battānī

Al-Battānī, an astronomer of great reputation, lived at the turn of the ninth and tenth century: he was born about the middle of the ninth century and died in AH 317 (AD 929). Originally from Ḥarrān, like Thābit, he lived for most of his life in Raqqa, on the Euphrates in the north of present-day Syria, where for more than thirty years he made many high-quality observations from a private observatory. He wrote a survey of his work in a monumental book called 'The Sabian tables' (*al-Zīj al-Ṣābī*).[37] This had a great influence on the astronomy of the Latin West in the Middle Ages and at the beginning of the Renaissance, because it is the only complete treatise on Arabic scientific astronomy of this era to have been translated in its entirety into Latin in the twelfth century (and then directly into Spanish in the thirteenth century), naming the author Albategni or Albatenius. His was therefore the only majorly important work of eastern astronomy in the Arabic tradition that was known and studied until relatively recent times, which is why al-Battānī is so renowned and has been hailed as 'the greatest Arab astronomer' by successive authors of most of the books on the history of astronomy.

He was indeed a great observer, but his work in theoretical astronomy is not of major importance; it depends almost entirely on his immediate Arab predecessors, who are never explicitly cited although al-Battānī frequently refers to Ptolemy. He recalculates certain parameters and compares the results of his own observations with some preceding theories without criticizing them or making any notable additions.

His important contribution, then, lies in the area of pure observation. He measures the obliquity of the ecliptic with a high degree of accuracy (23; 35); he finds that the apogee of the sun's orb on the ecliptic is at 22; 50; 22 of Gemini, which is much closer to the true position in his day than the value recorded in the 'Book on the solar year', and he thus confirms the mobility of this apogee. He calculates the length of the tropical year and finds it equal to 365; 14, 26, a value slightly less accurate than that in the 'Book on the solar year'. Having checked it but without naming his source, he accepts the value of the precession constant given in the 'Verified table' (*al-Zīj al-mumtaḥan*) (1° every sixty-six years), and from this he recalculates the figures in the catalogue of fixed stars in the *Almagest*, reducing their number by slightly more than half (489 stars instead of 1,022).

His most famous observation, and deservedly so, is that of the variation of the apparent diameter of the sun and the moon, on the basis of which he comes to the conclusion, for the first time in the history of astronomy, that annular eclipses of the sun are possible as long as the apparent diameter of the moon at its minimum is slightly less than that of the sun. He finds that the apparent diameter of the moon, when in conjunction with the sun, can vary from 0; 29, 30 to 0; 35, 20 (real variation from 0; 29, 20 to 0; 33, 30), and that the apparent diameter of the sun can vary from 0; 31, 20 to 0; 33, 40 (real variation from 0; 31, 28 to 0; 32, 32), whereas Ptolemy had considered that the apparent diameter of the sun would remain equal to 0; 31, 20 – curiously disregarding the difference in its distance from the earth during its movement on the eccentric – and that this value was the same as the minimum apparent diameter of the moon, thereby eliminating the possibility of annular eclipse.[38]

In conclusion, to attempt a brief summary of the study of astronomy under the Abbasids in the ninth century, we can say first that original research took place in this domain as soon as the basic resources became available to scholars, whether those resources were Indian, Persian, Syriac or above all, Greek. Translation into Arabic of earlier sources and pure scientific research went hand in hand, for astronomy as for all the exact sciences, right from the beginning and during the whole of this century (Rashed 1989).

The work of astronomical research really got underway with the establishment of a collective programme of continuous observations under al-Ma'mūn a little before 830, and the caliph strongly encouraged this fundamental research, many of his successors doing the same. It is clear that right from this period astronomers stressed the precision of the instruments and the necessity for continuous and repeated observations – for the sun and the moon at first in Baghdad and Damascus, and then for the rest of the heavenly bodies – whereas the sources transmitted only observations that were isolated in space and time; this programme was continued and developed throughout succeeding history.

The collective aspect of the work should also be emphasized even outside a purely observational framework for in addition to the existence of communal structures such as the observatories at Baghdad and Damascus financed by central power, numerous traces of scientific correspondence between astronomers are to be found cited in ancient Arabic bio-bibliographic works concerning this era. We can therefore speak of the founding of a real 'school of Baghdad' in astronomy of the ninth century.

The constant movement back and forth between theory and observation, markedly more systematic than in the astronomy of Hellenistic tradition, allowed an early and sometimes vigorous critique of parts of Ptolemy's theories or results, but still exclusively within the framework of the system and the geometrical models that he had proposed.

During this century the progress of spherical trigonometry, considered at the time to be just an 'auxiliary science' to astronomy, enabled much more rigorous and elaborate geometrical reasoning concerning the arcs of the celestial sphere, due to the systematic utilization of sine and cosine and the introduction of tangents and cotangents (see vol. II, chapter 15); finally, the research begun by Thābit to apply to astronomy the results achieved by mathematicians, who were often also astronomers, was continued by most of his great successors, consequently making astronomy increasingly 'scientific'.

Thus the subsequent developments in Arabic astronomy had already taken root in the ninth century, especially in Baghdad, for the programme and methods of work which reached a high state of organization there would be followed without notable change, at least in basic principles, for several centuries.[39]

ASTRONOMY IN THE TENTH AND ELEVENTH CENTURIES UNTIL AL-BĪRŪNĪ

We saw in Chapter 1 that it was between the tenth and eleventh centuries that decisive progress was made with regard to the conception and organization of permanent large-scale observatories in Baghdad or in Persia, and the chapter on trigonometry (vol. II, chapter 15) shows the importance of the results achieved during the tenth century for the development of this science, on which the accuracy of astronomical calculations partly depends.

However, few but fragmentary or incomplete texts of theoretical astronomy have come down from this era, and it is paradoxically more difficult to sketch the evolution of eastern Arabic astronomy in the tenth century than in the preceding century. We shall therefore simply take three examples of scientists from this period, who seem to have worked in greater isolation than their predecessors of the ninth century, and then turn our attention to the line that, from master to student, leads to al-Bīrūnī, who lived partly in the tenth and partly in the eleventh century and who stands at the summit of this first period of eastern astronomy.

Abū Jaʿfar al-Khāzin, ʿAbd al-Raḥmān al-Ṣūfī and Ibn Yūnus

Abū Jaʿfar al-Khāzin was a brilliant mathematician; originally from Khurāsān, he spent part of his life in Rayy and he died between AH 350 and 360 (AD 961 and 971). He composed several treatises of theoretical astronomy, but only some fragments of his *Commentary on the Almagest* – mainly on trigonometry – remain; however, the allusions to this work made by certain later authors, notably al-Bīrūnī, demonstrate its importance for his successors. He had studied the motion of the sun and, unlike al-Battānī, he used Ptolemy's observation of the constant value of its apparent diameter, therefore necessitating a fixed distance from the earth. He thus proposed a new model for the motion of the sun: not on an eccentric but on a circle concentric with the earth, the uniform motion occurring around an eccentric point in an analogous way to the movement of the epicycle around the 'equant point' in the Ptolemaic model of the upper planets.[40] This is currently the only point which reveals that he had made a critical evaluation of Ptolemy's models.

He also wrote the *Kitāb fī sirr al-ʿālamīn* ('Book on the secret of the worlds'), which is lost in its entirety and in which he proposed a new global conception of the universe based on Ptolemy's results in the *Planetary Hypotheses*.[41] Although we cannot yet determine its precise measure, the work of Abū Jaʿfar al-Khāzin had an undoubted influence, a century later, on the work of Ibn al-Haytham, *al-Shukūk ʿalā Baṭlamyūs*, relating to his criticism of the Ptolemaic system, which is frequently based on arguments of a cosmological nature (see the following chapter).

ʿAbd al-Raḥmān al-Ṣūfī (AH 291–376 (AD 903–86)) was born in Rayy and worked in Shiraz and Isfahan. Several of his observations, on the obliquity of the ecliptic and on the motion of the sun or the length of the solar year, have been reported, but he is most famous for his *Kitāb ṣuwar al-kawākib al-thābita* ('Book concerning the constellations of the fixed stars');[42] this is a reworking of the catalogue of fixed stars from the *Almagest*, written around 965. In his introduction, al-Ṣūfī defines his position in relation to the Arab astronomers of the preceding generation who dealt with the fixed stars or to the makers of celestial spheres, criticizing the way in which one or another constellation was handled, and he chooses the value of the precession constant calculated by the authors of the 'Verified table' under al-Ma'mūn – 1° in sixty-six years – instead of the 1° per century stated by Ptolemy. This work is not just an adaptation of the catalogue of the *Almagest* achieved by modifying the longitude of each star with the aid of the correction corresponding to the precessional movement between the second and tenth centuries, because al-Ṣūfī also made many verifications by observation, for the magnitude of the stars as well as for their ecliptic longitude – he states that he preserved the value for the latitudes given by Ptolemy – and introduced notes on the apparent colours of the principal stars. Al-Ṣūfī's book was widely read in Arabic, and from the twelfth century was translated and disseminated in Latin – the name of its author being transcribed as 'Azophi' – resulting in many stars being given names of Arabic origin in the West.

The work describes each of the forty-eight constellations according to a unique format: first a presentation of the constellation concerned, listing all its stars and the different Arabic names under which they could be known; then a table giving their ecliptic co-ordinates and their magnitude. All copies of the book, from the earliest, contain miniatures of the mythical figures representing each of the constellations with the positions of its different stars, always sketched twice in symmetrical fashion – 'as seen in the sky' and 'as seen on the sphere' (i.e. on a representation, in wood or metal, of the celestial sphere) – thus enabling easy location of the constellations even by a beginner. The author intended a double purpose for his work, at once theoretical and practical, for example in orientation on land or sea, and this was part of the reason for its success. A number of illustrations printed here show the quality and diversity of the representations of constellations in the manuscripts of this famous work.

Ibn Yūnus (d. AH 399 (AD 1009)) was a great Egyptian astronomer, and above all an observer, who worked in Cairo during the first period of the Fatimids and probably had his observatory on Mount Muqattam, east of

Cairo. His most important work is *al-Zīj al-Ḥākimī al-Kabīr* ('The great Hakemite table') – from the name of the Fatimid sultan al-Ḥākim who reigned in Cairo from AH 386 (AD 996) to AH 411 (AD 1021) – a monumental book in eighty-one chapters of which only a little more than half is preserved.[43] Ibn Yūnus set out to produce a complete treatise of astronomy including the greatest possible number of previous observations, critically reviewed and analysed and enriched with the results of his own numerous observations. His work is therefore a means of gaining access to much scientific material of the ninth and tenth centuries which is known only through his quotation of it.

There is very little theoretical reasoning in this work of Ibn Yūnus; it is a 'zīj' in the strict sense of the term, i.e. a work that concentrates exclusively on the compilation of tables of the movements of heavenly bodies, with calculations of their various parameters and details of how to use them. Since his results became available in translation at the beginning of the nineteenth century, the accuracy of his observations has been exploited by modern scientists to gain, for example, a better knowledge of the secular acceleration of the moon.

Al-Bīrūnī

Al-Bīrūnī was born in AH 362 (AD 973) in the Khwārizm and died around AH 442 (AD 1050), probably in Ghazna (modern Afghanistan). He was the pupil of Abū Naṣr Manṣūr b. ʿIrāq, himself the pupil of Abū al-Wafāʾ al-Būzjānī; he explicitly recognized these two scholars as his masters, and he worked in Rayy with al-Khujandī. These three contributed toward al-Bīrūnī's becoming simultaneously a mathematician, an astronomical theorist and an observer.

Abū al-Wafāʾ al-Būzjānī, mathematician and astronomer, who was born in AH 328 (AD 940) at Būzjān in Persia, and died at Baghdad in AH 388 (AD 998), represents a return to the tradition of astronomical research of the 'Baghdad school', which had been so strong in the preceding century, as we have seen, for he gained his scientific training in this environment and worked in Baghdad thereafter. For his astronomical research Abū al-Wafāʾ used the large observatory built under the patronage of Sharaf al-Dawla in the gardens of the royal palace in Baghdad. He entitled his principal astronomical treatise *Almagest*, but only part of this text has been preserved and this mainly concerned with questions of trigonometry, a science considerably developed by this author.[44] We thus know little about his contributions to theoretical astronomy, but al-Bīrūnī makes numerous allusions to his studies concerning the motion of the sun and the value of the precession constant.[45]

We have less information about al-Bīrūnī's immediate 'master', Abū Naṣr Manṣūr b. ʿIrāq, who died around AH 427 (AD 1036) at Ghazna. We know that he was a student of Abū al-Wafāʾ al-Būzjānī, and his surviving works consist mainly of important texts on trigonometry, written partly at al-Bīrūnī's request when the latter was puzzling over specific problems (Samsó 1969). Al-Khujandī, who died *c.* AH 390 (AD 1000), devoted a great deal of study to the question of observational instruments, about which he wrote several books, and he was responsible in particular for the building of the great sextant at Rayy described in the preceding chapter.

Al-Bīrūnī is a scholar of exceptional stature, who wrote about 150 works on all the known sciences of his time. These included thirty-five treatises of pure astronomy of which only six have survived; his other works – on India, for example, or about chronology – contain numerous references to astronomical matters. His major survey in this area is *al-Qānūn al-Masʿūdī* ('Tables Dedicated to Masʿūd'), a work consisting of eleven treatises written around AH 426 (AD 1035) which comprises 1,482 pages in the original edition (Boilot 1955).

His mother tongue was Persian, but his chief working language was Arabic and he also knew Sanscrit fluently because he used it and he made several translations of scientific texts from Sanscrit to Arabic. He thus had direct access to all the sources of Indian scientific astronomy, to which he constantly refers alongside the Greek sources or the works in Arabic of his predecessors, whereas since the transmission of Sanscrit texts at the end of the eighth century, those predecessors seem only to have had access to a few Indian astronomical documents or to secondary sources, while the Greek scientific tradition was much more widespread. Al-Bīrūnī could therefore bring together and study directly the entire astronomical heritage of his day, from the Greek world, the Indian world and the Arab world, and all his work does in fact tend toward the goal of a rigorously conducted synthesis. Rather than attempting to present all the astronomical work of al-Bīrūnī, which would be particularly difficult, we shall consider some aspects of his method.

In the first treatise of *al-Qānūn al-Masʿūdī*, al-Bīrūnī states some general principles of astronomy and sets out the bases of chronology in different cultures, including that of China. In Chapter 2 he deals with the position of the heavens in relation to the earth, and considers the hypothesis of the rotation of the earth about itself to explain diurnal movement.[46] He states that this hypothesis was supported in India by Aryabhata and his disciples but that it is not compatible with one of Ptolemy's arguments according to which a body in free fall will not fall vertically if the earth has a rotational movement; al-Bīrūnī then asserts that a 'great scholar' (whom he does not name) contends that Ptolemy's argument is not valid in so far as every terrestrial body is carried by this rotation along the vertical line through which the body falls. After setting out this argument, which he appears to find consistent, al-Bīrūnī returns to the question, considers the problem of horizontal motion, calculates the speed of a point on the earth in the hypothesis of its rotation about itself, and concludes that this great speed could only be added to or subtracted from the other movements of terrestrial bodies from East to West, which cannot be verified, and therefore, according to him, it is not possible for the earth to rotate about itself.

In general, al-Bīrūnī deals with a specific astronomical problem according to the following scheme: first, he outlines some general principles about the problem in question; then he gives the various solutions proposed by the Indians, by Ptolemy and by the Arab astronomers, all presented and critically analysed on the basis of the general principles stated at the beginning; then, where relevant, he details the earlier observations that are most important or most noteworthy for the phenomenon in question, and describes his own observations; finally he selects one of the preceding solutions, or proposes his own solution based on all the foregoing material. As an example, we shall take the question of the visibility of the crescent as set forth in *al-Qānūn al-Masʿūdī*.[47]

Treatise VI of this work concerns the motion of the sun, treatise VII concerns the motion of the moon and treatise VIII concerns the observable phenomena on the connection between the motions of the sun and of the moon, i.e. the question of eclipses of one or other of these 'two luminaries' and that of the visibility of the crescent. Chapter 13 of treatise VIII is devoted to the morning and evening twilight, with a description of this phenomenon as the approach of the horizon to the limit of the cone of shadow created on the earth by the sun, and al-Bīrūnī states that 'the astronomers' (without citing them) have determined that the beginning of the morning twilight in the East, or the end of the evening twilight in the West, occurs when 'the arc of depression of the sun below the horizon' is 17° or 18° – without choosing between the two values. Chapter 14 then deals with the visibility of the crescent.

General principles

The ability of the eye to see the crescent depends on several factors: first, the distance between the moon and the sun, which determines the portion of the moon's surface which is illuminated; then the earth–moon distance on which its apparent luminosity for the same amount of illumination depends; then the luminosity of the atmosphere on the horizon depending on the inclination of the ecliptic on the horizon, and therefore on both the position of the sun on the ecliptic and the latitude of the place; finally the position of the setting moon on the horizon, more or less close to the 'brightest point on the horizon', and thus to the vertical of the sun's position below the horizon.[48] Al-Bīrūnī concludes that all these parameters must be carefully taken into account.

Earlier solutions

Ptolemy did not study this question because the problem did not arise in his culture. Four Arab astronomers, al-Fazārī, Yaʿqūb b. Ṭāriq, al-Khwārizmī and al-Nayrīzī, used an Indian method, taking the difference in time between the setting of the sun and of the moon, but this criterion did not allow for the inclination of the ecliptic on the horizon, and it was therefore not valid; al-Nayrīzī, however, did a little better than the other three, because unlike them, he took account of the correction of the lunar parallax. Following several corrections, al-Battānī took into consideration the distance between the sun and the moon both on the equator and the ecliptic but did not pay sufficient attention to the inclination of the ecliptic on the horizon. Finally, Ḥabash used as his principal criterion 'the arc of depression of the sun below the horizon', which can only be calculated from all the other parameters.

Conclusion

Ḥabash's method must be chosen. Al-Bīrūnī does not offer a personal solution, and concludes his chapter by describing the means of finding the lunar crescent on the horizon with the aid of the observation tube described in the preceding chapter.

The problem of the motion of the sun according to al-Bīrūnī has been studied by Hartner and Schramm (1963). As well as all the stages of the preceding scheme, al-Bīrūnī includes large numbers of observations of the sun in addition to his own, and a mathematical study of the apparent motion on the eccentric which recalls that of Thābit b. Qurra described earlier. Following a critical analysis of the findings of previous authors, al-Bīrūnī establishes in definitive manner the motion of the apogee of the sun, recalculates all the parameters and draws up tables of its movement.

This type of astronomical work did not disrupt the overall system of astronomy as perceived by al-Bīrūnī, because he remained faithful to the model of epicycles and eccentrics defined by Ptolemy. However, al-Bīrūnī reviewed everything in detail, continuing, for example, the movement toward the mathematization of astronomy begun by Thābit a century and a half before him,[49] and rigorously taking stock of the current state of the science in all its aspects. In so far as such a comparison is possible, this work is analogous to that carried out by Ptolemy in the *Almagest* eight centuries earlier: establishing a rigorous scientific tradition, but without major global innovation, using all the preceding research and all the mathematical tools available to the astronomer at the given time.

Al-Bīrūnī accomplished this synthesis brilliantly; it was the crowning achievement of the first period of Arabic astronomy, remaining within the general framework erected by Ptolemy. It was his contemporary Ibn al-Haytham who began to break free of that framework, a development that might not have been possible without the precise contribution of al-Bīrūnī.

NOTES

1 See al-Qifṭī.
2 See the Indian sources referred to in Chapter 1.
3 Cf. al-Bīrūnī, *Kitāb fī Taḥqīq*, pp. 351–2. Al-Bīrūnī is usually very reliable when reporting traditions of a scientific nature, particularly with regard to India, and the tradition recounted here is probably based on historical fact, but certain elements are lacking, without which we cannot be absolutely convinced of the authenticity of everything that has been reported concerning the episode: the various Arabic sources disagree on an exact date; who was the Indian astronomer and in which language did the exchanges between him and his questioners take place; was this the translation of a text in the proper sense of the term – and if so, which text, since the expression *Zīj al-Sindhind* may be purely generic – or was it simply a transmission of results in the form of tables? And so on . . . .
4 For al-Fazārī, see Pingree (1970). For Yaʿqūb b. Ṭāriq, see Pingree (1968: 97–125).
5 Latin text edited by Suter (1914); translation with commentary by Neugebauer (1962a).
6 The four texts by Euclid were translated by Ḥunayn and Thābit; the three texts by Theodosius were translated by Qusṭā; the two texts by Autolycus were translated by Isḥāq and by Qusṭā respectively; the text by Aristarchus and that of Hypsicles were translated by Qusṭā; and the book by Menelaus was translated by Ḥunayn or his son Isḥāq.
7 For the transmission of the *Almagest* in Arabic, see Kunitzsch (1974).
8 Ibn al-Ṣalāḥ, Arabic text, p. 155, ll. 12–18.
9 Only one part of these two versions has been published: the star catalogue of the *Almagest*. See Ptolemy, *Almagest: Der Sternkatalog*, edited and translated into German by Kunitzsch.
10 See Leiden, or. ms. 180, fol. 1a.
11 al-Masʿūdī, *Kitāb al-Tanbīh*, pp. 15–16.
12 In his book *Hay'at al-aflāk*, ms. Oxford, Bodl., Seld. 3144.
13 See al-Kindī for the edition of the text and Rosenthal for his analysis.
14 See al-Farghānī.

15 We no longer possess his commentary, which numbered 200 folios.
16 Thābit Ibn Qurra, treatises 1 and 2. For Qusṭā's text see note 12.
17 The Arabic manuscript Escurial 927 carries the explicit title 'Table verified according to the observations of al-Ma'mūn', but the text contains many elements dating from after the ninth century; see the analysis by Vernet (1956) and Kennedy (1956: 145–7).
18 They are regrouped in tabular form in al-Hāshimī, pp. 225–6.
19 Cited in Thābit, treatise 2, p. 22, lines 4–5, and in al-Farghānī, pp. 50–3.
20 The Arabic text of this treatise can be found in Thābit, pp. 26–67; see the introduction, pp. XLVI–LXXV, and the complementary notes, pp. 189–215, where the arguments condensed here are detailed.
21 Thābit, treatise 3, p. 61.
22 Thābit, treatise 3, p. 49.
23 Recalculated results for the period (year 830) are given in parentheses.
24 For details of the reasoning see Thābit, pp. LX–LXIII.
25 See the commentary on this point in Neugebauer (1962b: 274–5).
26 The contents of this manuscript have been analysed in detail by Debarnot (1987).
27 See Kennedy (1965, 1968), reprinted in Kennedy (1983: 151–63).
28 See the detailed exposition in Thābit, pp. XXVI–XXX.
29 His works on astronomy preserved in Arabic have been edited and annotated (see Thābit); the account that follows is summarized from that study.
30 Treatise entitled 'Ralentissement et accélération du mouvement apparent sur l'écliptique selon l'endroit où ce mouvement se produit sur l'excentrique'; cf. Thābit, pp. LXXVI–LXXIX, 69–82, 216–21.
31 Ptolemy, *Almagest*, Heiberg, vol. I, p. 220; Toomer (1984), p. 246.
32 Treatise entitled 'Clarification d'une méthode rapportée par Ptolémée, à l'aide de laquelle ceux qui l'avaient précédé avaient déterminé les divers mouvements circulaires de la lune, qui sont des mouvements uniformes' or 'Le mouvement des deux luminaires'; cf. Thābit, pp. LXXX–XCII, 84–92, 222–9.
33 Ptolemy, *Almagest*, Heiberg, vol. I, pp. 272–5, and Toomer (1984), pp. 176–8.
34 See Thābit, pp. XCIII–CXVII, 94–116, 230–59, for details of the description summarized here in an attempt to reconstruct the reasoning of the author.
35 For the explanation of this hypothesis see Thābit, pp. CXIII–CXV.
36 Ptolemy, *Almagest*, Heiberg, Vol. I, p. 524; Toomer (1984), p. 308.
37 The full name of this author is Abū ʿAbd Allāh Muḥammad b. Jarīr b Sinān al-Battānī al-Ṣābī al-Ḥarrānī; see the bibliography under al-Battānī.
38 For these various observations see al-Battānī (translation and commentary, vol. 1; Arabic text, vol. 3; tables, vol. 2): the obliquity of the ecliptic, translation p. 12, commentary pp. 157–62, Arabic p. 18, l. 14; apogee of the solar orb, translation p. 72, Arabic p. 107, l. 23 to p. 108, l. 7; tropical year, translation p. 42, commentary pp. 210–11, Arabic p. 63, l. 22 to p. 64, l. 1; precession, translation p. 128, Arabic p. 192, ll. 1–5; catalogue of fixed stars, translation pp. 144–86, Arabic pp. 245–79; apparent diameters of the sun and the moon, translation p. 58, commentary pp. 236–7, Arabic p. 88, ll. 3–15.
39 This topic is discussed further in Morelon (1994), under the following areas: the link between theory and observation, the 'mathematization' of astronomy, and the link between 'mathematical' astronomy and 'physical' astronomy.

40 See al-Bīrūnī, *al-Qānūn al-Masʿūdī*, pp. 630–2. Al-Bīrūnī (ibid., p. 1312) also refers to a *Book on the Sizes and the Distances of Heavenly Bodies* by the same author.

41 Al-Kharaqī al-Thābitī, an author from the twelfth century, refers to al-Khāzin while mentioning similar work by Ibn al-Haytham, in the introduction to his book of cosmology: *Muntahā-l-idrāk fī taqāsīm al-aflāk* ('The farthest point of knowledge for the divisions of celestial spheres') – manuscript: Paris, B.N., Ar. 2499.

42 See al-Ṣūfī, ʿAbd al-Raḥmān.

43 See Ibn Yūnus for the edition in French translation of the first chapters of this work.

44 The Paris manuscript, B.N., Ar. 2494, is very incomplete; it was studied by Carra de Vaux (1892). The use by this author of a particular term sparked a controversy, begun by L. A. M. Sédillot, concerning the discovery by Abū al-Wafā' of the motion of lunar variation, by showing that this was not what was referred to in the text.

45 See al-Bīrūnī *al-Qānūn al-Masʿūdī*, pp. 640–77.

46 al-Bīrūnī *al-Qānūn al-Masʿūdī*, pp. 42–53; see Pines (1956).

47 al-Bīrūnī *al-Qānūn al-Masʿūdī*, pp. 950–65.

48 See Figure 2.3 and the discussion of it, including a description of the different methods. Note that al-Bīrūnī was obviously unaware of Thābit b. Qurra's method, described above, which uses all the parameters cited in a more complete way than that of Ḥabash.

49 For the complexity of the interpolation methods employed by al-Bīrūnī for the use of the tables, see Rashed (1991).

3

Arabic planetary theories after the eleventh century AD

GEORGE SALIBA

This chapter takes its starting date as the eleventh century AD for several reasons. First, one can argue that it was in the eleventh century that Arabic astronomy was finally 'acclimatized' within the Islamic environment and from then on it began to be coloured with whatever prerequisites that environment demanded. From this perspective several works began to be characterized by original production, and were no longer mere repetitions of problems that were discussed in the Greek tradition. Figures such as Abū Sahl al-Kūhī, Abū al-Wafā' al-Būzjānī, Bīrūnī, Manṣūr ibn Naṣr ibn 'Irāq, etc., who lived just around the turn of the previous century, were setting the grounds for this new production in astronomical research. This work could still be considered as a continuation of that of Ḥabash al-Ḥāsib, Thābit Ibn Qurra, Khwārizmī and others of the previous ninth century.

Second, the eleventh century witnessed a series of works all characterized by a genuine interest in the philosophical basis of Greek astronomy. As a result, these works led to a new school of writers on astronomical subjects whose main concern was to point out the problems that were inherent in the Greek astronomical system. One should recall the works of Ibn al-Haytham in his *Shukūk*, Abū 'Ubayd al-Jūzjānī in his *Tarkīb al-Aflāk* and the anonymous Spanish astronomer in his *Istidrāk*. The problems raised by these astronomers were later taken up by 'Urḍī, Ṭūsī, Quṭb al-Dīn al-Shīrāzī and Ibn al-Shāṭir, among others. The last four astronomers have been referred to in the literature as the 'School of Marāgha', mainly because of the association of the first three with the observatory built by the Ilkhānid monarch Hūlāgū in the city of Marāgha in northwest modern Iran in AD 1259. If one were to take their works alone, one could show that within the thirteenth century, when the first three lived, there occurred a real revolution in astronomical research and a definite change in attitude towards astronomical presuppositions. This tradition, started in the eleventh century, reached a sophisticated maturity during the thirteenth century, and climaxed with the works of Ibn al-Shāṭir in the fourteenth, but lingered on well into the fifteenth and sixteenth centuries if one takes into consideration the works of a student of Ulugh Beg, ʿAlāʾ al-Dīn al-Qushjī (d. 1474), and *al-Hayʿa al-Manṣūrīya* of Manṣūr ibn Muḥammad al-Dashtāghī (1542).

By defining this genre of writing as the main motivating direction of astronomical research after the eleventh century, one has to accept the fact that from this perspective the works of someone like Jamshīd b. Giyāth al-Dīn al-Kāshī in the fifteenth century, especially in his *zīj-i Khāqānī*, represent a return to the older tradition as represented by such works as those of Khwārizmī and Bīrūnī, where the main concern is mathematical and computational and not at all concerned with theory of science and philosophy.

Other important figures of the fifteenth and the sixteenth centuries such as Abū ʿAlī al-Birjandī seem to have taken it upon themselves to write commentaries on earlier works, mainly the works of Ṭūsī, and have produced very little original material that could be classified either way. Works of people like Jaghmīnī and Qushjī were elementary indeed, and of the two authors only Qushjī seems to have understood the originality of the Marāgha School, as we shall see later.

In what follows, it will be shown that the Marāgha School astronomers not only produced original mathematical astronomy, but also left their imprint on later astronomical research, mainly in the Latin West, and may perhaps have laid the foundation for Copernican astronomy itself.

This chapter will introduce the problems that were the main concern of the New School, discuss their various solutions by various authors and conclude with an evaluation of their possible relationship to Copernican astronomy.

THE CONTROVERSIAL PROBLEMS

The most outstanding problems in the Ptolemaic astronomy as expounded in the *Almagest* and the *Planetary Hypotheses* were seen to be (1) the problem of prosneusis, (2) the problem of inclination and deviation of the spheres of Mercury and Venus, (3) the problem of the equant in the model of the superior planets, and (4) the problem of the consistency of the planetary distances as they were perceived to be within nested shells.[1] But this list could be enlarged depending on the seriousness with which one took the various lists compiled during the later centuries such as the list compiled by al-Akhawayn sometime during the latter years of the fifteenth century and the early years of the sixteenth. The list of problems – called *ishkālāt* – that was the subject of the treatise of al-Akhawayn will be reproduced here as an example of the kind of comprehensive coverage that these problems received.

According to al-Akhawayn, the general *ishkālāt* in astronomy were classified as follows:

Ishkāl 1 Speeding up, slowing down and average motion are inappropriate motions in the heavens and need a special solution. Such an *ishkāl* is easily solved, in the case of the sun for example, by the adoption of either the eccentric model or the epicyclic one.
Ishkāl 2 Planetary bodies appear sometimes to be larger than at other times. One such problem involves the explanation of the reason why there is a total solar eclipse when the sun is in the middle of its slow motion sector, while there is only an annular eclipse when it is on the opposite side in its fast moving sector, knowing that in both cases the sun is covered by the same-size body, i.e. the moon. The solution would obviously follow from the kind of model one would adopt for the solution of the first *ishkāl*, i.e. if one were to take the eccentric model then it is easy to see why the sun would look smaller when it is on the far side of the eccentric than when it is on the near side.
Ishkāl 3 Stationary, retrograde and forward motion of the planets are phenomena which seem to deny the assumed regular motions of the planets. Here again, the adoption of the epicyclic model could explain the three phenomena without any inconsistencies with respect to the general principles of uniform circular motions as being the only appropriate motions for the heavenly bodies.

All of the three problems mentioned above could have been solved by relying on principles already enunciated by Ptolemy in the *Almagest* and without recourse to any conditions contrary to the general principles.

Ishkāl 4 Uniform motion is measured around a point which is not the centre of the body that generates this motion. This is the general problem of the *equant* directly applicable to the model of all the planets, but in a special application includes the motion of the moon which is uniform with respect to the earth rather than the centre of its deferent.

It is this problem which generated a great amount of research, for it implied a contradiction in the Ptolemaic theory between the physical assumptions and mathematical ones. In what follows, we shall discuss in great detail the various solutions that were applied to this problem.

Ishkāl 5 The uniform motion is around a point, in spite of the fact that the moving body draws sometimes nearer to that point than at other times. The solution of this problem required the development of a mathematical theorem – now called the 'Ṭūsī Couple' – that became an integral part of most astronomical research after its discovery.

Ishkāl 6 A problem which results from requiring in the motion of a sphere that its diameter be slanted away from the centre of the carrying sphere that moves it. This *ishkāl* will be made clear in the discussion of the prosneusis problem mentioned above. The most notable illustration of this problem is in the model of the moon as it was perceived by Ptolemy.

Ishkāl 7 The problem of incomplete circular motions being part of the motion of the heavenly bodies. The best illustration of this *ishkāl* is the assumed motion of the epicyclic diameter in the Ptolemaic latitude theory of the lower planets. This is the same problem referred to above as the problem of the inclination and the deviation.

PTOLEMAIC THEORY OF THE MOTION OF THE PLANETS IN LONGITUDE

In order to begin to appreciate the seriousness of these problems, and the nature of the criticism as well as the solutions applied to them, it is best that we review very quickly the Ptolemaic theory for the motion of the planets.

The motion of the sun

The motion of the sun is described in *Almagest* III by Ptolemy in either one of two models, namely the eccentric model or the epicyclic model. The equality of these models had already been shown by Apollonius (Neugebauer 1959; re-edited in 1983), and was incorporated by Ptolemy as an integral part of the *Almagest* terminology. In Figure 3.1, the observer is supposed to be at point O, the centre of the ecliptic. The sun could either move along the eccentric ABCD with uniform speed so that it will appear to the observer on the earth to be moving fast when in the lower half of the eccentric BCD and to slow down when in the upper half DAB – of course it would seem to be at its slowest point when it is at the apogee A – or the whole motion could equally be described if one assumed the sun to be moving on an epicycle with centre E in the direction contrary to the order of the signs (i.e. in the direction indicated by the arrow, which is also called 'forward', while the opposite motion is called 'to the rear' or 'backward'),[2] while the centre E itself is moved on a concentric circle (the broken one in the figure) in an equal and contrary motion to that of the epicycle. The resulting motion will obviously be the same as the one anticipated by the eccentric model. The equivalence of the two models, and thus the resulting motions, is best described in *Almagest* III, 3.

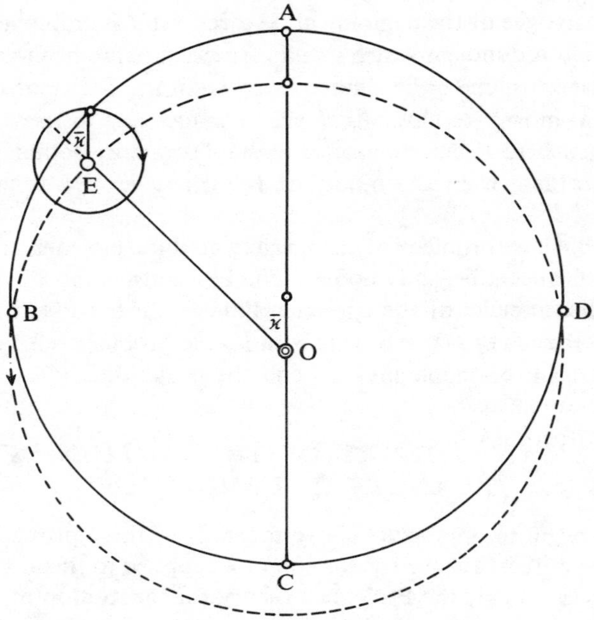

Figure 3.1

Although the motion of the sun could appear to the novice as implying a contradiction with the original principles of uniform motion, the Ptolemaic double-model explanation was found to be perfectly satisfactory, in that all the motions were indeed taking place around the centre of a sphere which was different from the position of the observer, in the first case, but could also be explained as identical with it in the epicyclic model and still be a combination of motions all uniform and all around centres of spheres – hence satisfying the primary principles.

The motion of the moon

In the case of the moon, however, the situation is quite different, for the motion is much more complex. In *Almagest* IV, Ptolemy first tried the Hipparchian model, which was essentially an extension of that of the sun, but soon realized that it did not describe all the motions properly. After a lengthy description that seems like a change of mind, he finally adopted a rather complicated model, in *Almagest* V, to account for the lunar movements.[3] In Figure 3.2 the observer is supposed to be at point O, the centre of the ecliptic. An encompassing sphere, called the sphere of the nodes, would move uniformly forward around the centre of the universe carrying with it the apogee of the deferent A. The deferent itself moves in the opposite direction around its own centre F, such that angles $\bar{S}$OA and $\bar{S}$OC are equal and opposite. This is obviously *ishkāl* 4 of al-Akhawayn's list mentioned above, for we have a deferent moving in a non-uniform motion around its centre F but moving uniformly around another point O. The epicycle carrying the moon is supposed in this model to be at point C, moving backward. The moon itself moves with its epicycle in the forward direction but measuring the forward angle from the extension of the line that connects point N – a point diametrically opposite to F from the centre of the universe called the prosneusis point – to the centre of the epicycle C, and extended to the mean apogee H at the epicycle's circumference. Since point N is always in motion, to remain opposite to the moving point F, it was thought to be a non-stable point for one to begin the description of motion from it, thus giving rise to the *ishkāl* of prosneusis mentioned above.

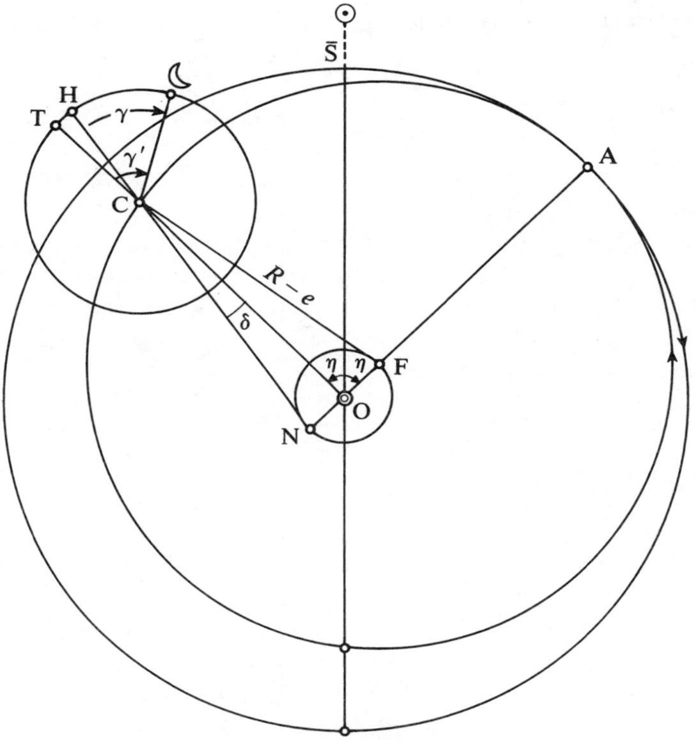

Figure 3.2

To summarize therefore, in the Ptolemaic lunar model one had to accept inconsistencies that gave rise to serious problems when one thought of the heavenly spheres as actual solid spheres. For it was impossible to move these spheres uniformly with respect to a centre other than their own, or with respect to a moving point that is not a fixed reference point for equal motion. All the criticism and the reformulations of Ptolemaic astronomy were really centred around these two points.

The motion of the upper planets (Saturn, Jupiter, Mars) and Venus

The motion of the upper planets as described by Ptolemy was relatively simpler than that of the moon, and involved the following elements. In Figure 3.3 the observer is taken to be at point O. The centre of the sphere that carries the epicycle of the planet – i.e. the deferent – in a backward motion is at point T. The epicycle of the planet itself moves backward around its centre, point C. The planet, P, is carried by the epicycle in its backward motion at a uniform speed measured by the angle of the anomaly. The reference point for the measurement of the anomaly, however, is located along the extension of the line joining the centre of the epicycle C to a point

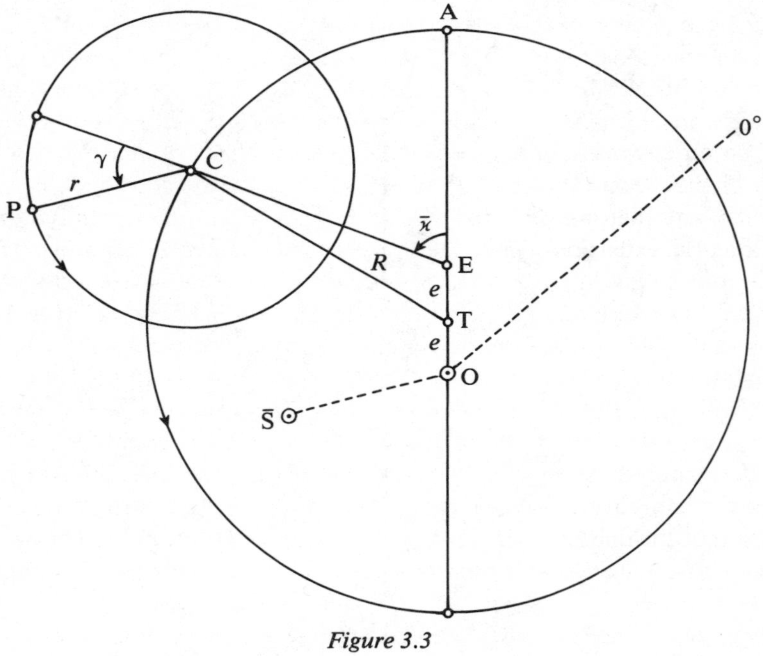

Figure 3.3

E, which is located along the line of centres OTA and is at such a distance from O that T bisects the line OE.

The difficulty in this model is in the motion of the deferent. For, as it was described by Ptolemy, the deferent carried the epicycle in a backward motion but did not do so in such a way that point C, the centre of the epicycle, would describe equal arcs in equal times around the centre of the deferent T, but rather around point E, the so-called *equant*. In essence then, the deferent, assumed by Ptolemy to be a physical sphere in the *Planetary Hypotheses*, is here being forced to move uniformly around a point different from its own centre. Put differently, the condition seems to require a sphere to move uniformly around an axis that did not pass through the centre of that sphere, which is an impossibility (*muḥāl*).

The motion of Mercury

Because Mercury is difficult to observe on account of its proximity to the sun and on account of its relatively fast motion, the Ptolemaic model for this planet was conceived to involve very complicated motions, which could not be included in the models described above. Moreover, this planet was unlike any other planet, in that Mercury was thought to have two perigees instead of one like all the others. These perigees were supposed to be symmetrically located with respect to the line of centres at angles equal to 120° from the apogee.

For an observer at point O, the centre of the universe, the motion of Mercury could be described as resulting from the following motions.[4] Let there be an encompassing sphere, analogous to the one that carried the nodes of the moon (Figure 3.4), and let it move forward around centre B in such a way that it carries the deferent's apogee with it. Let that apogee fall along the extension of line BG. Now let the deferent itself move backward around its own centre G, thereby carrying the epicycle's centre to point C, and making angle AEC equal to angle ABG. The epicycle itself moves backward around centre C, thus moving the planet M with the anomaly motion, which is measured from the extension of line EC. This model will allow the centre of the epicycle C to come close to the earth – i.e. to reach the perigee – twice per revolution, namely when ABG = 120° and 240° approximately. In both of these situations line GC will pass through E. Moreover, since the mean anomaly is always measured from the extension of the line EC, point E therefore will always act as the *equant* for the planet Mercury in an analogous fashion to the *equant* of the upper planets.

It is obvious that this model of Mercury suffers from difficulties similar to the ones encountered so far in connection with the models of the moon

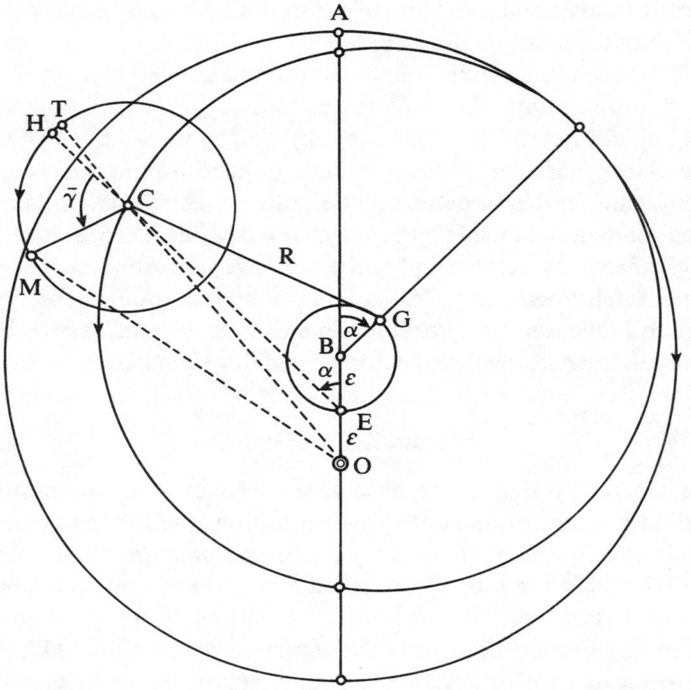

Figure 3.4

and the upper planets. In the first instance, the crank-like mechanism suggested here, with which the deferent is moved in one direction around a centre different from its own, while it moves on its own centre in the opposite direction, is very similar to the type of mechanism used in connection with the moon. The main difference between the two models is that in the case of the moon the mean anomaly was measured from a line that connected the centre of the epicycle with a moving point called the prosneusis point N, while here, in the case of Mercury, the analogous point is taken to be the midpoint of line OB, and thus a fixed *equant*, like that of the upper planets. In both models though, the deferent was supposed to move around its own centre at a non-uniform speed, while its uniform motion is supposed to occur around another centre, the centre of the universe O, in the case of the moon, and around the *equant* E in the case of Mercury.

It is not surprising therefore to find that the objections that were raised against Ptolemy's lunar model – especially in regard to the prosneusis point – and the model of the upper planets, in connection with the *equant* problem in particular, were also raised against the Ptolemaic model for

Mercury, for it seemed to have had the disadvantages of both of the other earlier models.

THE MOTION OF THE PLANETS IN LATITUDE

The previous description of the Ptolemaic models assumed that their motion in latitude is either negligible or that if it existed it did not affect the motion in longitude, which is not true. The facts are such that the planets are rarely seen to coincide with the plane of the ecliptic where the longitude motion is really measured, and that the latitude component is sometimes considerable and has to be taken into consideration. But in a typical Ptolemaic approach, this latitude component is thought of as an adjustment to the longitude and was left to be described in a separate section by itself.

The models accounting for the latitude theory were described in the *Almagest* in three separate models, i.e. one for the moon, one for the upper planets Saturn, Jupiter and Mars, and one for the inferior planets Venus and Mercury. This order also represents their order of complexity.

The latitude of the moon

For the moon, the model is rather simple, on account of the fact that the lunar orbit does pass through the earth, and thus for a geocentric observer the calculation of the lunar latitude is not too difficult. In fact, since the lunar orbit is at a fixed angle with respect to the ecliptic, and since the observer is at the centre of the ecliptic, the computation of the lunar latitude is very similar to the computation of the solar declination with respect to the celestial equator.

But because the lunar orbit is inclined at a fixed angle with respect to the ecliptic, the angle being about 5°, this means that the maximum latitude of the moon will also reach around 5° which it does according to observations. On the other hand, the observations have also shown that the moon's maximum latitude is not reached at any specific position on the ecliptic, but that it rather 'moves' around. This, coupled with the fact that solar eclipses also happen at various points on the ecliptic, meant that the line of intersection between the lunar orbit and the ecliptic, the nodal line, was also moving. That could only happen if one assumes the existence of a sphere that carries the whole configuration around so that the sphere whose cincture (*minṭaqa*) is the lunar orbit itself, or in Ptolemaic terms the deferent, is also moved by this assumed sphere. This last sphere is called the sphere of the parecliptic (*mumaththal*), or that of the nodes (*jawzahar*), and was supposed to move at about three minutes per day in the direction opposite to the order of the signs (i.e. forward in the sense used above).

To summarize, the lunar model, in its complete form, included the following spheres: (1) a sphere called the sphere of the parecliptic (*mumaththal*) which moves the nodes and everything else in a forward direction; (2) an inclined sphere (*al-mā'il*), which also moves in the same forward direction and also accounts for the lunar latitude, and whose *minṭaqa* has the same plane as the deferent; (3) the sphere of the deferent, which moves in the backward direction in its own motion; and finally (4) the sphere of the epicycle which carries the body of the moon itself and is itself carried by the deferent sphere.

We have seen above the two objections raised against this model in terms of its performance when describing the lunar motion in longitude. These objections do not apply to the motion in latitude, for in this case all the motions that are needed to account for the motion in latitude are performed by spheres moving around their own centre, in this case the centre of the universe as well.

Latitude of the upper planets

For the upper planets, the situation is more complex, just because the actual orbit of these planets does not pass through the earth, the centre of the universe, but rather through the sun. For an observer on the earth, the transformation of the motion in latitude to geocentric co-ordinates involves more complicated procedures than the ones used in the lunar latitude which was just described.

To use the analogy of the lunar model, the upper planets' deferents (Figure 3.5) were also thought to be inclined, at a fixed inclination i_1 with respect to the ecliptic. The line along which the deferent intersects the ecliptic plane is also called the line of nodes, where the point at which the

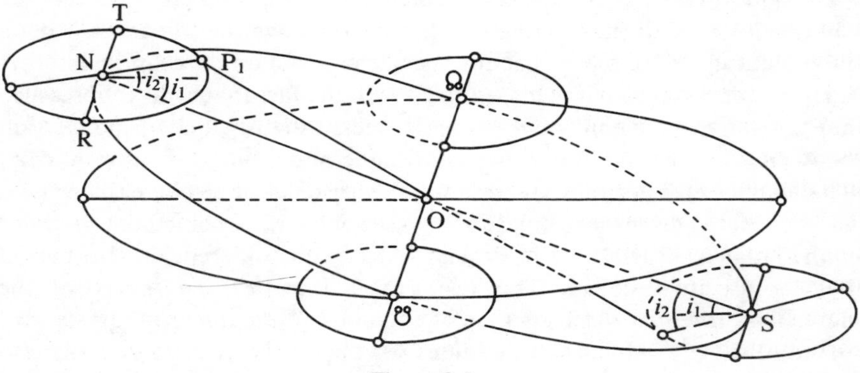

Figure 3.5

epicycle ascends from the south to the north is called the ascending node (or the head) and the diametrically opposite one is called the descending node (or the tail). The line issuing from the observer's point, perpendicular to the line of nodes, defines the top of the deferent N when it intersects the northern circumference of the deferent, and the bottom of the deferent S when it intersects the southerly one. This line is not in general the same as the apsidal line for it only passes through the centre of the ecliptic O, not through the centre of the deferent and the *equant* as the apsidal line does.

But unlike the lunar model, the epicycles of the upper planets do not lie in the plane of the deferent, as they were assumed to do when the component of the longitude was being considered on its own. Instead the plane of the epicycles themselves becomes inclined with respect to the plane of the ecliptic by an angle i_2 when the epicycle moves from the nodes. This last angle of inclination (called a deviation) reaches a maximum northerly inclination when the epicyclic centre is at the top of the deferent. The same deviation reaches a maximum southerly value, a value larger than the northerly one in absolute terms, when the epicycle is at the bottom of the deferent. This situation occurs because the portion of the deferent that is to the north of the ecliptic is larger than the southerly portion, which implies that the bottom part of the deferent is closer to the observer, and thus subtends a larger angle.

But when the centre of the epicycle is along the line of nodes, the plane of the epicycle is supposed to go back to lie in the plane of the ecliptic. Both angles of latitude, that of the inclined deferent and that of the deviation of the epicycle, will be equal to zero.

In effect, therefore, the plane of the epicycle seems to undergo a see-saw motion about an axis, RNT, which is perpendicular to the line joining the real apogee to the real perigee of the epicycle, and is always parallel to the plane of the ecliptic. This result is in itself an awkward one, for it involves a see-saw kind of motion in a portion of the heavens that was supposed to allow only complete circular motions. To account for it, Ptolemy suggested in *Almagest* XIII, 2, that small circles be attached to the tip of the see-sawing diameter P_1 of the epicycle such that the radius of the small circle would be equal to the maximum angle of deviation and the plane of the small circle would be perpendicular to the plane of the deferent from which the deviation is measured. With the insertion of such small circles, the line connecting the real apogee of the epicycle to the real perigee will no longer see-saw, but would have its tip moving along these small circles. But here again, since the time the epicycle takes to move along the larger northerly portion of the deferent is in general greater than the time it takes to cover the southern portion of the same deferent, and since the period along the small circle for the tip of the epicyclic diameter is equal to the period of the epicyclic motion along the deferent, the motion of the tip of the epicyclic diameter along the small circle is not uniform, and like the epicyclic centre has to be moving uniformly along its own *equant*.

This must have proved embarrassing to Ptolemy, since he begs the reader not to consider this arrangement as over-complicated, 'for it is not appropriate to compare human [constructions] with divine, nor to form one's beliefs about such great things on the basis of very dissimilar analogies' (*Almagest* XIII, 2). He goes on to say that he accepted that only because it yields a simple representation of the motions of the heavens.

It was this specific point that was objected to above (*ishkāl* 7) and was thought to be a violation of the accepted premises of the discipline of astronomy. As we shall see, the invention of what later came to be called the 'Ṭūsī Couple' would allow for the solution of this problem. In fact there is enough evidence to support the claim that the 'Couple' was specifically invented by Ṭūsī to solve this inconsistency, and was only applied later to produce a linear motion as a combination of two circular motions. Moreover, the 'Couple', being composed of two circular motions, allows the tip of the epicyclic diameter to oscillate back and forth in the same plane, without violating the circular motion principle, and thus allows the longitude component not to be disturbed.

Latitude of the lower planets

The Ptolemaic model for the lower planets is still more complicated, and assumes, in the case of Venus for example, that the inclination of the deferent plane is not fixed, but that it oscillates back and forth; that the deviation of the epicycle, like that of the upper planets, also undergoes a see-sawing motion about an axis that passes through the centre of the ecliptic; and finally, that the epicyclic plane also see-saws about another axis perpendicular to the first, and thus undergoes two see-sawing motions of its own. In the case of Mercury, all these motions are also required except that they are taken to be in the opposite direction.

To illustrate the case of the model for Venus, we take (Figure 3.6) the eccentric deferent to be inclined with respect to the plane of the ecliptic at an angle i_0, and let it intersect the plane of the ecliptic along the nodal line that passes through the point of the observer at the centre of the ecliptic. Unlike the case of the upper planets, in this model the apsidal line itself is perpendicular to the nodal line. But the inclination of the deferent is no longer fixed as it used to be in the case of the upper planets and the moon. In this case, the inclination of the deferent i_0 is coupled with the motion of the epicycle so that the plane of the deferent would coincide with the plane

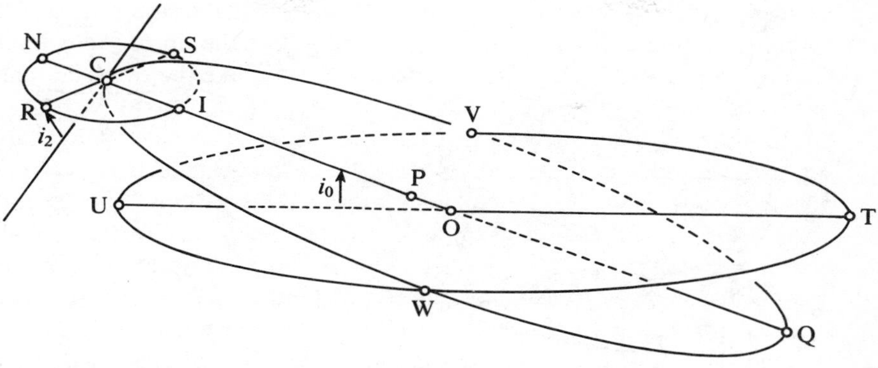

Figure 3.6

of the ecliptic when the epicycle is along the ascending node. As the epicycle begins to move towards the north, the inclination of the deferent begins to increase also in the northerly direction to reach a maximum inclination i_0 when the epicycle reaches the apogee. The inclination will then decrease as the epicycle moves from the apogee to the descending node, to get back to the plane of the ecliptic as shown in Figure 3.7. But as the epicycle moves from the descending node towards the perigee, the inclination of the deferent will increase again in the northerly direction, as shown in Figure 3.8, to reach another maximum value i_0 when the epicycle reaches the perigee. As the epicycle goes back to the ascending node, the plane of the deferent goes back to its original position in the plane of the ecliptic as in Figure 3.7. This is the first see-sawing motion in the model for Venus.

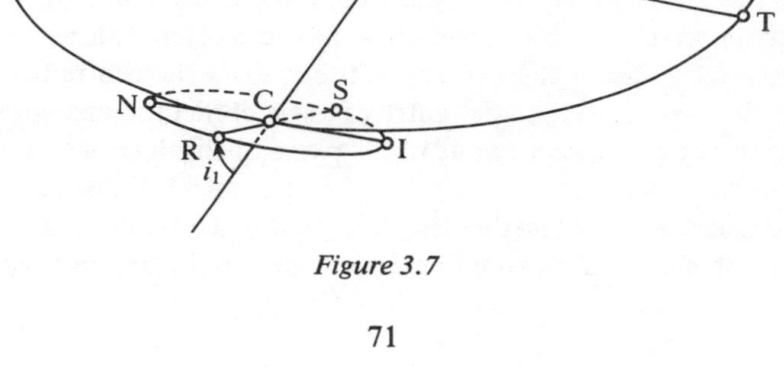

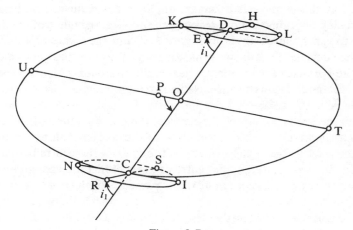

Figure 3.7

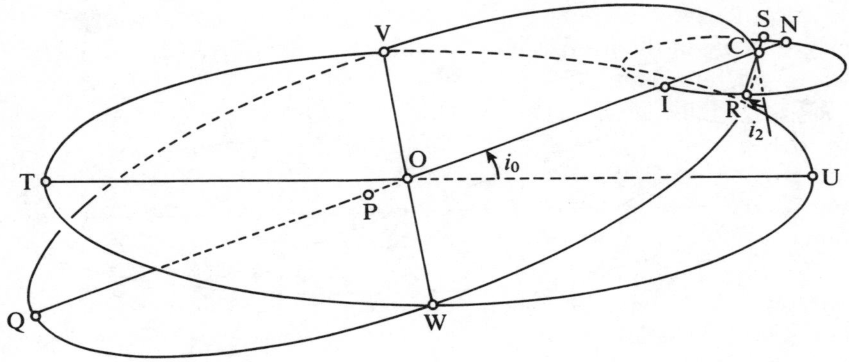

Figure 3.8

The second see-sawing motion is called a 'slant'. To explain it, Ptolemy assumed the plane of the deferent to coincide with the plane of the ecliptic when the epicycle is at the ascending node as in Figure 3.7. The line COD defined by the intersection of the plane of the ecliptic with the perpendicular plane produced by the line joining the apparent epicyclic apogee R, H and perigee S, E and the centre of the ecliptic is taken to be the first axis about which the slanting motion will take place. The line perpendicular to this last axis KDL, NCI, and which passes through the centre of the epicycle D, C (the mean diameter), is then taken to be the second axis about which the deviation motion will be described.

When the epicycle is at the ascending node, the epicyclic mean diameter KDL lies in the plane of the ecliptic, and thus has no latitude component for that slant. But at that position the plane of the epicycle is 'deviated' in such a way that it reaches its maximum deviation i_1 just at that position. As the epicycle begins to move towards the apogee, the plane of the deferent moves towards the north as in Figure 3.6, and the plane of the epicycle begins to 'deviate' back from its maximum position to reach zero deviation when the epicycle reaches the apogee, while the slant which was equal to zero at the node begins to increase to reach a maximum slant i_2 when the epicycle is at the apogee.

When the epicycle reaches the apogee, the plane of the deferent will be at its maximum inclination i_0, the plane of the epicycle will be slanted such that the eastern side of the epicycle will be northerly at a maximum slant i_2, while the line connecting the centre of the ecliptic to the apparent perigee and apogee of the epicycle will lie in the plane of the deferent having a zero deviation.

As the epicycle moves to the descending node as in Figure 3.7, the plane of the deferent goes back to coincide with the plane of the ecliptic, while the epicyclic plane reaches its maximum deviation i_1 with the apogee of the epicycle being in the northerly direction, and at that position the epicycle will have a zero slant.

But when the epicycle moves to the perigee of the deferent as in Figure 3.8, the inclination of the deferent increases to bring the perigee of the deferent to a maximum northerly direction i_0. The plane of the epicycle is slanted in that position to bring the easterly side of the epicycle to its maximum northerly slant i_2, as it did when the epicycle was in the apogee of the deferent. Here again, the deviation of the epicycle will be zero.

In the case of Mercury, the inclination of the deferent, the slant and the deviation of the epicycle are all in the direction opposite to that of Venus. When the epicycle is at the nodes, the deviation is northerly for Mercury when it was southerly for Venus and vice versa. At the apogee the inclination of Mercury's deferent moves to its maximum southerly inclination when that of Venus moves to its maximum northerly direction. Similarly at the apogee the slant is southerly for Mercury where it was northerly for Venus.

If the deviation in the case of the upper planets had been an embarrassment for Ptolemy in terms of his having to use small circles to account for the motion of the deviation of the epicycles of the upper planets, the inclination, the deviation and the slant of the inferior planets, all of them requiring such small circles to allow them to see-saw about their various axes, are a triple embarrassment, and it should not be surprising that such models were considered to be inconsistent with the original premises of the discipline of astronomy. Here again, the 'Ṭūsī Couple' could be efficiently used to account for all those linear motions of the tips of the various axes as resulting from circular motions.

This then is a brief description of the main features of the Ptolemaic theory of latitude. And as we have seen, it was easy to find many faults with it in spite of its observational base and its ability to predict positions of specific planets at specific times. The main problem that permeates the whole theory at all levels is the one referred to above as *ishkāl* 7, and could be simply described as the problem of admitting oscillating linear motion among the heavenly motions which are supposed to be circular. Once those oscillating motions could be produced by circular motions, as in the case of the 'Ṭūsī Couple', the problem would then be reduced to adjusting the periods of motion so that the circles of the 'Couple' will themselves move at a uniform speed, which is not an easy matter.

THE REFORM OF THE PTOLEMAIC PLANETARY MODELS

It was said above that serious criticisms of the Ptolemaic models started, as far as we know, sometime during the eleventh century. In that century, two main lines of research seem to have been developed simultaneously; namely, the line of criticism that was limited to identifying the main defects of the Ptolemaic models, and that of finding alternative models to replace the defective Ptolemaic ones.

The first line of research, that devoted to criticism, was represented by Ibn al-Haytham (*c.* AD 965–1039) in his work *al-Shukūk ʿalā Baṭlamyūs*, and an anonymous astronomer whose work *al-Istidrāk ʿalā Baṭlamyūs* has not yet been located.[5] From Ibn al-Haytham's work, we know that the criticism was not only limited to Ptolemy's planetary models, but that it also included other works of Ptolemy such as the *Optics*. This means that the whole work was probably motivated by considerations that were much more general than the astronomical ones. One could argue that this genre of writing was in all likelihood in the same tradition as the work of the tenth-century physician Abū Bakr al-Rāzī (d. *c.* AD 925) who wrote a similar work against Galen (second century AD), which he called *al-Shukūk ʿalā Jālīnūs*. The astronomical contents of Ibn al-Haytham's work will be summarized in the following section. The work of the anonymous astronomer, on the other hand, seems to have been devoted to astronomical issues. For in his surviving treatise, whenever he reaches one of the difficult points in the Ptolemaic models that were noted above, he says that the point is rather difficult to accept and is explained in his work *al-Istidrāk*.

Contents of Ibn al-Haytham's *Shukūk*[6]

The book begins with an introduction in which Ibn al-Haytham sets down the principles that he intends to follow in his work. After admitting the excellence of Ptolemy's works, he goes on to say that, in this book, he will only mention those problems (*shukūk*) that cannot be explained away, and are in direct contradiction with the accepted original principles.

Apparent size of the sun

The rest of the book is divided into three main parts; each is devoted to the contradictory positions in one of the three works of Ptolemy, i.e. the *Almagest*, the *Planetary Hypotheses* and the *Optics*. In the first part, following the order of the *Almagest*, Ibn al-Haytham begins with the problem in Book I,3, which is the problem of the apparent size of the sun, as it appears bigger when the sun is closer to the horizon than when it is in the middle of the heaven. Here Ibn al-Haytham uses Ptolemy's own results from the *Optics* against his statement in the *Almagest*.

Directions from the centre of the world

In regard to chapter 5 of Book I of the *Almagest*, Ibn al-Haytham requires of Ptolemy more precision in his use of his own concepts, and objects to Ptolemy's use of the earth being 'higher' or 'lower' with respect to the centre of the world when there are no such directions with respect to the centre of the universe, since all directions from the centre are in the 'higher' direction. This kind of mistake (*ghalaṭ*), Ibn al-Haytham identifies as a mistake in conceptualization (*taṣawwur*) rather than a contradiction. Similarly, when Ptolemy uses the terms east and west to describe the position of the earth, he would be committing the same conceptual error.

The value of the chord of 1°

In the same vein, Ibn al-Haytham objects to Ptolemy's use of a value being bigger and smaller than another quantity, as a proof of its being equal to that quantity. He would have forgiven Ptolemy had he said at that point that the value of the chord of 1° was approximately equal to that quantity and that it differed from it by some small number rather than being at the same time greater than and smaller than that quantity.

The inclination of the ecliptic

Ibn al-Haytham also objects to Ptolemy's method for determining the inclination of the ecliptic, for he said that he observed the sun along the meridian circle and found the difference between the highest position of the sun at the summer solstice and its position at the winter solstice to be 47° and a value greater than $\frac{2}{3}^\circ$ but less than $\frac{1}{2}^\circ$ and $\frac{1}{4}^\circ$.

The reason he objects to that is that the solstices need not occur when the sun is crossing the meridian circle of that specific locality, and Ptolemy knew that. But he agreed to take the approximate value, when he ought to have explained how such a value could be determined with precision. Moreover, he also knew that the sun will never return to the same point on the meridian circle in an integral number of days in the years that follow. But in spite of that he still said that he observed the sun cross that point of the solstice year after year, which could not be true. Because several parameters are connected to this measurement, Ibn al-Haytham concludes that neither the solar year, nor the solstice point, nor the declination, nor the equinoctial points are known from Ptolemy's statements.

The proof that Ptolemy did not really determine these parameters is that modern astronomers have found them to be different. They found the declination to be different and the solar apogee to be moving when Ptolemy had determined that it was fixed.

The prosneusis point

The objection about the prosneusis point is the same as the one mentioned by al-Akhawayn under *ishkāl* 6. It derives from the Ptolemaic lunar model in which it was required that the mean epicyclic apogee be determined by the extension of a line joining the centre of the epicycle and the prosneusis point, which was itself defined as being diametrically opposite to the centre of the deferent with respect to the centre of the universe. Such an apogee is not only imaginary to Ibn al-Haytham, but it could not be a reference point from which one could measure motion. Ibn al-Haytham's real concern, however, is expressed in the following terms:

> The epicyclic diameter is an imaginary line, and imaginary lines have no perceptible motion that produces an existing entity in the world. Similarly, the plane of the ecliptic is also an imaginary plane, and imaginary planes do not exhibit an observable motion. And nothing moves in an observable motion, which produces an existing entity in this world, except the bodies that do [indeed] exist in this world.
>
> (*Shukūk*, p. 16)

Moreover, even if one were to accept the existence of such an imaginary line, and thus the existence of a mean apogee defined by it, one still could not explain according to the accepted principles the motion of this line, for it seems to oscillate back and forth producing positive and negative angles within a period of half a lunar month. None of these motions seemed to have been produced by full revolutions of spheres moving at uniform speed as they ought to be. Ibn al-Haytham concludes this section with a tirade of criticisms, exhausting all possible excuses for Ptolemy, and finally rejecting the existence of such lines or bodies that could move these lines in this manner. 'If such bodies were then found to be impossible [to exist] [*muḥāl*] then it is impossible for the diameter of the epicycle to move in such a way that it would be in line with the assumed prosneusis point' (p. 19).

The later attempts by other astronomers to modify the Ptolemaic lunar model included in one way or another a statement about the problem of the prosneusis point and an effort to avoid using it.

Limits of eclipses

In this section Ibn al-Haytham objects to Ptolemy's apparent use of an approximate method when he determined the limits of eclipses. The brunt of the objection centres around Ptolemy's use of an arc – equal to the sum of the solar and lunar radii – perpendicular to the orbit of the moon rather than the ecliptic, as Ibn al-Haytham would have preferred. Ibn al-Haytham then argues that Ptolemy's choice of that procedure does not allow him to compute the beginning of the eclipse, nor its middle nor its end, and 'his assumption that this arc could determine the limits of an eclipse in longitude and in latitude is an obvious error without any doubt'.[7]

The equant problem

This section is by far the most important of Ibn al-Haytham's criticisms of Ptolemaic astronomy. It deals with the problem referred to above as *ishkāl* 4, which simply states that a sphere could not possibly move uniformly around an axis which does not pass through its centre as Ptolemy assumed. To construct his argument, however, Ibn al-Haytham started by showing that Ptolemy was quite aware that he was violating his own premises in regard to the *equant* problem.

Ibn al-Haytham begins by pointing to *Almagest* IX, 2, where it is clearly stated that the upper planets are supposed to move uniformly[8] just like the other planets that he had already described. This section is then contrasted with *Almagest* IX, 5, where Ptolemy clearly states that in the model for the upper planets 'We find, too, that the epicycle centre is carried on an eccentre which, though equal in size to the eccentre which produces the anomaly, is not described about the same centre as the latter'. (Toomer 1984: 443) Later on, in *Almagest* IX, 6, Ptolemy describes his model for the upper planets in more detail. It is there that Ptolemy defines the *equant* (to use the later medieval term) as simply the point around which the centre of the epicycle moves uniformly. Without any proof, Ptolemy also states in this chapter that the centre of the deferent lies midway between the centre of the universe and that of the *equant*.

To this Ibn al-Haytham says: 'What we have reported is the truth of what Ptolemy had established for the motion of the upper planets; and that is a notion [*maʿnan*] that necessitates a contradiction' (p. 26). The proof of the contradiction is then constructed as follows. Since Ptolemy accepted the principle of uniform motion, and since he had shown in the case of the sun that if any body moves uniformly around one point it must necessarily move non-uniformly around any other point, therefore, Ptolemy must have contradicted himself by stating that the centre of the epicycle moves uniformly around the *equant* for then it means that it does not move uniformly around the centre of its own deferent, which is impossible.

In the details of the response, Ibn al-Haytham states quite clearly that his objection is actually based on the fact that these motions are supposed to be motions of real bodies, not imaginary ones, 'since imaginary circles do not move by themselves in any perceptible motion' (p. 28). Moreover, Ibn al-Haytham makes the obvious remark that if a body is supposed to move uniformly around a point, it also means that the body must be always equidistant from that point. In effect, if the bodies that were described by Ptolemy were supposed to be real physical bodies, then a sphere could move uniformly only around an axis that passes through its centre.

Ibn al-Haytham then extends his criticism to the Ptolemaic model of Mercury, *Almagest* IX, 9, for it suffers from the same contradiction. He goes on to conclude this section by casting doubt about the method in which Ptolemy determined the eccentricities of the planets.

In order to clinch his argument, Ibn al-Haytham quotes Ptolemy, *Almagest* IX, 2, which proves that even Ptolemy himself had already admitted that he was using hypotheses that were contrary to the accepted principles (*khārija ʿan al-qiyās*). Since Ptolemy

> had already admitted that his assumption of motions along imaginary circles was contrary to [the accepted] principles, then it would be more so for imaginary lines to move around assumed points. And if the motion of the epicyclic diameter around the distant center [i.e. the equant] is also contrary to [the accepted] principles, and if the assumption of a body that moves this diameter around this center is also contrary to [the accepted] principles for it contradicts the premises [*al-uṣūl*], then the arrangement, which Ptolemy had organized for the motions of the five planets, is also contrary to [the accepted] principles. And it is impossible for the motion of the planets, which is perpetual, uniform, and unchanging to be contrary to [the accepted] principles. Nor should it be permissible to attribute a uniform, perpetual, and unchanging motion to anything other than correct principles, which are necessarily due to accepted assumptions that allow no doubt. Then it becomes clear, from all that we have shown so far, that the configuration, which Ptolemy had established for the motion of the five planets, is a false [*bāṭila*] configuration, and that the motions of these planets must have a correct configuration, which includes bodies moving in a uniform, perpetual, and continuous motion, without having to suffer any contradiction, or be blemished by any doubt. That configuration must be other than the one established by Ptolemy.
>
> (*Shukūk*, pp. 33–4)

Motion in latitude

Ibn al-Haytham's objections to the theory of latitude begin with a long quotation from *Almagest* XIII, 1, which treats the motion of the inferior planets in latitude. He then follows that with his own summary of Ptolemy's statement, and concludes by saying that

> This is an absurd impossibility [*muḥāl fāḥish*], in direct contradiction with his earlier statement about the heavenly motions – being continuous, uniform and perpetual – because this motion has to belong to a body that moves in this manner, since there is no perceptible motion except that which belongs to an existing body.
>
> (*Shukūk*, p. 36)

Moreover, since the motions of the inclined plane in which the deferent lies are contrary in direction, Ibn al-Haytham concludes that Ptolemy did indeed commit a great error in allowing the same body to have two different natures, thus signalling a possibility of change in the heavens which is contrary to the accepted principles.

Conclusion

The concluding section of the critique of the *Almagest* is a long reflective statement by Ibn al-Haytham on the reasons why Ptolemy said what he said. He admits that there are places where these contradictions might have occurred as a result of negligence from which no human is free. At these places, Ptolemy could be excused. But at the places where he intentionally falls into contradiction, he has no excuse whatsoever. To prove that Ptolemy did indeed intend to accept the contradictions, Ibn al-Haytham quotes the famous passage of the *Almagest* IX, 2, in which Ptolemy says that he was obliged to employ devices that were contrary to the accepted principles (*khārija ʿan al-qiyās*) and that he demonstrated his proof by using imaginary circles. Ibn al-Haytham then isolates the main problem with Ptolemy's configuration for the upper planets as being exactly that; i.e. that he had demonstrated the motion of these planets in reference to imaginary circles and lines. But once the existence of real bodies was assumed, the contradiction then became clear.

Nor would Ibn al-Haytham accept the statement of an apologist who would say that these configurations are all imaginary, and that they would not affect the behaviour of the real planets, because one need not assume a contradictory configuration to describe the motions of existent bodies. Nor could Ptolemy be excused for saying as he did (*Almagest* IX, 2) that he reached a correct description of the motion of the planets without being able to describe the method by which he reached those conclusions. He should rather have admitted first that the configuration that he was describing was not the real one, and that he had not yet come to understand the correct one. Only then would he be excused.

This section is followed by Ibn al-Haytham's summary of Ptolemy's models for the planets, a straightforward rendering of the models described in the *Almagest* (*Shukūk*, pp. 39–41). He concludes by saying that Ptolemy

> had gathered together all the motions that he could verify from his own observations and from the observations of those who have preceded him. Then he sought a configuration of real existing bodies that could exhibit such motions, but could not realize it. He then resorted to an imaginary configuration based on imaginary circles and lines, although some of these motions could possibly exist in real bodies. He resorted to this method simply because he could not devise another one. But if one imagines a line to be moving in a certain fashion according to his own imagination, it does not follow that there would be a line in the heavens similar to the one he had imagined moving in a similar motion. Nor is it true that if one imagined a circle in the heaven, and then imagined the planet to move along that circle, that the [real] planet would indeed move along that circle. Once that is accepted, then the configuration assumed by Ptolemy for the five planets is a false configuration [*hay'a bāṭila*], and he established it knowing that it was false for he could not devise anything else. But the motions of the [real] planets have a correct configuration in [real] existing bodies that Ptolemy did not comprehend, nor could he achieve. For it is not true that there should be a uniform, perceptible, and perpetual motion which does not have a correct configuration in existing bodies.
>
> (*Shukūk*, pp. 41–2)

Doubts concerning the Planetary Hypotheses

In his doubts engendered by the text of the *Planetary Hypotheses*, Ibn al-Haytham starts by enumerating the points of variation between that text and the text of the *Almagest*. He enumerates, for example, the number of motions attributed to the planets in the *Almagest*, which were found to be thirty-six, with the number of motions of those in the *Planetary Hypotheses*, which were found to be only twenty-six.

Then, in the description of the movements of the epicyclic spheres in the first book of the *Planetary Hypotheses*, Ibn al-Haytham finds Ptolemy's text wanting in that it did not include the 'small circles', referred to in the *Almagest*, that carried the epicycles in latitude, nor did he find any account of how the planet is supposed to be moved in latitude (*Shukūk*, pp. 43–4).

He then concludes that Ptolemy's statements in the first book of the *Planetary Hypotheses* not only describe an erroneous configuration (*hay'a fāsida*), but are in fact contrary to what is found by observation – in terms of the latitudinal motion of the planets – and to what is found in the *Almagest* itself.

In analysing the causes (*ʿilal*) for the planetary movements, Ptolemy proposes in the first book of the *Planetary Hypotheses* that each of these planets has two movements: 'One that is voluntary [*irādīya*], and the other that is by compulsion [*yuḍṭarru ilayha*]' (Goldstein 1967b: 26, ll. 16–18). In the second book of the *Planetary Hypotheses*, he goes on to say that 'each of these various movements that vary in quantity and kind must have a body that produces it by moving about some poles . . . in such a way that it undergoes no forcing or compelling from outside' (*Shukūk*, pp. 45–6).

Ibn al-Haytham finds these two statements to be contradictory, for how could a body be compelled to move in one case and in the other accept no compulsion from an outside agent?

Then he attacks Ptolemy for using the idea of spherical shells (*manshūrāt*) instead of spheres, saying that instead of solving the problems under discussion, the planetary shells suffer from the same disadvantages and introduce some of their own in addition.[9]

This brings Ibn al-Haytham back to the theory of latitude for the inferior planets, and the 'small circles' which were assumed, in the *Almagest*, to move the epicycles of the inferior planets along two perpendicular axes. These 'circles' are not mentioned in the *Planetary Hypotheses*. To this Ibn al-Haytham says:

> If one were to explain them in the same manner as before [i.e. in the case of the *Almagest*] then they would produce the same impossibilities, but if not, then one has to assume that Ptolemy had made a mistake in their regard [by not mentioning them here] or that he had made a mistake by mentioning them in the *Almagest*.
>
> (*Shukūk*, p. 58)

Similarly, Ptolemy did not mention in the *Planetary Hypotheses* the oscillating motion of the inclined planes of the inferior planets, as he did in the *Almagest*.

Moreover, Ptolemy dropped from consideration the motion of the prosneusis point while describing the spheres of the moon, which he had included among the movements of the moon in the *Almagest*.

At the end of the second book of the *Planetary Hypotheses*, Ptolemy seemed to have accepted the fact that planets could move by themselves, without the need for any bodies to move them. But to this Ibn al-Haytham says that it necessitates the existence of void in the heavens, by allowing the planet to 'empty' one place and 'fill' another. Then he goes on to object to the motion as being one of rolling (*tadaḥruj*), and he concludes by saying:

> If Ptolemy could find it permissible that a planet could move by itself, without the need for any body to move it, then all the spherical shells, which he had proposed for the planets, as well as the spheres themselves would be invalid.
> (*Shukūk*, p. 62)

This section is concluded like the one concerning the criticism of the *Almagest* by saying that Ptolemy

> either knew of the impossibilities that would result from the conditions that he assumed and established, or he did not know. If he had accepted them without knowing of the resulting impossibilities, then he would be incompetent in his craft, mislead in his attempt to imagine it and to devise configurations for it. And he would never be accused of that. But if he had established what he established while he knew the necessary results – which may be the case befitting him – with the reason being that he was obliged to do so for he could not devise a better solution, and [on top of that] he went ahead and knowingly fell into these contradictions, then he would have erred twice: once by establishing these notions that produce these impossibilities, and the second time by committing an error when he knew that it was an error.
>
> When all is considered, and to be fair, Ptolemy would have established a configuration for the planets that would have been free from all these impossibilities, and he would not have resorted to what he had established – with all the resulting grave impossibilities – nor would he have accepted that if he could produce something better.
>
> The truth that leaves no room for doubt is that there are correct configurations for the movements of the planets, which exist, are systematic, and entail none of these impossibilities and contradictions, but they are different from the ones established by Ptolemy. And Ptolemy could not comprehend them, nor could his imagination come to grips with them.
> (*Shukūk*, pp. 63–4)

As if that condemnation was not enough, Ibn al-Haytham then reminds the reader once more that Ptolemy neglected to mention in the *Planetary Hypotheses* the 'small circles' that he had used in the *Almagest* to account for the latitudinal motion. Ibn al-Haytham then guesses that Ptolemy did not do so either because he knew of the contradictions it would lead to if he adopted the model of the spherical shells, or because he wanted to avoid the cumbersome additional spheres if he adopted the model of the spheres. 'He then saw that it was better to remain silent about this (latitudinal) motion than to fall into these contradictions that it entailed' (*Shukūk*, p. 64).

Contents of *al-Istidrāk ʿalā Baṭlamyūs*

We know very little about the author of this work, or about the work itself, which has not yet been located. But whatever we can glean from the existing treatise by the same author, called *Kitāb al-Hay'a*, now kept at the Osmania University Library in Hyderabad (Deccan, India), seems to indicate that the author of *Kitāb al-Hay'a* had lived in Spain sometime during the eleventh century; he spoke of the famous Spanish astronomer al-Zarqāel (or al-Zarqāllu) (d. 1099) as his personal friend. The author also mentioned that in one of his works he described the instrument used for the observations that were conducted at Toledo, without specifying the year.

This author of *Kitāb al-Hay'a* tells us that he had found some of the statements of Ptolemy to have been objectionable, and states quite explicitly that he did not want to interject his objections in this elementary text which he was writing, for he had already devoted a special book to such objections which he called *al-Istidrāk ʿalā Baṭlamyūs* ('Recapitulation Regarding Ptolemy').

The manner in which he refers to this work is quite revealing of the subject matter that the book must have included. While speaking of the inaccuracy of the instrument which 'was found in Toledo, in al-Andalus', he says that 'the instrument was set in accordance with (the position) designated by the man who actually used it for observations, Abū Isḥāq Ibrāhīm b. Yaḥyā known as al-Zarqael (*sic*) as he himself had told me' (fol. 15^{v}). On folio 16^{r}, the author says that he had composed a book that he called *al-Istidrāk ʿalā Baṭlamyūs*. And while discussing the solar apogee, the author says: 'It was during the time of al-Ma'mūn at twenty degrees of Gemini and about two thirds of a degree. These matters ought to be better mentioned in the book *al-Istidrāk*' (fol. 41^{v}).

When the author discussed the motions of the moon, he had the following to say: 'I may object to Ptolemy in these matters in several ways, but I ought to mention that in what is simpler (?) than this book, and I will mention it in *al-Istidrāk* if God wills it' (fol. 48^{r}).

Finally, while discussing the planetary apogees, the author had the following to say: 'And Ptolemy found that the motion of the (origin of the) longitudes of these five planets takes place at the rate of one degree every one hundred years, while the more recent (astronomers) found it to be one degree in about sixty six years. We will mention the reasons for this variation in *Kitāb al-Istidrāk*' (fol. 68^{r}).

ALTERNATIVES TO PTOLEMY'S PLANETARY MODELS

The two critical works that were cited above represent what is now known of this type of literature. But that does not mean that the scope of the critical activity was limited to those two, nor does it mean that other criticisms were not as influential as those. From the recovered works that were written in the later centuries we can assert that the criticism of Ibn al-Haytham was taken very seriously by astronomers, and more than one astronomer took it upon himself to find an alternative set of models that was free of the contradictions that had bedevilled Plotemaic astronomy.

At this point, and in the interest of space and time, it is useful to divide the response to such criticisms – which found expression in the attempts to construct planetary models which were construed as alternatives to the Ptolemaic ones – into two schools; namely, the Andalusian School and the Eastern School.

The Andalusian school

The anonymous Andalusian astronomer who wrote *al-Istidrāk* may have been the forerunner of a later school of astronomers who continued his work and added some of their own criticisms; they all attempted to reformulate the Ptolemaic models. Names such as those of Jābir ibn Aflaḥ (d. *c.* middle of the twelfth century), Al-Biṭrūjī (fl. *c.* 1190) and Averroes (d. 1198) are but a few of those whose works have been critical of Ptolemy's models and have been subjected to some study.[10]

Considering Jābir ibn Aflaḥ's work, *Iṣlāḥ al-Majisṭī* ('Correction of the *Almagest*'), the main contribution of that work is that it lists some ten to fifteen problems – called 'errors' by Jābir – through which the reader is led step by step to realize the difficulties and the problems in the Ptolemaic text. One such major difficulty, for example, is the treatment of the planetary distances in the *Almagest*, for, according to Ptolemy's values, at least Venus would have to be placed above the sun.[11] Noting this difficulty, Jābir argues[12] that his computations required that both Venus and Mercury ought to be above the sun.

The main arguments of Jābir for placing both Venus and Mercury above the sun are as follows. (1) Ptolemy admits that the sun exhibits a parallax of about three minutes of arc, and that Venus and Mercury do not exhibit any observable parallax. This, according to Jābir, could only mean that they were further away than the sun, and hence above the sun in the arrangement of the heavenly spheres. (2) Jābir uses Ptolemy's values for the ratio between the epicyclic radii of Venus and Mercury and their respective deferents, and proves that if these values were to be taken seriously then Venus and Mercury would have to exhibit a parallax of about six to seven minutes, almost twice as much as that of the Sun. Since none of that takes place then they must be above the sun.

After citing the full text of Ptolemy in regard to the relative distances of the spheres of the planets, Jābir remarks in the following manner: 'I am extremely amazed at this man, and quite perplexed by him, because he appears to contradict himself without even knowing it' (fol. 78^v).

Since the absolute distances of the planets could not be determined with any certainty, this problem of the relative order of the planetary spheres remained a challenging one throughout the Middle Ages, and was taken up, as we shall see below, by al-Biṭrūjī and by Mu'ayyad al-Dīn al-ʿUrḍī (d. 1266) among others.

For al-Biṭrūjī, as for Ibn al-Haytham, the main problem with Ptolemaic astronomy was that it was not sufficiently Aristotelian. But unlike Ibn al-Haytham, who understood motion along an eccentric as being acceptable in the Aristotelian sense, al-Biṭrūjī did not even tolerate eccentrics and epicycles if these were to be understood in the traditional Ptolemaic sense. His main concern was that the universe must have only one point around which all other points must revolve, and that point had to be fixed and must coincide with the centre of the earth. This purist Aristotelian attitude is supposed to have been first championed by al-Biṭrūjī's teacher Ibn Ṭufayl (d. 1185), who promised that he would produce a book in which such an astronomy would be described, but does not seem to have done so.

These attempts were then followed by al-Biṭrūjī's book *Kitāb al-Hay'a* ('Book on Astronomy') which was written especially to develop such an astronomy, and later by the work of Ibn Rushd (mainly his commentary on the Aristotelian *Metaphysics*), who only recorded his objections in a qualitative manner.

All this activity remained limited in its applicability and scope, simply because the new proposed configurations – such as the one proposed by al-Biṭrūjī – were not successful enough in their ability to duplicate the Ptolemaic observational and analytical results. There was a real need therefore for a set of new models that would avoid the Ptolemaic shortcomings, with the provision that these models had also to conform to Ptolemy's valid observations, and save the same phenomena that were saved by Ptolemy's models.

The real progress in that regard was achieved in the eastern part of the Islamic world, where generations of astronomers, beginning in the eleventh century and continuing well beyond the fourteenth, had achieved several results first by isolating the Ptolemaic problems and then by applying new techniques to them which rendered them consistent with the original principles of the Aristotelian cosmos.

The Eastern School

The Eastern School discussed here has been referred to in the literature as the 'Marāgha School', [13] simply because all but one of the astronomers who were then known to have discussed non-Ptolemaic astronomical models had worked at one time or other during the latter part of the thirteenth century at the Marāgha Observatory in northwest modern Iran. But since we now know more about this activity, and we know that it was not restricted to the environs of the Marāgha Observatory, nor to the thirteenth century, we chose a term that contrasts the activities that were carried out in this region of the Islamic Empire with those that have been described as belonging to the Andalusian Revolt.

Luckily, the activities in the Eastern School have some cohesion, and could therefore be characterized as belonging to the same tradition. The general attitude of the astronomers belonging to this School towards Aristotle and Aristotelian cosmology was distinctly different from the attitude of their Western colleagues in Andalus. As the Andalusians were concerned with the inadmissibility of eccentric and epicyclic motion, for it violated the principle of the Aristotelian centre of the universe which marked the centroid around which all circular motion had to take place, the Eastern astronomers did indeed realize that this problem was only a pseudo-problem. For as Ibn al-Shāṭir put it:

> The existence of small spheres such as the epicyclic spheres, that do not surround the earth, is not impossible, except in the ninth sphere. The proof of that is that [it is similar] to the existence of a planet in each sphere and a multitude of stars in the eighth sphere where each of them [i.e. the planets or the stars] is greater than some of the epicycles of some planets, and the planet [or star] is different from the body of the [heavenly] sphere. Thus it is not inadmissible that there be epicycles and such things. From this, it is understood that the [heavenly] spheres have some sort of composition [*fīhā tarkīb^{un} mā*] and that the only one which is absolutely simple [*basīṭ muṭlaq*] is the ninth [sphere] for it is not possible to imagine in it the existence of a planet [or star] or such similar thing. [14]

Ibn al-Shāṭir expresses the same opinion later on when he says:

> They [i.e. the astronomers] have disagreed about the motion of the small spheres which do not encompass the center of the universe, such as the sphere of the epicycle and the like. They have all accepted that it could move towards any assumed direction, for they indicated that the epicyclic sphere has an upper and a lower half. So if it moves in the direction of the signs in the upper

> half, it would move in the direction opposite to the signs in the lower half, and vice versa. Its motion would not be by compulsion [*qasrīya*] nor by accident [*ʿaraḍīya*], but would rather be natural [*ṭabīʿīya*]. They have also agreed that it was permissible to have epicycles in other than the ninth [sphere], because we see planets in the spheres. And the existence of a planet in a sphere signifies some sort of composition [*tarkīb^{un} mā*]. And whoever says that the spheres are all simple, and thus no epicycles could exist in them, and every motion that is not a motion around the center could not be a simple motion, I would say that the motion of the epicycles has been proven to exist. And if that could ever be proven with a conclusive proof [*burhān qaṭʿīy*], then the composition and the lack of simplicity in the spheres would be proven. My opinion is that it is composed of simple [essences] and not of elements [*ʿanāṣir*], except for the ninth [sphere]. And only God knows best.[15]

The problem for the Eastern School astronomers was therefore a problem of devising models that would preserve Ptolemy's observations, save the phenomena and be consistent mathematically as well as physically. In other words, their main concern was to find a set of models that would describe the motion of the physical spheres that carry the various planets by using geometrical mathematical terms, without having the mathematical statements contradict the physical assumptions.

The general thrust of the research that was initiated at the Eastern School is usually described in the literature as being of a philosophical nature, for it had accepted in general all the observational results of Ptolemy and only expressed some philosophical objections to the Ptolemaic models.

In a separate article, I have argued that Ibn al-Shāṭir's model for the sun was, as far as we can tell, the only model that seems to have been motivated by philosophical as well as observational considerations (Saliba 1987a). In that article, I discussed in great detail Ibn al-Shāṭir's attitude to observations, and the manner in which his solar model was definitely conceived to accommodate his observational results and that it was not proposed only as a philosophical objection to Ptolemy's model which, as we have seen, was in no way philosophically objectionable. In fact I know of no other astronomer who considered Ptolemy's solar model as objectionable, or who had offered any alternative to that model.

In order to trace the general activities of the astronomers of the Eastern School, however, I shall single out Ibn al-Shāṭir's solar model, since it is unique in its conception and the only model for the sun, lay down the main outlines of the argument against the Ptolemaic solar model and then proceed to give a short description of the model itself. For the sake of economy, and to avoid duplication of material, the models for the other planets will be treated in a thematic fashion, by giving, in a chronological progression, all the known alternative models that were proposed for each planet.

Ibn al-Shāṭir's solar model

The Ptolemaic model for the sun (Figure 3.1) was conceived as being either an eccentric or an epicyclic model, both alternatives being philosophically acceptable for they could actually describe the behaviour of physical bodies. But from other considerations of the solar theory, Ptolemy had assumed, for example, that the apparent disk of the sun always had the same size, 0; 31, 20°, at all distances of the sun, and was therefore equal to the size of the lunar disk when the moon is at its maximum distance from the earth. Naturally, this assumption implies, on the one hand, that the solar eccentricity has at best a negligible effect on the size of the apparent diameter of the sun, and denies, on the other hand, the possibility of annular eclipses. The first statement is obviously an approximate one, and the second is simply contradicted by observations.

We unfortunately do not have an explicit statement by Ibn al-Shāṭir in which he objects to this Ptolemaic assumption. But from his remarks throughout his *Nihāyat al-Sūl* ('The Final Quest'),[16] we know, for example, that, in contradistinction to Ptolemaic theory, Ibn al-Shāṭir did admit the possibility of annular eclipses (fol. 38^r). From his other observational results, which are only quoted in the *Nihāya*, we also know that he considered the apparent size of the solar disk to be variable, again contrary to Ptolemaic theory. Ibn al-Shāṭir referred the reader to another one of his texts, namely *Taʿlīq al-Arṣād* ('Discourse on Observations'), in which these observations were supposed to have been analysed in detail. Unfortunately, the *Taʿlīq* itself has not yet been identified, and is presumed to be lost.

In the *Nihāya*, in at least two places (fols 12^v, 41^r), Ibn al-Shāṭir gives the apparent size of the solar diameter to be

at apogee	0; 29, 5°
at mean distance	0; 32, 32°
at perigee	0; 36, 55°

which proves beyond doubt that he must have been reporting only observational results as he himself says in many different expressions such as '*taḥarrara bi-l-raṣd*' and '*ḥaqqaqtu dhālika bi-l-raṣd*', both meaning 'to verify by observation'.

In a different context (fol. 3^r), Ibn al-Shāṭir also says that he observed the sun in the middle of the seasons and found that the maximum equation of the sun, a function of the eccentricity, was different from the one given by Ptolemy; Ibn al-Shāṭir's maximum equation was 2; 2, 6°, which implies an eccentricity of about 2; 7 parts instead of 2; 30 parts as given by Ptolemy.

As long as we do not know the details of Ibn al-Shāṭir's observational methods, we refrain from making any comments on the plausibility or the accuracy of these reports. We simply state that Ibn al-Shāṭir must have convinced himself that his results were indeed more accurate than those of Ptolemy, and thus require a new model that accommodates them for they could not be accommodated by the Ptolemaic model. What he had to do, therefore, was to devise a model that has the effect of a smaller eccentricity than that of Ptolemy, to accommodate the smaller maximum equation, but at the same time allow the sun to move much closer to the earth to appear at an angle of $0;36,55^\circ$, and farther from the earth to appear at an angle of $0;29,5^\circ$. The ratio of the maximum size to the minimum one should therefore be approximately the same as $0;36,55/0;29,5 = 1.26934$.

To do that, he assumed (Figure 3.9) the following orbs for the solar model. (1) An orb of radius 60 parts, which he called the parecliptic, concentric with the observer at point O, the centre of the world, and moving in the direction of the signs at the same speed as the daily mean motion of the sun, namely $0;59,8,9,51,46,57,32,3^\circ$ per day. This parecliptic orb carries another smaller one (2), called the deferent, of radius 4; 37 parts in

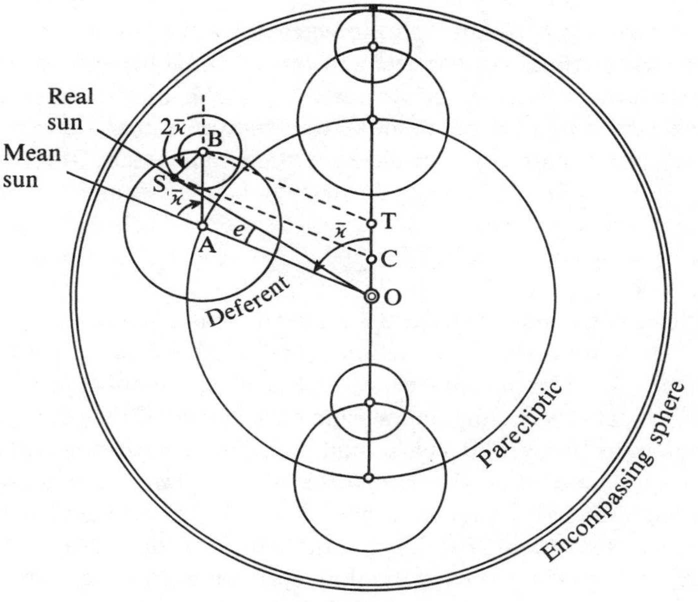

Figure 3.9

the same units that make the radius of the first orb 60 parts. The second orb moves on its own centre at the same speed as the first, but in the opposite direction, thus keeping line AB always parallel to OCT and having the same effect as transferring the eccentricity OT to an epicycle with centre A (as in Figure 3.9). (3) The third orb, called the director, of radius 2; 30, is carried by the deferent in a direction opposite to that of the signs, but moves on its own centre in the opposite direction at twice the speed of the first orb. This third orb carries the body of the sun S, which now seems, according to ʻUrḍī's lemma (discussed later), to be moving at uniform speed around point C. Finally, all of these orbs are embedded within a final orb (4) called the encompassing one (*al-shāmil*), that moves at the same speed as the solar apogee, in the direction of the signs, which was found to be 1° per 60 Persian years.

The effect of this model is to allow the sun S to move uniformly around point C, i.e. eccentricity $OC = 4;37 - 2;30 = 2;7$ which is smaller than the Ptolemaic eccentricity of 2; 30, and thus achieve longitudes close to those of Ptolemy, to be later corrected for the maximum equation. But unlike the Ptolemaic model, that of Ibn al-Shāṭir allows for a variation in the apparent size of the solar disk in the magnitude of

$$\frac{67;7}{52;53} = 1.26914$$

which is very close to the value predicted by the observations of the apparent size of the solar diameter. As an additional benefit, Ibn al-Shāṭir adds that his model yields a further advantage in that all the angles for mean motion are measured around point O, the observer, rather than around the centre of the eccentric, as would have been required by the Ptolemaic model.

The lunar models

As we have seen above (Figure 3.2), the Ptolemaic model for the moon suffered from two main contradictions: (1) the impossibility (*muḥāl*) of the motion of the deferent sphere, as it seemed to describe, according to Ptolemy, equal arcs in equal times around the centre of the universe rather than around its own centre as it should; and (2) the unaccountability for a mechanism that could allow the diameter of the epicycle that connects the mean apogee and the epicyclic centre to be always directed towards the prosneusis point rather than towards the centre of the deferent.

The astronomical reforms of the thirteenth century included several suggestions for alternative lunar models. One such alternative was proposed by the Damascene astronomer Mu'ayyad al-Dīn al-ʿUrḍī (d. 1266) sometime before 1259.[17]

ʿUrḍī's lunar model

In order to avoid the first impossibility, ʿUrḍī required that the direction of the motion of the Ptolemaic inclined sphere be reversed so that it now moves, according to ʿUrḍī, in the same direction as that of the signs instead of the reverse. In this new arrangement (Figure 3.10) the apogee of the deferent will be carried in the same direction as that of the order of the signs, say to point B. ʿUrḍī further required that the motion of the inclined sphere be in absolute value three times as much as the motion required by the Ptolemaic model. Since the inclined sphere is concentric with the centre of the universe, it meant that angle $\bar{S}OB$ would be three times as large as angle $\bar{S}OA$.

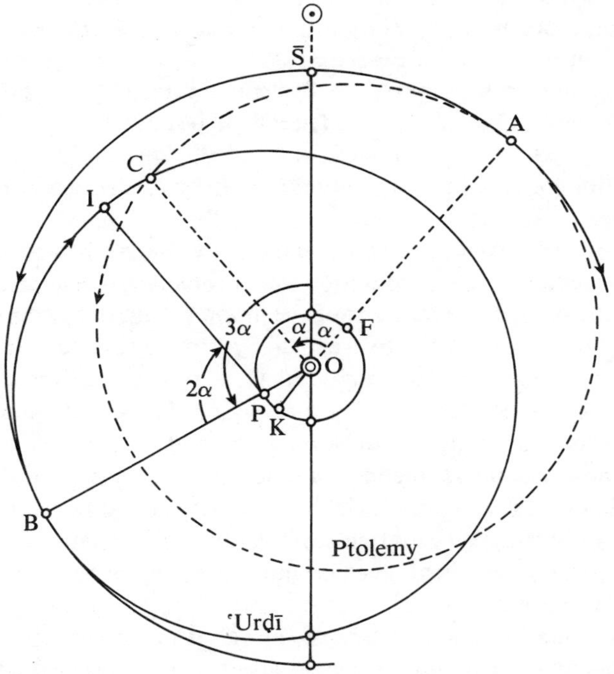

Lunar model

Figure 3.10

Once the whole deferent is carried with the inclined sphere in the same direction as that of the order of the signs, then the apogee that would have reached point A in the Ptolemaic model would now be carried to point B. Now ʿUrḍī requires that the deferent itself move around its own centre P in the direction opposite to that of the order of the signs, by an absolute amount equal to twice the motion required by Ptolemy. This would require that point B would be carried back to point I, thus making line PI, parallel to OC, the original direction of the Ptolemaic epicyclic centre from the observer's position O. All of these motions described so far are performed by spheres that move around their own centres, and thus do not contradict the principles of uniform motion. ʿUrḍī notes at this point that his model is only describing mean motions, just like the Ptolemaic model, and thus the direction PI should be taken as equivalent to the direction OC since it is parallel to it. With this configuration the epicyclic centre could then be carried to the same position required by the Ptolemaic model without having to fall into the same contradiction as the first one mentioned above.

The new model also solves the second contradiction, i.e. that of the prosneusis point, since now one should notice that line PI passes, in general, very close to point N, K in Figure 3.10, thus making it appear at point I as if it is coming from point N, the Ptolemaic prosneusis point. The mean apogee will therefore be, in this model, a fixed point defined as the common point of tangency between the deferent and the epicycle, falling naturally at the extremity of the line connecting the centre of the deferent to that of the epicycle.

By reversing the order of the motions, and by changing their magnitudes, ʿUrḍī managed, therefore, to retain the Ptolemaic observations and to reproduce the predictable motions of the moon without any compromise on the physical principles that were accepted by Ptolemy himself. He was quite aware of this major step that he had taken, and of the differences between his model and that of Ptolemy. But that did not bother him greatly, for he tells his reader that Ptolemy could only demand that his observations be taken for facts, not the mathematical methods – such as the directions and amount of motions – that he used to account for these facts. These, ʿUrḍī claims, are only guesses (*ḥads*) on the part of Ptolemy, and no one should be held responsible for them, because anyone's guesses could be as good as those of Ptolemy.

ʿUrḍī then takes up the problem of the variation between his model and that of Ptolemy, and computes the variations between the equation due to the prosneusis point as predicted by his model and the one predicted by Ptolemy's model. After a lengthy argument he concludes that the variation between the two models is less than two and a half minutes of arc, which he considers as quite permissible since Ptolemy himself had accepted variations from the facts of up to four minutes of arc and said that such variations could escape even the best of observers. For that reason he felt quite satisfied with his model, and urged the reader to accept it and to reject that of Ptolemy, since the latter had been shown to have been riddled with contradictions.

In ʿUrḍī's words, the alternative would be to accept that there are spheres that move irregularly on their own centres, and

> if we were to accept that there is a sphere that moves around its own center, sometimes speeding up and other times slowing down, then there would be no need for all the efforts expended in regard to this astronomy, and the final quest would then be the knowledge of the equations to be applied to the motions, even if those were based on false notions.
>
> (ʿUrḍī, *Kitāb al-Hay'a*, p. 135)

Ṭūsī's lunar model

Ṭūsī discusses in Chapter 7 of his most famous astronomical work *al-Tadhkira fī ʿilm al-hay'a* the lunar model according to Ptolemy. At the difficult points, however, he mentions that there were problems with that model and that he intended to treat them later on. In fact, after he finishes surveying the remaining models for the upper planets and for Mercury, he devotes a special chapter to the solution of all those difficulties that had been encountered so far. The strategy of that approach becomes obvious when one considers that the model which Ṭūsī finally proposed for the lunar motion was also applied to the motion of the upper planets, and hence placing it at the end meant that he could take advantage of combining both models under one solution.

Ṭūsī's understanding of the difficulties in the Ptolemaic model for the moon seem to have been centred around the inability of that model to allow the centre of the epicycle to approach the centre of the universe and to draw away from it without having to incorporate the crank-like mechanism of Ptolemy. If one could, by some method, keep the centre of the deferent sphere at the centre of the universe, and only allow the line joining the centre of the deferent to that of the epicycle to be shortened at quadratures and be elongated at conjunction and opposition, then that could allow the deferent to move uniformly around the centre of the universe, and, at the same time, account for the great observable variation in the equation due to the epicyclic radius.

Once the problem was conceived as such it was reduced to having to devise a mechanism that would allow a vector-like magnitude to be shortened and elongated as a result of circular motion only. Put differently, the problem could be solved once the tip of a vector line could be made to oscillate back and forth as a result of uniform circular motion. The same problem was faced when Ptolemy had to consider the oscillating planes of the latitude model for all the planets with the exception of the moon. It was in that context that Ṭūsī proposed a new mechanism in one of his other books called *Taḥrīr al-Majisṭī* ('A Redaction of the Almagest', composed in 1247), with the help of which he managed to get the tip of the oscillating diameters to be mounted on a pair of circles – discussed elsewhere as the 'Ṭūsī Couple' – and thus made that tip to move in an oscillating linear motion which was produced by circular motion. All that Ṭūsī had to do was to generalize that solution which he had proposed for the latitude model and apply it to the special requirement of the lunar model, and then use the same technique for the model of the upper planets as well.

It is not surprising, therefore, that Ṭūsī would begin the chapter in which he proposed his alternative models with a statement of the lemma that came to be called the 'Ṭūsī Couple' lemma and a formal proof of the same lemma. That was done in Chapter 11 of his *Tadhkira* mentioned above.

The lemma is stated in the specific case first, i.e. the plane case, and afterwards generalized to include the spherical case.[18] In the plane case a paraphrase of that lemma states the following: Let there be two circles (Figure 3.11), such that one of them is tangent to the other from the inside, and whose diameter is half the diameter of the other encompassing circle. If we then assume that the smaller interior circle moves in a direction opposite to the direction of motion of the exterior encompassing circle, and at twice the speed, then the point on the diameter of the larger circle and the circumference of the smaller circle, namely, the initial point of tangency, will oscillate back and forth between the extremities of the diameter of the larger circle.

Once that result was proved, Ṭūsī remarked that instead of these circles one could take two spheres whose diameters and positions have the same relationship to each other as these two circles. If that were true, then those spheres could be taken to be of an appropriate thickness to encompass other spheres such as the epicycle of the moon in the Ptolemaic model. In fact, Ṭūsī replaces the initial point of tangency with the initial position of the centre of the lunar epicycle which is now embedded inside two such spheres. What that did was to allow the centre of the lunar epicycle to oscillate back and forth along the diameter of the larger sphere. Once that was achieved, there was no longer any need for the eccentric deferent of Ptolemy, nor for his crank-like mechanism, both of which were originally required to bring the lunar epicycle closer to the earth at quadrature and farther away at conjunction and opposition.

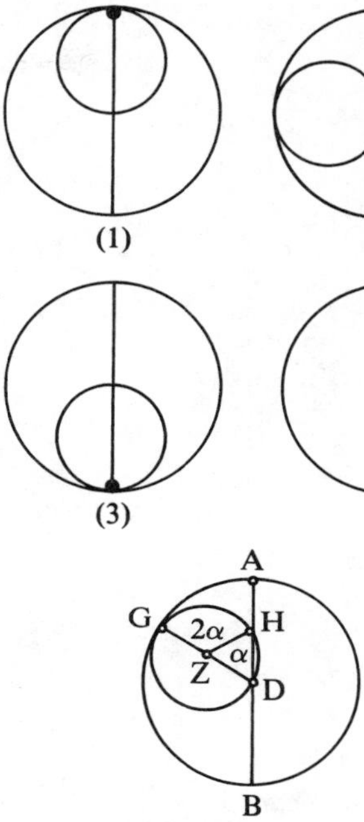

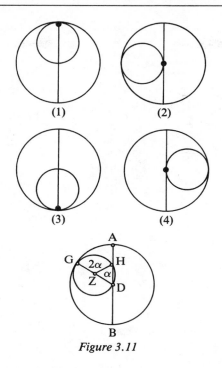

Figure 3.11

If one ascribes the appropriate motions to these spheres in such a way that they would match the Ptolemaic observed motions, then one could devise a model (Figure 3.12) whereby the deferent of the moon could move uniformly around the centre of the universe, to solve the first difficulty of the Ptolemaic model, and the epicycle could be drawn closer to the earth at quadratures and farther at conjunction and opposition in order to approximate the maximum equations observed by Ptolemy. For the prosneusis point, Ṭūsī adopts a spherical 'Couple' which, like the plane one, allows the tip of the epicyclic diameter to oscillate back and forth covering in each direction an angle equal to the maximum equation of Ptolemy.

With that resolved Ṭūsī goes on to show that the resulting path of the centre of the epicycle around the earth is not a circle, although it looks like one.

But once the advantage of the 'Couple' was realized, Ṭūsī goes on to use it in the model for the upper planets, discussed below, and in the latitude theory, as was mentioned above.

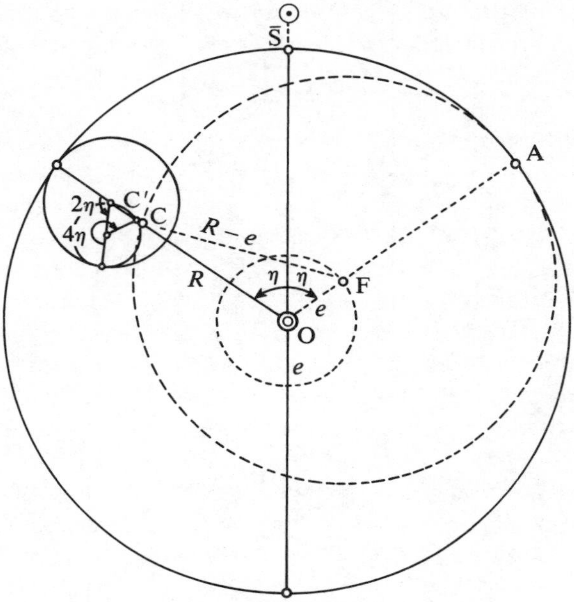

Figure 3.12

The lunar model of Quṭb al-Dīn al-Shīrāzī (d. 1311)

In his *Nihāyat al-Idrāk*,[19] Shīrāzī begins the discussion of the lunar model, on folio 54ʳ, with a general survey of the conditions for the Ptolemaic model, concluding in effect that the Ptolemaic model answers very well for a description of the observational phenomena. After giving the observational facts that would account for the existence of the spheres of the lunar model, he lists in detail the number of spheres that the lunar model must have in order to account for all the observational variations. The following section is then devoted to the various motions of these spheres and to the way in which such motions could be combined to produce the observational variations, giving in each case the mean motion for each of these spheres. That synopsis is immediately followed by a description of the variations that are observed between the mean motions produced by these spheres and the actual motions of the moon. Here, in this section, he gives the maximum equations, which, like the mean motion parameters quoted above, are strictly Ptolemaic.

On folio 60ʳ, he recapitulates and summarizes the objections that have been raised to this Ptolemaic model that he had just finished describing. In effect, he gives the two famous objections quoted so far, namely, the inadmissibility of the motion of the deferent around its centre while it describes equal arcs in equal times around the centre of the universe, and the prosneusis point.

Then he quickly says that such objections could be answered. One of those answers, which had to do with the objection to the uniform motion around the centre of the universe rather than around the deferent centre, had already been given by the principle of 'the large and the small (spheres)' – an obvious reference to the 'Ṭūsī Couple'. Moreover, from the description that he gives for this principle, and the way in which it responds to the objections against the Ptolemaic model, it becomes very clear that he is only summarizing Ṭūsī's solution that was given in Chapter 11 of the *Tadhkira*, and which was described above. Even the terminology used is transparently that of Ṭūsī, and, at best, one could say that the model offered in the *Nihāya*, so far, was nothing other than a paraphrase of Ṭūsī's model.

As for the objection concerning the prosneusis point, 'that is a matter of some subtlety' (*maḥall naẓar*). He asserts that it was difficult to achieve, and without repeating the statements of Ṭūsī in this regard moves on to say that it could be answered by using the ninth principle – referring to a principle that he had described earlier in the text – which he now calls the principle of inclination '*aṣl al-mayl*'. On the other hand, Shīrāzī does not offer a detailed description of how that principle, which was mainly applied to the latitude theory problem, could be applied here for the prosneusis point. Nor is it clear as to how Shīrāzī intended to combine this principle with Ṭūsī's model. He then continues to describe the behaviour of the Ptolemaic model that necessitated the assumption of the prosneusis point.

Then, without any introduction, Shīrāzī juxtaposes a long quotation from ʿUrḍī's text *Kitāb al-Hay'a*, beginning it simply with the statement 'one of the learned men of the moderns here, who is versed in this discipline had said' (*qāla baʿḍ afāḍil al-muta' akhkhirīn min ahl al-ṣināʿat hāhunā*).[20] That is followed by a detailed paraphrase of ʿUrḍī's lunar model which apparently had been accepted by Shīrāzī as the preferable solution, for he ends this section with the following statement:

> This is the configuration for the spheres of the moon, the magnitudes of their motions, and the manner in which it could be taken according to the chosen method [*al-wajh al-mukhtār*] that suffers from none of the objections [*al-ishkālāt*] and which conforms to the principles and the observations. The method is only different from that which has been accepted by common opinion, but that should not be detrimental to it once it is the true one. For the truth is beloved, and the teacher is beloved, but the truth is lovelier still.[21]

To summarize, therefore, Shīrāzī, who had promised in the introduction of his *Nihāya* to offer an anthology of solutions that would answer the objections raised against the Ptolemaic models, had offered, in the case of the moon, two models – one by Ṭūsī, which he did not think answered both objections to the lunar model, and one by ʿUrḍī, which he seems to have preferred.

But in the *Tuḥfa*, which Shīrāzī had written later, he proposed a model of his own. The model consists of vector connections that will ultimately allow the centre of the epicycle to move uniformly around the centre of the universe as a result of a combination of two other uniform motions. Instead of the regular eccentric sphere of Ptolemy, Quṭb al-Dīn proposed (Figure 3.13) an eccentric of his own, DHK, which has only half the Ptolemaic eccentricity. He then allowed this new eccentric sphere to move in the direction of the order of the signs at a speed equal to twice the speed of the inclined sphere of Ptolemy, ABG, which carried the apogee D in the direction opposite to that of the order of the signs. Then at the circumference of the cincture of that eccentric, Quṭb al-Dīn required the introduction of another small sphere with centre H whose radius is equal to half the Ptolemaic eccentricity. He required that the smaller sphere should move in the same direction as the new eccentric, and at the same speed. This allowed the epicyclic centre E, which is placed at the cincture of this sphere, to be

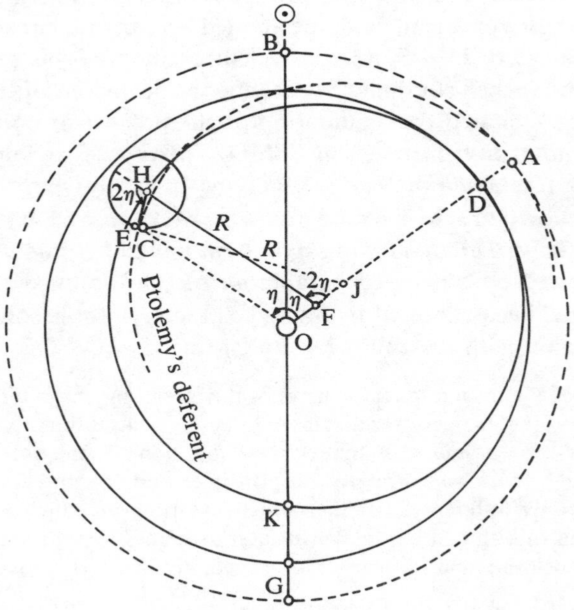

Figure 3.13

brought very close to the old Ptolemaic epicyclic centre C, and to move uniformly around the centre of the universe.

The improvement in the new model is that one could show that the new position of the epicyclic centre will appear as if it is moving uniformly around O, the centre of the universe, when in reality it is moving uniformly around H, the centre of its own carrier – the smaller sphere – which in turn is moving uniformly around F, the centre of its own carrier as well – the centre of the new eccentric proposed by Quṭb al-Dīn. To prove that such a relationship actually exists, Quṭb al-Dīn used a theorem that was first proposed by Mu'ayyad al-Dīn al-ʿUrḍī which will be discussed below as ʿUrḍī's lemma. What the new model does to the Ptolemaic one is to remove the first objection that was raised in connection with the Ptolemaic model, namely that of having a sphere move uniformly around a point which is not its own centre.

But what it does not do is solve the second objection, namely that of the prosneusis point. Quṭb al-Dīn remains silent on this second issue, in Chapter 10 of the *Tuḥfa*, and takes it up again at the end of Chapter 12 of the same book. But, even then, Quṭb al-Dīn did not seem to have succeeded in responding to the second objection. In the words of a later astronomer by the name of ʿUbaydallāh b. Masʿūd b. ʿUmar Ṣadr al-Sharīʿa (d. AH 747 AD 1346/7), who attempted to solve this specific problem of Quṭb al-Dīn's model,[22] he said that the author of the *Tuḥfa* 'had spoken profusely about the prosneusis point, without any apparent success, for the import of his statement was that the motion of the eccentric alone was sufficient to exhibit the difference between the two (epicyclic) apogees. But there is no doubt that that would not do so.'[23] The work of Ṣadr al-Sharīʿa himself, which is encyclopedic in nature, has now been studied by Dallal (1995a). What he seems to have done (Figure 3.14) is to suggest the addition of yet another sphere – of radius r_1 – to be carried at the tip of the epicyclic radius, whose own radius is equal to 0; 52 parts in the same units that make the radius of the inclined sphere 60 units. This additional sphere is supposed to move at the same speed as the deferent and in the same direction, i.e. in the direction opposite to that of the epicycle. The effect of this small additional vector is to increase the anomaly by an amount proportional to the first equation at the intermediate points between syzygies and quadratures, and to leave it as is, i.e. have the effect of a zero equation, at syzygies and quadratures. This small epicyclet could also allow the radius of the epicycle to appear larger at quadrature, and smaller at syzygies as required by the Ptolemaic observations.

A more successful astronomer, and a better known one, a contemporary of Ṣadr al-Sharīʿa, was a Damascene by the name of Ibn al-Shāṭir (d. 1375),

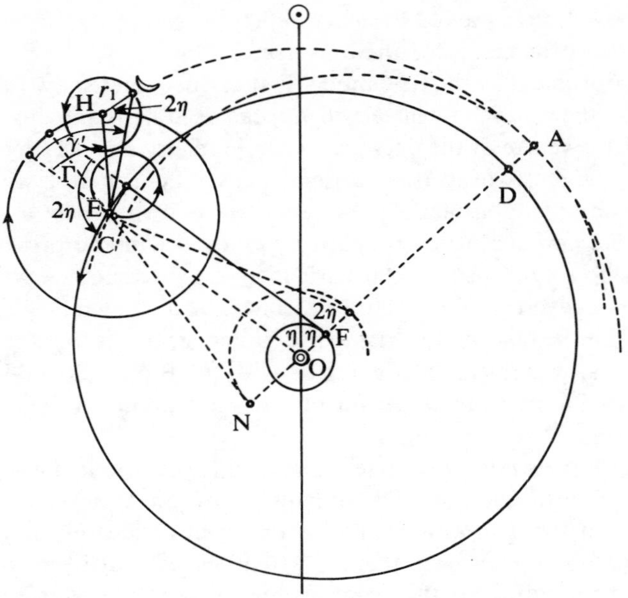

Figure 3.14

who proposed a new set of models that did not suffer from the same difficulties as the Ptolemaic models, and that were apparently very close – even at times identical, as in the case of the lunar model – to those proposed by Copernicus some two hundred years later.

The lunar model of Ibn al-Shāṭir[24]

Ibn al-Shāṭir's approach to the lunar model, like his approach to the remaining planets, as we shall see below, centres around his interest in doing away with eccentric spheres altogether. As a result of that strategy, he felt that he could not tolerate Ptolemy's crank-like mechanism which caused the problems in the first place, despite the fact that it did explain the phenomenon of the variation of the lunar equation between the two positions of the moon, i.e. when the moon is in conjunction with the sun and when it is at quadrature with it.

To solve the problem of the lunar model, Ibn al-Shāṭir proposed a new configuration. Let the lunar model (Figure 3.15), not to scale, have the following spheres. (1) A parecliptic sphere (*mumaththal*) concentric with the sphere of the zodiacal signs, whose centre is naturally the centre of the universe O and whose radius is 69 parts.[25] (2) An inclined sphere whose

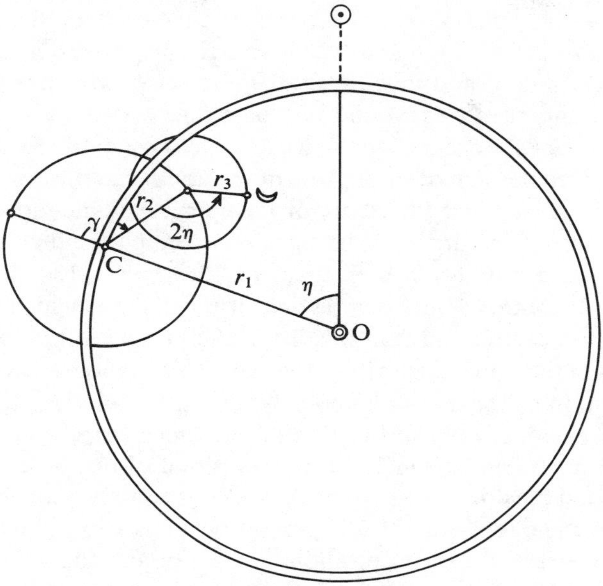

Figure 3.15

equatorial plane is inclined with respect to that of the parecliptic by a fixed amount having a maximum value of 5°. The centre of this second sphere coincides with the centre of the universe O, which is at the same time the centre of the parecliptic, and its radius r_1 is assumed to be 60 parts. The cincture (*minṭaqa*) of this sphere intersects that of the parecliptic at two points that are called the nodes. The concave surface of this sphere has a radius of 51 parts.[26] (3) A third sphere whose radius r_2 is 8; 16, 27 parts (i.e. eight parts, sixteen minutes and twenty-seven seconds)[27] is then assumed to be embedded within the inclined sphere, and let it be called the epicyclic sphere.[28] (4) A fourth sphere whose radius r_3 is 1; 41, 27 parts is assumed to be embedded within the sphere of the epicycle, and let it be called the sphere of the director (*al-mudīr*). The moon itself is embedded in the sphere of the director, and its diameter is 0; 32, 54 parts.

Since the fourth sphere is embedded in the third one and the moon whose radius is 0; 16, 27 is embedded in the fourth, the circular representation of these spheres will have the following dimensions. The radius of the third circle will be 6; 35 parts, that of the fourth will be 1; 25 parts and that of the moon will be 0; 16, 27 parts.

The motions of these spheres are as follows. (1) The parecliptic sphere moves around the centre of the universe in a direction opposite to that of the order of the signs at a speed equal to the speed of the nodes, i.e. 0; 3, 10, 38, 27° per day. Obviously this sphere carries with it all the other spheres of the moon, and thus causes them to move with this motion. (2) The inclined sphere moves around the same centre as the first sphere, but in the direction of the order of the signs, at a speed equal to 13; 13, 45, 39, 40 which is equal to the sum of the motion of the nodes and the mean longitude of the moon. As a result the centre of the epicycle of the moon is moved in the direction of the order of the signs by an amount equal to the mean longitude of the moon which is equal to 13; 10, 35, 1, 13. (3) The third motion, 13; 3, 53, 46, 18° per day, is that of the epicycle which takes place around its own centre, and is in the direction opposite to that of the order of the signs in the upper part of the epicycle. This motion was called in the past the motion of the moon in anomaly, and it begins from the apparent apex of the epicycle. (4) The fourth motion, which carries the moon with it along the cincture of the inclined sphere, is that of the director, a simple motion around the director's own centre in the direction of the order of the signs, and is equal to 24; 22, 53, 23° per day which is also equal to twice the elongation between the mean position of the sun and that of the moon.

This model will in effect answer the two objections that were raised against the Ptolemaic model, for it now allows the variations in the observed positions of the moon to be accounted for while all the motions are motions of spheres around their own centres. When the moon is in conjunction with the sun (Figure 3.15) all centres will be in line with the sun, or as Ibn al-Shāṭir would put it they would be in the direction of what he calls the position of the apogee. As the inclined sphere moves in the direction of the order of the signs, the epicyclic sphere will move in the opposite direction. The two motions will therefore satisfy the phenomena of elongation of the moon and the lunar anomaly. To account for the evection, the director is then made to move at a speed equal to twice that of the inclined sphere, thus leaving the moon at the perigee of the director, i.e. towards the earth, but in the direction of the apogee when the moon is in conjunction and at the director's apogee when the moon is in quadrature. This will allow the lunar equation to increase from the observed Ptolemaic value 5; 10 at conjunction (Ibn al-Shāṭir has 4; 56) to the maximum value of 7; 40 at quadrature.

But more importantly, this model allows the lunar distance to vary between 1, 5; 10 and 54; 50 at syzygies and between 1, 8; 0 and 52; 0 at quadrature in the same parts that make the radius of the inclined sphere equal to 60 parts. This model is then a vast improvement over that of Ptolemy for in the latter model the moon was allowed to come as close as 34; 7 parts to the earth, which would have had to make the moon look twice as big at quadrature than when it is at syzygies which is contrary to the observed facts. Although this result was probably one of the main aspects that made this model attractive to Copernicus, for he used the same dimensions and configuration in his *De Revolutionibus*, it was only mentioned in passing by Ibn al-Shāṭir who definitely knew of the advantages of his own model (*Nihāya*, fol. 3^r).

The model for the upper planets

In the case of the upper planets, the Ptolemaic model, as described above (Figure 3.3), suffered from one major difficulty, namely that of the *equant*. In brief, this problem is essentially the same as having to force a sphere to move uniformly around an axis that does not pass through its centre, a true physical impossibility as long as these spheres were thought of as real physical bodies, which they were. The Arabic-writing astronomers proposed several models to get around the *equant* problems of Ptolemy.[29]

Abū ʿUbayd al-Jūzjānī (d. *c.* 1070)

The first astronomer–philosopher that we know of who left us a treatise that purports to reform Ptolemy's astronomy, namely to solve the problem of the *equant*, is Abū ʿUbayd al-Jūzjānī, the student and collaborator of Avicenna (see Saliba 1980). In it he tells us that Avicenna had also made the incredible claim that he too had solved the same problem, but he was not going to tell his student about it, for he wanted him to work it out for himself. In a mixture of cynicism and wit, Abū ʿUbayd continues: 'I suspect that I was the first to have achieved these results' (*ibid.*: 380).

A summary of Abū ʿUbayd's solution of the *equant* problem is given in Figure 3.16. He clearly thought that he could replace the Ptolemaic deferent by the *equant* sphere itself – shown in broken lines – and thus transfer the motion of the epicycle from point H on the deferent to point B, now carried on a secondary epicycle of radius e equal to the Ptolemaic eccentricity of the planet. The advantage of such a model would obviously be that B, the epicyclic centre, would now move uniformly around H, while H itself also moves uniformly around T, thus not violating the uniform motion requirement. Moreover, if the secondary epicycle, with centre H, is made to move at the same speed as the Ptolemaic deferent, but in the opposite direction, then point B, the planet's epicyclic centre, will look as if it is moving uniformly around D, the *equant*, as required by observation. As such the new model seems to satisfy both uniform motion and observation.

All this would have been acceptable if the distance of point B, the planet's epicyclic centre, from the observer at Q were not also determined by observation, and thus could not be easily replaced. The laborious and lengthy

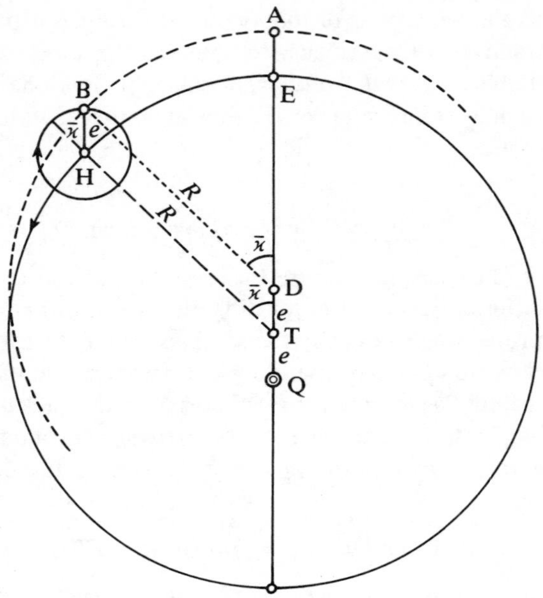

Figure 3.16

computations in Book X of the *Almagest* were specifically carried out to determine the relative dimensions of the model of each planet in such a way as to satisfy the observational data that Ptolemy was trying hard to save.

Moreover, if Abū ʿUbayd's model were to work, Ptolemy would have been the first to adopt it, for it only seems to replace an eccentric sphere, the deferent, by a concentric one and a secondary epicycle. This equation was very well known to Ptolemy who further attributed it to Apollonius in Book XII, 1, of the *Almagest*, and used it efficiently in Books III, 3, and IV, 6 (See Neugebauer 1959). It would be naive, therefore, to assume with Abū ʿUbayd that the observational problem of the *equant* could be solved simply by replacing the eccentric hypotheses by the epicyclic hypotheses as Ptolemy would have called this transformation.

The problem, therefore, was still to find a model that preserved both the Ptolemaic deferent distance and the effect of the *equant*, and would still be the result of the motion of spheres that move uniformly around their own centres.

Mu'ayyad al-Dīn al-ʿUrḍī[30]

Taking advantage of the fact that one could transfer motion on an eccentric to a motion along a concentric with an epicycle – the Apollonius equation referred to above – ʿUrḍī's problem was to devise such a motion so that point B (Figure 3.17) in Jūzjānī's model could be brought closer to Ptolemy's deferent, if possible to coincide with Z. This does not necessarily mean that ʿUrḍī was trying to emend the model of Jūzjānī directly, for he does not mention Jūzjānī at all, and he could have been working directly with the Apollonius equation. But it was a stroke of genius to realize that one does not have to transfer the whole eccentricity TD = BH to the secondary epicycle, but instead accept a compromise and transfer only half of that eccentricity KD = NB. To do so, and approximate Ptolemy's deferent as closely as possible, ʿUrḍī found that the epicyclet BOH must revolve in the same direction and by the same amount as the new deferent with centre K that he had just introduced. Only then will the combined motion of the deferent with centre K and the epicyclet with centre N produce a resultant path marked by point O which hugs very closely the Ptolemaic deferent EZH. Once this technique had been discovered by ʿUrḍī, it was used by every astronomer who came after him to adjust the Ptolemaic model in one way or another.

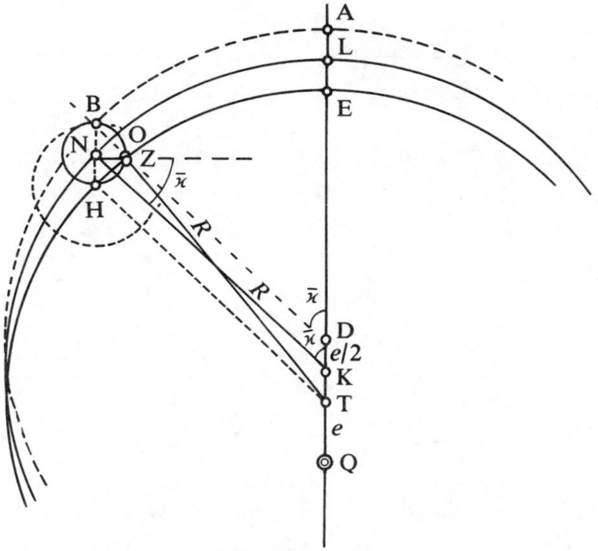

Figure 3.17

But to preserve the effect of the *equant* as well, ʿUrḍī had to show that the resultant motion of point O had to look as if it is itself uniform with respect to point D, the *equant*. This is tantamount to proving that under the stated conditions – namely, the epicyclet moving by the same angle $\bar{x}$ as the proposed deferent and in the same direction – the lines OD and NK will always be parallel.

To do so, ʿUrḍī stated the problem in the form of a general lemma, namely:

> Every straight line upon which we erect two equal lines on the same side so that they make two equal angles with the (first) line, be they corresponding or interior, if their edges are connected, the resulting line will be parallel to the line upon which they were erected.
>
> (*Kitāb al-Hay'a*, p. 220)

Figure 3.18 is taken from ʿUrḍī's text in which he shows that line GD is always parallel to AB in all the cases where AG and BD describe equal angles with line AB. It is also assumed that AG = BD. The proof is then straightforward both when the corresponding angles DBE and GAB are equal or the interior angles DBA and GAB are equal, for with the construction of line DZ parallel to AG, both cases become identical and require only *Elements* I, 27–33, to be proved.

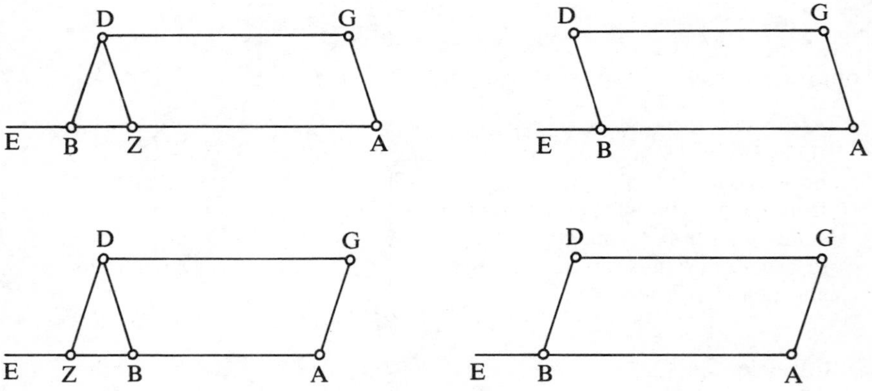

Figure 3.18

Now that the line OD (Figure 3.19) was shown to be always parallel to NK, point O could then be taken as the centre of the planet's epicycle and the Ptolemaic conditions would be very closely approximated. ʿUrḍī was quite aware of the fact that the path resulting from the motion of O coincides exactly with the Ptolemaic deferent only at the apogee E and the

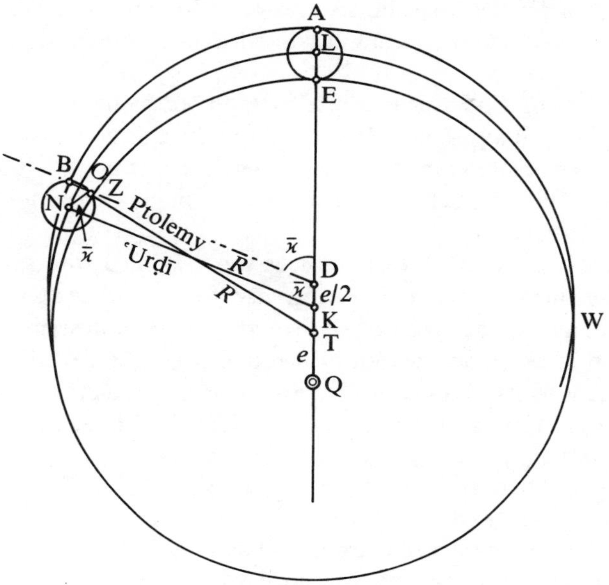

Figure 3.19

opposite perigee. To quote ʿUrḍī in full on this point, he says:

> As for the center of the epicycle, i.e. the point of tangency mentioned above [O in Figure 3.19], on account of the fact that the center of the epicycle will be on this circle [i.e. the deferent] at its two distances, i.e. its farthest distance from the eye and its closest distance to it, and since it is very close to its circumference at the remaining portions of its revolution, that has led Ptolemy to believe that the center of the epicycle is coincident with its circumference, and describes it with its motion.
>
> (*Kitāb al-Hay'a*, pp. 222–3)

Instead of calculating the variation between the resulting path of the model and the Ptolemaic deferent, which is very small indeed,[31] ʿUrḍī assumed confidently that his model was the true one, and that the burden of proof should be required of Ptolemy for it was Ptolemy who was confused about the true path when he assumed that it was along a circular deferent. This is exactly the same sentiment expressed by Maestlin some three centuries later when he explained this same point in Copernicus's astronomy to his student Kepler: 'For Copernicus shows (V, 4) that the path is not perfectly circular ... [and] that Ptolemy thought that this path of the planet ... was truly circular' (see Grafton 1973: 526). It is interesting to note that Maestlin also proves a specific case of the introductory lemma stated and proved by ʿUrḍī, without stating it in general terms (*ibid.*: 528).

In the words of Copernicus himself (V, 4), the same argument is made thus: 'Hence it will also be demonstrated that by this composite movement the planet does not describe a perfect circle in accordance with the theory of the ancient mathematicians but a curve differing imperceptibly from one' (*De Revolutionibus*, p. 743).

Therefore, both ʿUrḍī and Copernicus were satisfied with this new technique of bisecting the Ptolemaic eccentricity, for it allowed them to preserve Ptolemy's deferent, the effect of the *equant*, and still describe all motions in their respective models as uniform motions of spheres that revolve on their own centres, thus avoiding the apparent contradictions in Ptolemy's model. But to understand the possible relationship between the Copernican model for the upper planets and that of ʿUrḍī, we need to investigate the intermediary models that were described by Quṭb al-Dīn al-Shīrāzī (d. 1311), Ṣadr al-Sharīʿa (d. 1346/7) and Ibn al-Shāṭir of Damascus (d. 1375).

In an earlier article the present author has shown that the model of Quṭb al-Dīn was indeed identical with that of ʿUrḍī (Saliba 1979b), and was used as the basis for the model described by Ṣadr al-Sharīʿa. For these two astronomers, therefore, ʿUrḍī's model was quite sufficient to explain away the contradictions in the Ptolemaic model. As for Ibn al-Shāṭir, his main objection remains to be against the use of eccentrics. As in the case of the moon, he manages here too to devise a model that is perfectly geocentric, but incorporating ʿUrḍī's model as we shall see below.

Ibn al-Shāṭir's planetary model

Because of the historical importance of Ibn al-Shāṭir's model, and its possible relationship to the works of Copernicus, a full English translation of the short section in which Ibn al-Shāṭir describes his model for Saturn will be given here. The text comes from Ibn al-Shāṭir's *Nihāyat al-Sūl* in an unpublished edition of the present author, and varies from the text that describes the models for Jupiter, Mars and Venus only with respect to the actual dimensions. The general relationships that are applicable to the model of all the upper planets are summarized in Figure 3.20.

Chapter 12 of Ibn al-Shāṭir's *Nihāyat al-Sūl* begins thus:

> Concerning the Configuration of the Spheres of Saturn according to the Correct Manner.
>
> Of the spheres of Saturn, let there be a sphere that is represented [*mumaththal*] by the sphere of the zodiacal signs, occupying the same surface

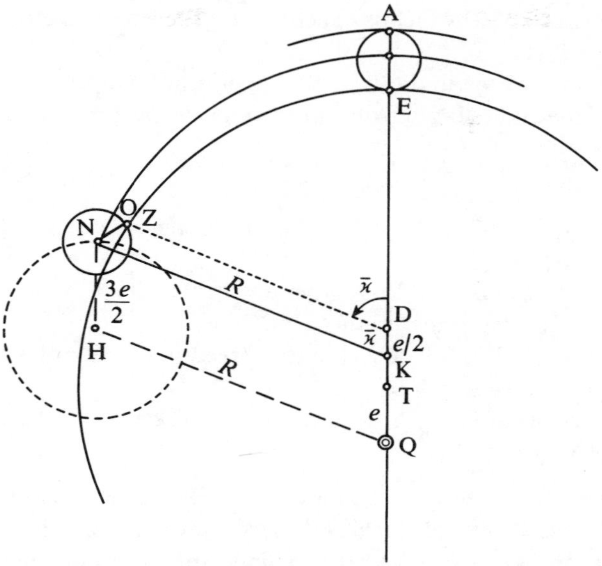

Figure 3.20

and having the same centre and the same poles [not represented in Figure 3.20 for simplicity].

One then imagines that there would be another sphere [represented by radius QH in Figure 3.20], inclined with respect to the first, at a fixed inclination of 2; 30 parts, intersecting it at two opposite points: one [of the points] is called the head and the other the tail.

Then let there be a third sphere [represented by the circle with centre H in Figure 3.20] whose centre is on the periphery of the inclined, and whose radius is equal to $5\frac{1}{8}$ parts, being in the same units that measure the radius of the inclined [R in Figure 3.20] as 60 parts. Let [this sphere] be called the deferent [*al-ḥāmil*].

A fourth sphere is imagined to have its centre on the periphery of the deferent [circle with centre N in Figure 3.20], and whose radius is 1; 42, 30 parts. Let it be called the director [*al-mudīr*].

A fifth sphere is imagined, with its centre on the periphery of the director [circle with centre O in Figure 3.22], and whose radius is 6; 30 of the same parts; it is called the sphere of the epicycle [omitted in Figure 3.20].

The centre of the body of Saturn is fixed to one point on the cincture [*minṭaqa*] of the epicycle.

From the dimensions given, we can verify the relationships $HN = 3\,e/2$ and $NO = e/2$, where e is the same as the Ptolemaic eccentricity, as applicable for all the other upper planets. In the case of Saturn, $HN = 5\frac{1}{8} = 5;7,30$ is indeed equal to $3\,NO = 3 \times 1;42,30$, and

HN + NO = 2 e = 6; 50 which is exactly twice the Ptolemaic eccentricity of 3; 25 parts.

The directions of the motions of the spheres in Figure 3.20 are to be surmised from the following set of values as given by Ibn al-Shāṭir.

Sphere 1 moves at 0; 0, 0, 9, 52° per day, in the direction of the order of the signs; not represented.
Sphere 2 moves at 0; 2, 0, 26, 17° per day in the direction of the order of the signs; radius QH.
Sphere 3 moves at 0; 2, 0, 26, 17° per day in the direction opposite to that of the order of the signs; radius HN.
Sphere 4 moves at 0; 4, 0, 52, 34 in the direction of the order of the signs; twice Sphere 2, radius NO.
Sphere 5 moves at 0; 57, 7, 43, 34, 22 in the direction of the order of the signs; not represented.

From these relationships, which are applicable to the other upper planets as well, it is clear that what Ibn al-Shāṭir calls the deferent, circle with centre H, moves at the same speed as the inclined sphere, represented by radius QH, but in the opposite direction. This, in effect, transfers the portion of the eccentricity QK from the centre to the periphery, using the same technique referred to above and used by Ptolemy in the *Almagest* III, 3. This allowed Ibn al-Shāṭir to make his model actually geocentric, for now the radius HQ revolves around the centre of the earth itself.

To adjust for the remaining portion of the eccentricity, and to retain the Ptolemaic deferent EZ, Ibn al-Shāṭir makes the epicyclet with centre N revolve in the opposite direction to the deferent with centre H, thus making angle HNO = $2\bar{\kappa}$. Since NH is parallel and equal to QK, lines NK and QH are also equal and parallel. Therefore, angle KNH is equal to $\bar{\kappa}$ = angle KNO.

But it was ʿUrḍī who proved earlier the general lemma (Figure 3.18) that if DK = NO and both lines describe the same angle with respect to KN, then OD, the line that connects their extremities, will be parallel to KN, and point O will be brought very close to Z on the Ptolemaic deferent.

What Ibn al-Shāṭir seems to have done, therefore, is to combine two results already available to him from previous research. First, he used the Apollonius equivalence to transfer the effect of QK to the periphery HN, and then used the result already reached by ʿUrḍī to draw point N back to O, by using ʿUrḍī's lemma. We do not need to speculate whether Ibn al-Shāṭir knew directly of the work of ʿUrḍī, for he explicitly tells us that he did, and he criticizes ʿUrḍī specifically for retaining eccentric spheres.

The net result is an orbit very close to the Ptolemaic deferent, and a geocentric model that is strictly concentric and free from the Ptolemaic contradictions. Figure 3.21 shows the relationship between Ibn al-Shāṭir's model – the model drawn in dashes – and that of Ptolemy – the continuous lines – with the dotted lines KN and DO as reminders of ʿUrḍī's model. I have intentionally exaggerated the distance between points O and Z just to make the point that they are not in general identical, but in no way to suggest that they could have been differentiated by any observational result. For Mars, the planet with the largest eccentricity, the value of OZ is in the order of 0.005 for a radius taken to be 60 units (see Swerdlow 1973: 469).

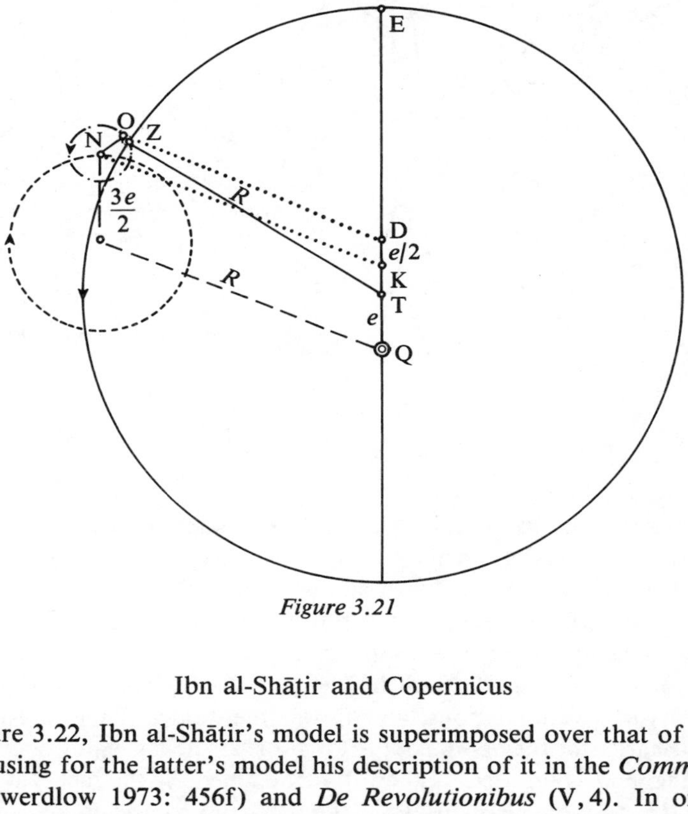

Figure 3.21

Ibn al-Shāṭir and Copernicus

In Figure 3.22, Ibn al-Shāṭir's model is superimposed over that of Copernicus, using for the latter's model his description of it in the *Commentariolus* (Swerdlow 1973: 456f) and *De Revolutionibus* (V, 4). In order to facilitate the transformation between the heliocentric Copernican model, the model drawn using broken lines, and the geocentric model of Ibn al-Shāṭir, drawn using continuous lines, the mean sun $\bar{S}$ in Ibn al-Shāṭir's model is held fixed and the other relationships and motions are allowed to remain the same. Once $\bar{S}$ was held fixed, Ibn al-Shāṭir's model, with all its dimensions, was translated to the model adopted by Copernicus. Since we

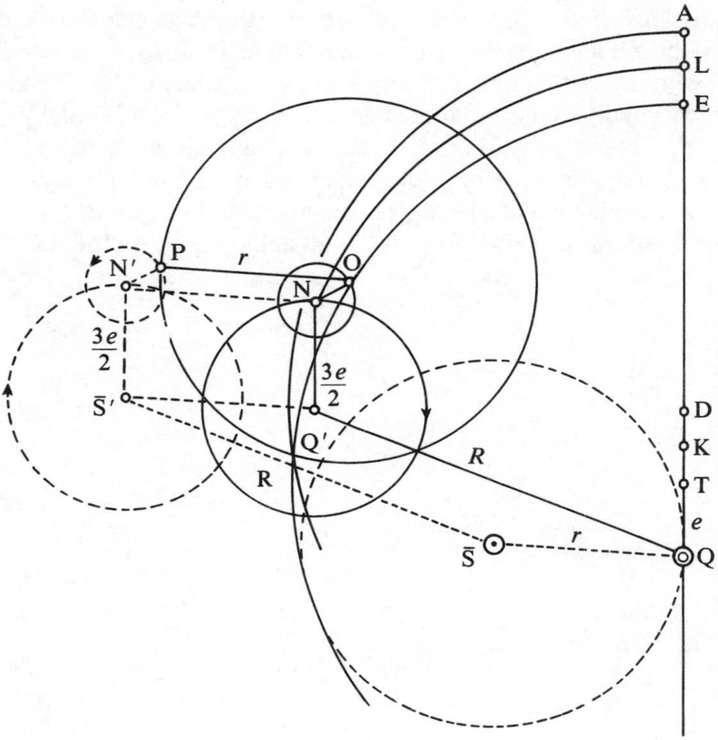

Figure 3.22

now know that the addition of vectors is commutative, it is not surprising to find that both models predict the same position for planet P, irrespective of whether the earth or the mean sun is taken to be fixed.

To conclude this section, all the four models discussed above, namely those of Ptolemy, ʿUrḍī, Ibn al-Shāṭir and Copernicus, are superimposed on the same Ptolemaic deferent in Figure 3.23. Jūzjānī's model was disregarded for obvious reasons, and so was the model of Quṭb al-Dīn and Ṣadr al-Sharīʿa for they both adopted that of ʿUrḍī. The equivalence of the remaining models, however, is best illustrated here by the fact that they all predict the position of planet P, without having to accept the Ptolemaic contradictions.

The historical relationships between ʿUrḍī and Ptolemy may have gone through the first attempt of Abū ʿUbayd, but may very well be a direct modification of the Apollonius equivalence, with the elegant and successful bisection of the Ptolemaic eccentricity. Once that result was achieved, Ibn al-Shāṭir, realizing its full significance, simply combined it with the

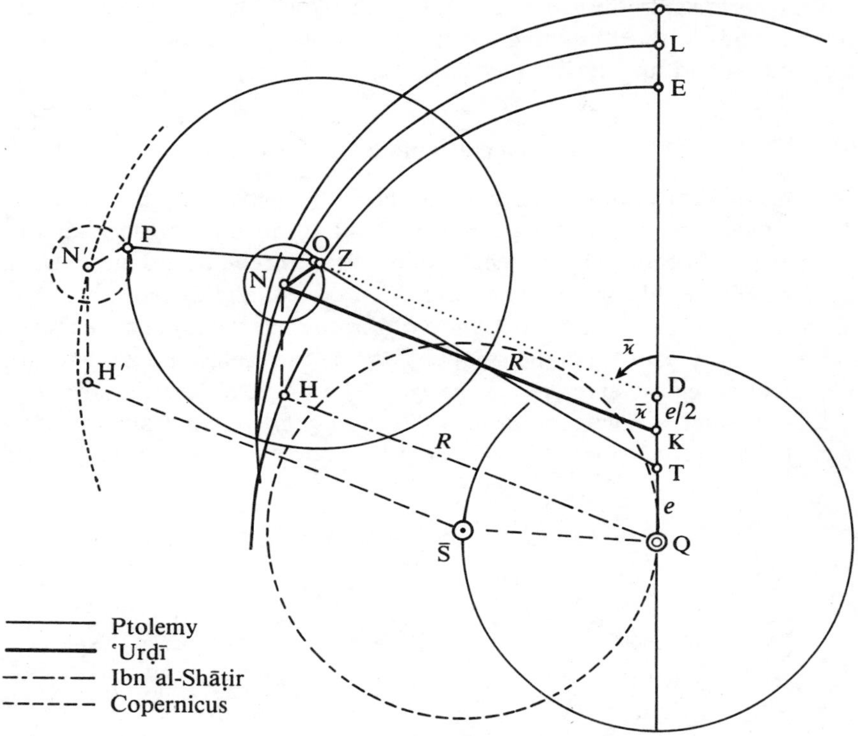

Figure 3.23

Apollonius equivalence to produce his own model. We noted that Ibn al-Shāṭir knew of the works of ʿUrḍī and took issue with him for retaining the eccentrics in his model. Then it is understandable that he did not feel obliged to prove the parallelism of OD and NK (Figure 3.23), for it was already proved by ʿUrḍī with the general lemma (Figure 3.18). Similarly, Copernicus did not prove that parallelism either, and it was Maestlin who explicitly proved it again in his letter to Kepler (see Grafton 1973: 528f).

The question of the explicit relationship between Copernicus and his Muslim predecessors, especially Ibn al-Shāṭir, remains open, and further research will have to be done before it can be decisively established one way or the other. What is clear, however, is that the equivalent model of Ibn al-Shāṭir seems to have had a well-established history within the results reached by earlier Muslim astronomers, and could therefore be historically explained as a natural and gradual development that had started some three centuries earlier. The same could not be said of the Copernican model. But some more research has to be done on the Arabic sources themselves before their inner relationships can be fully understood and exploited in this regard, and on the Byzantine sources for a possible connection between Copernicus and his Muslim predecessors.

The planetary model of Ṭūsī

In terms of its relationship to the Copernican model for the upper planets, Ṭūsī's model represents a tradition different from that of Ibn al-Shāṭir. Rather than bisecting the Ptolemaic eccentricity, in the tradition of ʿUrḍī, Ṭūsī generalizes his own lunar model (Figure 3.24) and allows a 'Couple' to move in such a way that the centre of the epicycle would also be moved closer to the equant when the epicycle is at the Ptolemaic apogee, and farther away when the epicycle is at perigee. The 'Couple' itself is carried by a deferent that is now concentric with the *equant*. All motions therefore would be uniform around the centres of the spheres concerned, and would

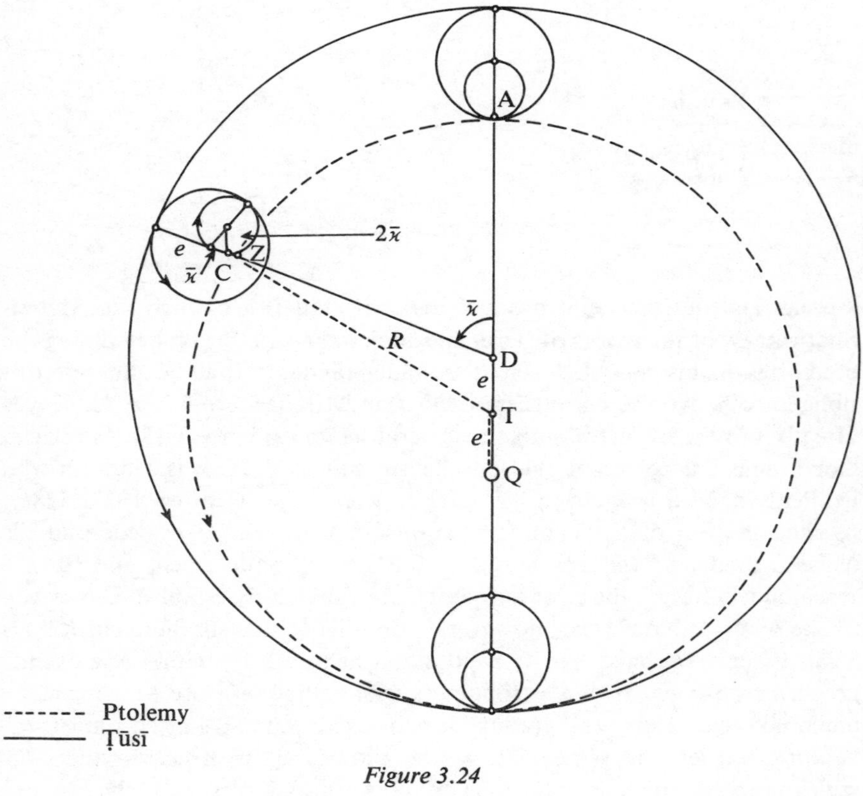

Figure 3.24

produce none of the contradictions that were assumed in the Ptolemaic model.

The model for the motion of Mercury

The Ptolemaic model for the motion of Mercury as described above (Figure 3.4) is very similar to that of the moon. In effect, Ptolemy uses the same crank-like mechanism that allows the planet to come close to the earth at two points instead of one, hence accounting for what Ptolemy observed to be the greatest elongations from the sun, and thus assumed the existence of two perigees for Mercury. The *equant* for Mercury, on the other hand, is now placed on the line of centres in between the centre of the universe and that of the eccentric, when the diameter of the eccentric is still in the direction of the apogee, instead of its being at twice that distance away from the centre of the universe as in the case of the upper planets. Unlike the lunar model, the model for Mercury requires the planet to move uniformly around the *equant* point instead of the centre of the universe as in the case of the moon.

The first astronomer known to have proposed an alternative model that would answer the objections to the Ptolemaic model is the same Mu'ayyad al-Dīn al-ʿUrḍī whose work we have seen above in connection with the models for the moon and the upper planets.

The Mercury model of ʿUrḍī

ʿUrḍī devotes two different chapters to the discussion of Mercury's model, in addition to the various remarks that he makes about it in connection with the other planets. Chapter 44[32] contains a straightforward description of the spheres of Mercury with brief remarks about the motions of these spheres. Whenever it was appropriate, ʿUrḍī would correct the Ptolemaic description to fit the new observations. He, at one point, reminds the reader that 'it is no longer necessary to add the conditions that were assumed by Ptolemy for these motions, for it was found that the solar apogee [assumed fixed by Ptolemy] indeed moves at the same rate as the apogee of the director (*al-mudīr*) which is in Libra'.

Chapter 48,[33] as its title *Iṣlāḥ Hay'at ʿUṭārid* ('A Reform of Mercury's Model') implies, is devoted to a reconstruction of Mercury's model in such a way that the two main problems in the Ptolemaic model are solved. These problems were, as in the case of the lunar model, (1) a deferent that moves uniformly around an axis that does not pass through its centre, and (2) an *equant* point which is neither the deferent's centre nor the centre around which the deferent describes equal motion.

In Ptolemy's model (Figure 3.25), the deferent is moved uniformly by the director around centre B in a direction contrary to the direction of the order of the signs, in order that it brings the apogee to point T. The deferent itself is made to move in the opposite direction around its own centre G, to carry the epicyclic centre to point C, but seems to be moving uniformly at equal and opposite speed around point E, the *equant*. It is necessary, therefore, that the deferent describe an irregular motion around its own centre G, a clear violation of the uniform motion principle.

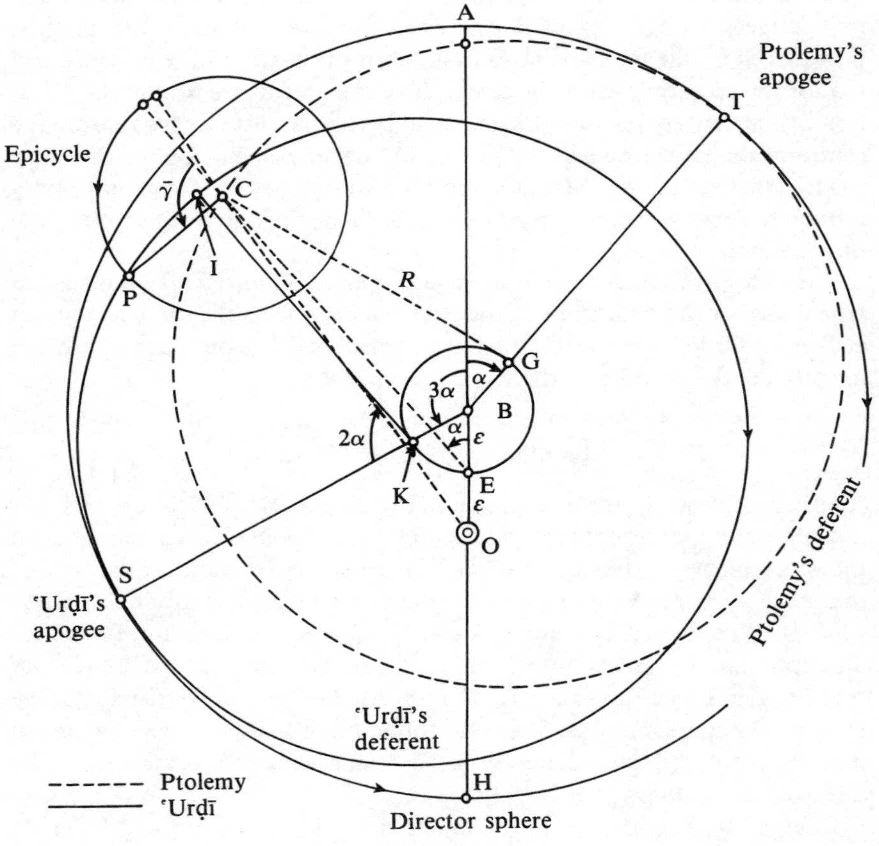

Figure 3.25

In response to that, ʿUrḍī states the following:

This total configuration resulted from several considerations. Among them are the observations, the proof that is based on the observations, the periodic

> movements, the model [*hay'a*] that he [i.e. Ptolemy] conjectured [*ḥadasa*], and the directions of these movements. No one should be critical of the observations, the proof, or the periodic movements, for there has not come to light anything to contradict them. As for the method of conjecture, he [i.e. Ptolemy] should have no priority over anyone else, especially that his error had been made evident. And if anyone were to find another proposition [*amr*] that agrees with the principles, and matches the particular movements of the planet that are found by observation, then that person should be deemed as closer to the truth [*awlā bi-iṣābat al-ḥaqq*].
>
> And now that we have seen his erroneous opinion and sought to emend it [*iṣlāḥ*] as we did in the case of the other planets, we found that we could do so if we reversed the directions of the two movements mentioned above, that is, the movement of the director and that of the deferent. Let us assume then, that the director moves in the direction of the order of the signs by as much as three times the mean motion of the sun, and that the deferent moves in the opposite direction [i.e. contrary to the direction of the order of the signs] by twice that motion. Then the resulting motion of the epicyclic center is in the direction of the order of the signs by as much as the mean motion of the sun, which is the same as in his [i.e. Ptolemy's] model.

Translated into the diagram (Figure 3.25) and superimposed, not to scale, over Ptolemy's model, ʿUrḍī's model describes the motion of Mercury by letting the director move uniformly, like Ptolemy's lunar deferent, in the direction of the order of the signs around centre B to carry the apogee to point S. Then the deferent should also move uniformly, but in the opposite direction, around its own centre K to bring the epicyclic centre back to I. The resultant motion of the epicyclic centre would then be parallel to that of Ptolemy and is very close to it, as in the figure. Moreover, ʿUrḍī's model will agree with the principles of uniform motion and will match the results of observations very closely, thereby, in ʿUrḍī's words, 'varying only slightly (from that of Ptolemy) by an amount that could escape the observer'. ʿUrḍī then continues to say: 'Our method, on the other hand, is free from doubt and contradiction, and is therefore clearly superior to any other'.[34]

The next astronomer to propose an alternative model for Mercury was Quṭb al-Dīn al-Shīrāzī, a student of Ṭūsī, for Ṭūsī himself clearly admitted in his *Tadhkira* that he had not yet devised a model for Mercury, and that he would describe it once he did.[35] The status of our present research does not indicate that he ever did.

The Mercury model of Quṭb al-Dīn al-Shīrāzī

A brief description of Shīrāzī's model has already been given by E. S. Kennedy (see Kennedy 1966: esp. pp. 373–5), and the following is mainly derived from his work and from the work of Shīrāzī in the *Tuḥfa*.

Shīrāzī proposed to replace the Ptolemaic model with one of his own (Figure 3.26), which was composed of six spheres: (1) a deferent with a radius r_1 equal to 60, eccentric to the centre of the universe by the same eccentricity as that of Ptolemy, and whose centre B does not move as in the Ptolemaic model, thus eliminating the need for a 'director'; (2–5) two sets of Ṭūsī Couples whose smaller spheres have radii $r_2 = r_3 = r_4 = r_5$ equal to half the Ptolemaic eccentricity; and (6) a final sphere of radius r_6 equal to the eccentricity.

The motions of these spheres as described by Kennedy, and by Quṭb al-Dīn in his *Tuḥfa*, are such that the deferent moves uniformly at the same

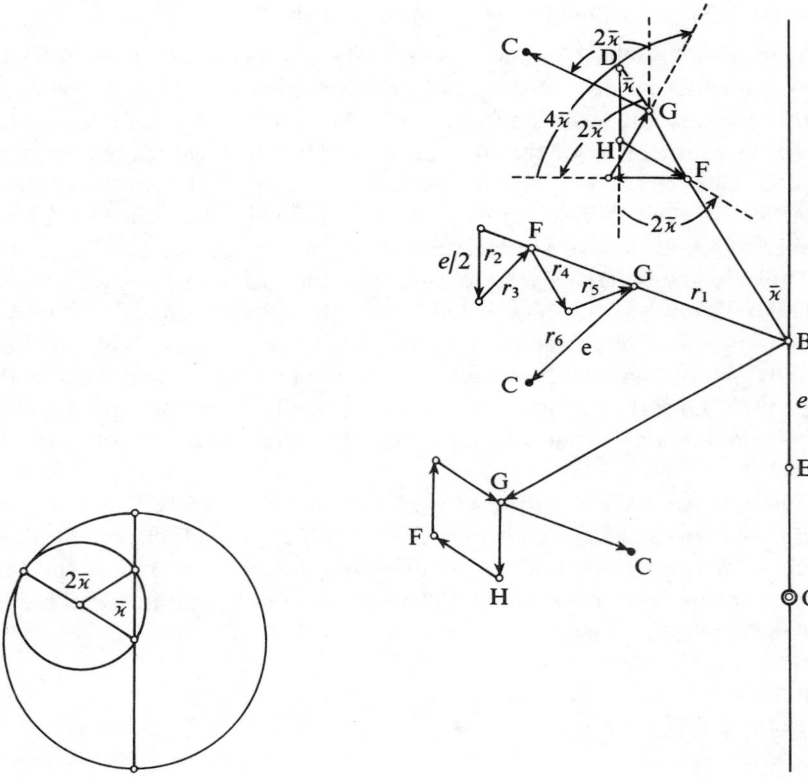

Figure 3.26

speed $\bar{x}$ as the mean motion of the sun, in the same direction as that of the order of the signs. This deferent carries with it all other spheres, the two Ṭūsī Couples and the sixth sphere of radius equal to the eccentricity. The first of the two Couples moves in such a way that the larger of the two spheres moves at the same speed as the mean motion of the sun but in the direction opposite to that of the order of the signs. This means that the smaller sphere will move at twice this speed but in the opposite direction, thus keeping the point of original tangency along the diameter of the larger sphere, which is in turn the radius of the deferent. This point F which has to oscillate along the radius of the deferent is taken to be the centre of the larger sphere of the second Couple. The second Couple will then take over and move in the opposite direction to the first Couple at twice the speed, which means that it will generate its own point G which will oscillate along the diameter of the larger sphere, which, in turn, is along the radius of the deferent. The effect of both Couples is to keep the centre of the sixth sphere G along the radius of the deferent, but to allow it to oscillate nearer to the earth and farther away from it. With this motion the radius of the sixth sphere $GC = r_6$ will, together with line BE, satisfy the conditions for ʿUrḍī's lemma, thus allowing the centre of the epicycle to describe a curve that looks like an egg-shape but pressed in at the waist, so to speak, when the centre of the epicycle is at the two perigees.

To describe these motions by using modern vector terminology, and if we assume that the deferent had already moved by an angle equal to $\bar{x}$, we could then take the radius of the deferent (Figure 3.26) to be a vector r_1 that has been moved by an angle $\bar{x}$, and vector r_2, the radius of the smaller sphere in the first Couple, to have been moved by the larger sphere in the opposite direction by an angle equal to $\bar{x}$. By the motion of the smaller sphere vector r_3 would have been moved in the direction opposite to r_2 by an angle equal to $2\,\bar{x}$. Now, in the second Couple, vector r_4 would be moved by the larger sphere through an angle equal to $2\,\bar{x}$, measured from the direction of r_1, and r_5 would be moved in the opposite direction to r_4 by the second smaller sphere through an angle equal to $4\,\bar{x}$ measured from the direction of r_4. Finally r_6 would be moved by its own sphere through an angle equal to $\bar{x}$, measured from the direction of r_1.

Perceived as such, the sum of vectors r_2, r_3, r_4 and r_5 will allow the centre of the sixth sphere G, i.e. the origin of vector r_6, to always oscillate along the radius of the deferent. In this model the centre of the deferent is fixed at a distance from the centre of the universe equal to twice the Ptolemaic eccentricity. Now, since vector r_6 will always move at an angle equal to that through which the deferent moves, and in the same direction, it means that the tip of that vector will seem as if it is always moving uniformly around the *equant* centre, as it would have been predicted by the lemma proposed by ʿUrḍī in the model for the upper planets, and as it would have been required by the Ptolemaic observational data.

What Quṭb al-Dīn seems to have done is to use the results reached by Ṭūsī and ʿUrḍī and to develop his own model by using both techniques that were developed earlier, namely Ṭūsī's Couple and ʿUrḍī's lemma.

Ibn al-Shāṭir's model for Mercury

To account for a uniform motion of Mercury with respect to the *equant*, and for the larger elongations from the sun at distances symmetrically placed at around 120° on both sides of the apogee, both facts supported by Ptolemaic observations, Ibn al-Shāṭir devised a model in which all these facts can be accommodated as resulting from uniform motions of spheres around their respective centres. Like Quṭb al-Dīn, he too, as we shall see, used the results reached by Ṭūsī and ʿUrḍī, namely the Ṭūsī Couple and ʿUrḍī's lemma.

Ibn al-Shāṭir used the same techniques he had used in the lunar and planetary models described above. And here too, he started constructing his model with the assumption that it should be strictly geocentric, so that he would not have to use eccentric deferents which he thought that others had erroneously used.[36] To make the model strictly geocentric he assumed (Figure 3.27) the existence of an inclined sphere, of radius $r_1 = 60$ parts, which is concentric with the centre of the universe O, and which moves in the direction of the order of the signs at the same speed as the mean motion of the sun. That inclined sphere is supposed to carry at its cincture (*minṭaqa*) another sphere, called the deferent (*al-ḥāmil*), whose radius r_2 is 4; 5 parts and which moves at the same speed as the inclined sphere but in the opposite direction. The deferent carries, in the same manner, a third sphere, called the director (*al-mudīr*), whose radius r_3 is 0; 50 parts and whose motion is like that of the inclined sphere in the direction of the order of the signs, but at twice the mean daily motion of the sun. The director then carries the epicycle whose radius r_4 is 22; 46 parts and whose motion is equal to that of the anomaly of Mercury. At the cincture of the epicycle there is a fifth sphere, called the encompassing sphere (*al-muḥīṭ* or *al-shāmil*), whose radius r_5 is 0; 33 parts and whose motion is equal to twice that of the daily mean motion of the sun and in the same direction as the order of the signs. In turn the encompassing sphere carries an identical sixth sphere, called the preserver (*al-ḥāfiẓ*), whose radius r_6 is the same as that of the fifth sphere and whose motion is four times as much as the daily mean motion of the sun, but in a direction opposite to the direction of the order of the signs. The planet Mercury is immersed at the cincture of this sixth sphere.

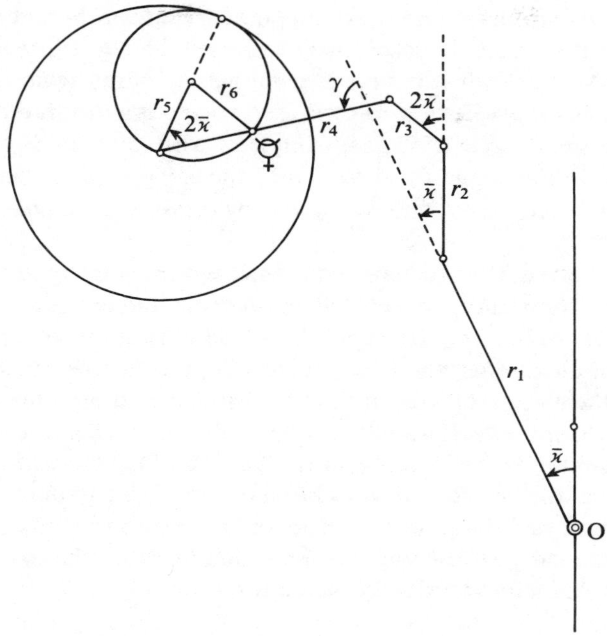

Figure 3.27

To use modern vector terminology, we let the radius of the inclined sphere be a vector r_1, 60 parts long. Its motion would then be equal to that of the daily mean motion of the sun in the same direction as the order of the signs. At the tip of this vector, another one, r_2, will represent the deferent and whose length will therefore be 4; 5 parts. Its motion will be equal and opposite to the motion of r_1. This means that r_2 will continuously be displaced in such a way that it will remain parallel to the line of apsides, and will in effect carry an amount of the eccentricity equal to 4; 5 from the centre to the periphery. The vector that represents the director, r_3, will move at twice the speed of r_1 and in the same direction. It can be easily shown by ʿUrḍī's lemma that the tip of r_3 will seem to move uniformly around a point on the apsidal line at a distance from the centre of the universe equal to 4; 5 – 0; 50 = 3; 15 parts. Since the tip of r_3 is actually the centre of the epicycle in the Ptolemaic model, this displacement will in effect produce the same effect as that of the motion of the centre of the epicyclic centre around the *equant* which is 3 parts away from the centre of the universe in the Ptolemaic model. Thus far the problem of the *equant* is resolved.

The last two vectors r_5 and r_6 are supposed to answer the second requirement of the Ptolemaic model, namely to create the effect of enlarging the epicycle of Mercury when the planet is about 90° away from the apogee. This will be achieved if we assume those two vectors to represent the two radii of the small circle of a Ṭūsī Couple,[37] whereby the diameter of the larger circle will be along the direction of the epicyclic diameter, and thus allowing the latter to be reduced in length by 0; 66 parts, or enlarged by the same amount.

With that solved, the two main requirements of the Ptolemaic model were met, and the contradictions were solved. As stated above, the model takes advantage of the two important results which were achieved by ʿUrḍī and Ṭūsī. Ibn al-Shāṭir was, therefore, unlike Copernicus who used the same model for the motion of Mercury – once without understanding it fully as in the *Commentariolus* (Swerdlow 1973: 504), and once in the *De Revolutionibus* (Swerdlow and Neugebauer 1984: 403f), where it was better described – a true heir to a long tradition of Arabic astronomy, which supplied him with such techniques that he only had to put them together, as he did in the model for the upper planets, and then add the requirement of making the whole model strictly geocentric.

The Mercury model of Ṣadr al-Sharīʿa

In his *Kitāb al-taʿdīl*, Ṣadr al-Sharīʿa describes the Ptolemaic model for Mercury (fols 32ʳ–33ᵛ) and concludes the section by a statement of the inadequacies of this model. He then repeats the statement of Ṭūsī in his *Tadhkira* where he admitted that he had not yet developed a model for the motion of Mercury. Ṣadr al-Sharīʿa then claims that, with God's help, he had been successful where Ṭūsī had failed. He goes on to describe a model which was essentially a modification of the lunar model proposed by Quṭb al-Dīn and is described above.

Ṣadr al-Sharīʿa proposes a new eccentric deferent whose centre F (Figure 3.28) is to be placed at a distance $e/2$ from the centre of the director, i.e. above the Ptolemaic *equant* in the direction of the apogee by one and a half times the Ptolemaic eccentricity, and whose motion is taken to be twice that of the director and in the opposite direction to that of the director, i.e. in the same direction as the order of the signs. He then uses ʿUrḍī's lemma, by affixing an epicyclet to the cincture of this deferent of a radius r_1 equal to $e/2$, and allows this epicyclet to move at the same speed and direction as the deferent. The actual epicycle of the planet is supposed to be carried at the cincture of this epicyclet. By ʿUrḍī's lemma, the centre of the actual epicycle H will seem as if it is describing equal arcs in equal times, i.e. moving uniformly, around the centre of the director B. Moreover, the

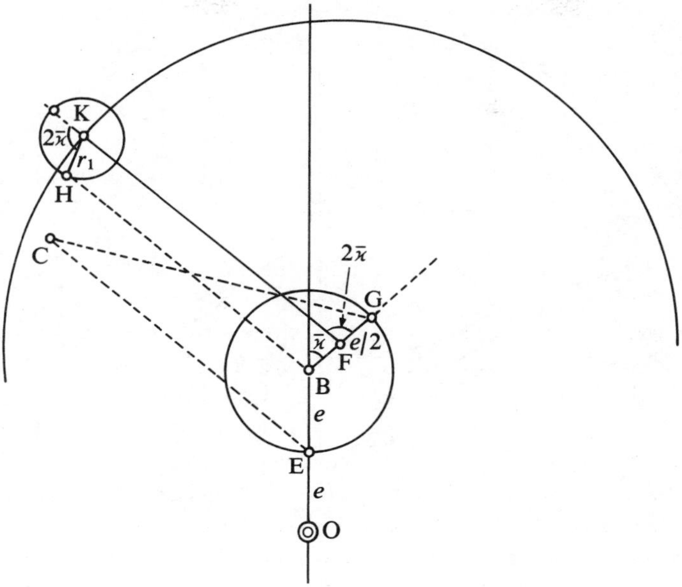

Figure 3.28

centre of the actual epicycle H would then be in the same direction from the centre of the director as the centre of the epicycle in the Ptolemaic model is from the *equant*. Since all these motions were described as mean motions, then to have the centre of the epicycle move in a direction parallel to the one that would have been anticipated in the Ptolemaic model must have satisfied Ṣadr al-Sharīʿa, for he claimed that he had found an equivalent model that did not suffer from the Ptolemaic inconsistencies.

ʿAlāʾ al-Dīn al-Qushjī (d. 1474)

In an anonymous treatise kept at the Asiatic Society Library in Calcutta (No. A1482), which, according to the present author, was written by Qushjī, we find yet another attempt to resolve the problem of Mercury.

After presenting the Ptolemaic model for Mercury, and criticizing it, Qushjī goes on to present his own solution of the problems entailed by the Ptolemaic model. He first assumes (Figure 3.29) the centre C (or G) of the Ptolemaic epicycle to be carried by an epicyclet with centre D, whose radius is half the Ptolemaic eccentricity, and which is in turn carried by another epicyclet, with centre B, of identical radius. Next he assumes that the epicyclet B is carried by a new deferent whose centre is point H, at a

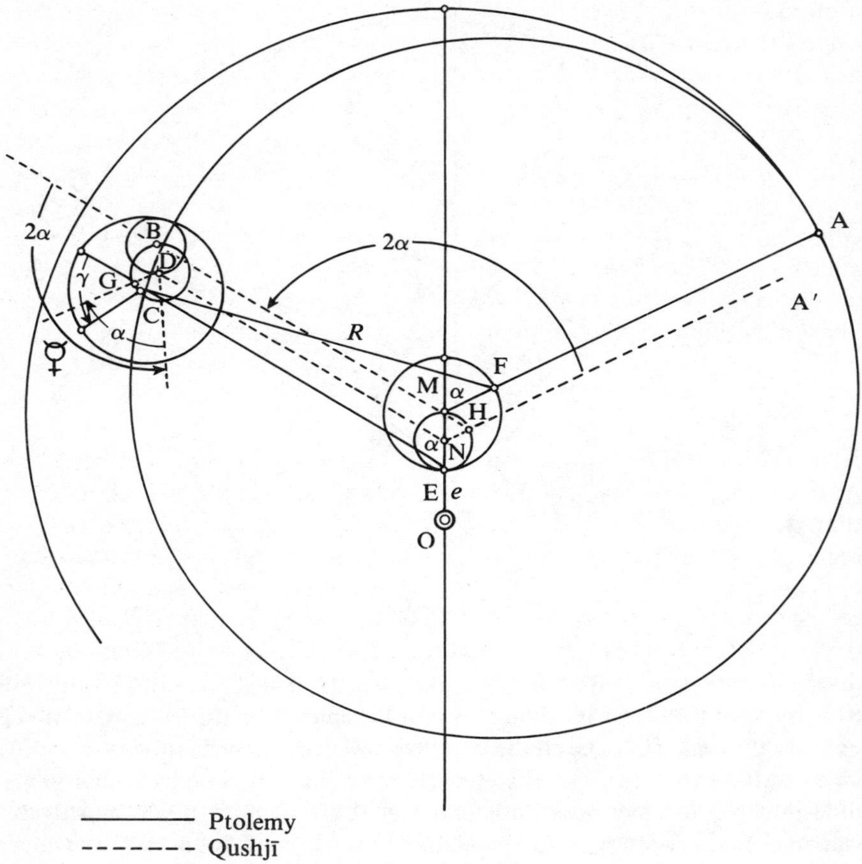

Figure 3.29

distance equal to half the Ptolemaic eccentricity from the centre of the director N after defining this new centre of the director to be at a distance from the centre of the universe equal to one and a half times the Ptolemaic eccentricity.

The motions of these spheres are then described as in the Figure to be as follows. The director carries the deferent with it in the direction opposite to that of the order of the signs at a speed equal to the mean daily motion of the sun to bring the apogee to point A′. The deferent moves in the opposite direction at twice that speed, bringing point B, the centre of one of the epicyclets, to the direction HB. The epicyclet with centre B moves in the same direction as the deferent and with the same speed, thus bringing point D, the centre of the second epicyclet, to look as if it is moving uniformly around centre N, the centre of the new director. The second epicyclet then carries the centre of the epicycle G by moving back in the same direction as the director and at the same speed as the director. This combination of motions will ensure that G will always be in line with C, and on the extension of the line connecting C to the *equant* E, thus making point G look as if it is always moving uniformly around the *equant* as it should.

Any close consideration of this model will immediately reveal its indebtedness to ʿUrḍī's lemma, used once to align D with N, and another time to align G with E, to Quṭb al-Dīn's lunar model, by retaining the crank-like mechanism of Ptolemy but also bisecting the eccentricity, and to Ṣadr al-Sharīʿa's more rudimentary form of the same model.

CONCLUSION

After this general review of the planetary theories which were developed by Arabic writing astronomers after the twelfth century, it has become clear that the two major achievements of this long tradition were, after disregarding motion in latitude and planetary distances for they are less important for that tradition, essentially two mathematical theorems: one that we referred to above as ʿUrḍī's lemma and the other being the so-called Ṭūsī Couple. With the help of these two theorems, and with the technique of dividing the eccentricities of the Ptolemaic models, it was possible to transfer segments of these models from the central parts to the peripheries and back. This freedom of movement not only allowed the retention of the effect of the *equant* in the Ptolemaic models, but also allowed the development of sets of uniform motions that would not violate any physical principles. The Ṭūsī Couple allowed, in addition, the production of linear motion as a combination of circular motions, and thus allowed someone like Ibn al-Shāṭir, and after him Copernicus, to create the effect of enlarging the size of the epicyclic radius and of shrinking it by using uniform circular motion only or combinations thereof.

The other result that has become clear from this overview is that the tradition of criticism of Ptolemaic astronomy became a well-established tradition after the thirteenth century, and very few astronomers could do any serious work without attempting some reform of Greek astronomy on their own. Ironically, this period of original production in Arabic is usually thought of as a period of decadence in Islamic science and little effort is spent to study it in any depth.

But recent scholarship on Copernican astronomy, especially that of Swerdlow and Neugebauer, has left no doubt that this Arabic tradition in astronomy must have had an impact on Copernicus himself, and only future research will reveal the exact nature of the channels of transmission from the East to the West that were responsible for this impact.

NOTES

1 For a complete statement of these problems and their solutions see the fuller discussion that follows.

2 For a description of the direction of motion and the problems associated with it, see Toomer (1984: 20 and 221) where he says that a point A moves 'clockwise' as being in 'advance [i.e. in the reverse order]of the signs'.

3 See, for example, Petersen (1969), Pedersen (1974: 167–95) and Neugebauer (1975: 68f).

4 For a geometric description of Mercury's model in the *Almagest*, see Toomer (1984: 444–5).

5 We know about this anonymous author from a treatise titled simply *Kitāb al-hay'a*, which seems to have survived in a unique copy at the Osmania Library in Hyderabad (India), and is summarized later in this chapter.

6 For the contents of this work I use the Cairo edition. There is a preliminary English translation of this text that was completed as a dissertation by Dan Voss at the University of Chicago under Noel Swerdlow's supervision (unpublished).

7 *Shukūk*, p. 23. For the clumsy derivation of the limits of eclipses by Ptolemy in the *Almagest*, VI, 5, see Pedersen (1974: 227f).

8 The actual statement of Ptolemy is: 'Now it is our purpose to demonstrate for the five planets, just as we did for the sun and the moon, that all their apparent anomalies can be represented by uniform circular motions, since these are proper to the nature of divine beings, while disorder and non-uniformity are alien (to such things)' (Toomer 1984: 420).

9 *Shukūk*, pp. 48–58. See also page 60 for the comparison between the conditions of the spheres and the shells of spheres.

10 The work of Jābir ibn Aflaḥ has not yet been fully analysed. That of al-Biṭrūjī was published by Goldstein (1971), and the work of Averroes in conjunction with that of al-Biṭrūjī was first analysed by Gauthier (1909) and more recently by Sabra (1984).

11 For a full description of the problem of the planetary distances in the Ptolemaic works, see Swerdlow (1968).

12 Escurial Ms. Arab. 910, fols 78^{v}–79^{r}.

13 Cf., for example, Kennedy *et al.* (1983), passim.

14 Ibn al-Shāṭir, *Nihāyat al-Sūl*, Bodleian Library Ms., Marsh 139, fol. 4^{v}.

15 *ibid.*, fol. 10^{r}.

16 The present author has completed a critical edition of this text of Ibn al-Shāṭir, which is now being prepared for the press. The references given here, however, are to the Bodleian Arabic manuscript Marsh 139.

17 For the date of 'Urḍī's work, see Saliba (1979a), and for the edition of the text, see al-'Urḍī, Mu'ayyad al-Dīn: *Kitāb al-Hay'a* (*Tārīkh 'ilm al-falak al-'arabī*).

18 A translation of that chapter, including the intended lemma, has been given, in French, by Baron Carra de Vaux (1893), and more recently in English by Faiz Jamil Ragep (1993), pp. 194–223.

19 We use for this study the Köprülü Ms 657, which is dated 20, Jumāda I, 681 AH (27 August 1282) within the lifetime of Shīrāzī (d. 1311).

20 Ms. fol. 61ᵛ.

21 Ms. fol. 66ʳ, the last sentence in the quotation is the same one quoted by ʿUrḍī, *Kitāb al-Hayʾa*, p. 136. See also p. 118 of the same text where ʿUrḍī states that he had gone against the opinion of all astronomers in matters relating to the directions of the motions of the lunar spheres and the magnitudes of these motions (*khālafnā fīhi jamīʿa aṣḥāb ʿilm al-hayʾa*). In a forthcoming article the present author will show the exact indebtedness of Shīrāzī to ʿUrḍī in regard to the lunar model.

22 For this astronomer, see Suter (1900: 165, n. 404). The work used for this study is Ṣadr al-Sharīʿa's *Kitāb al-Taʿdīl fī al-hayʾa*, British Museum Add. 7484, fol. 27ʳ sqq.

23 *Ibid.*

24 For a brief description of this model see Roberts (1957: esp. pp. 430–2).

25 In an earlier version of *Nihāyat al-Sūl* ('The Final Quest'), the book in which Ibn al-Shāṭir proposed his new astronomy, the radius of this sphere is taken to be sixty-seven.

26 This measurement was not given in the early version of the *Nihāya*.

27 Again, these measurements were not given in the earlier version of the *Nihāya*.

28 In the earlier version of the *Nihāya* he adds a note to the effect that this sphere should not be confused with the commonly known epicyclic sphere for they are not the same.

29 For a general survey of these solutions see Saliba (1984). The rest of this section depends heavily on this article.

30 For the edition of this author's work, see al-ʿUrḍī, *Kitāb al-Hayʾa*, and Saliba (1979a).

31 For the derivation of the greatest distance between the two, see Swerdlow (1973: esp. p. 469).

32 See ʿUrḍī, *Kitāb al-Hayʾa*, pp. 235–8. The following citation is found on p. 237.

33 *Ibid.*, pp. 246–57. The following citation is found on pp. 250–1.

34 *Ibid.*, p. 257.

35 In the *Tadhkira*, Leiden Ms. Or. 905, fol. 47ʳ, he says: 'As for [the model of] Mercury, I have not yet been able to imagine that in the proper manner. For it is difficult to imagine the cause for the uniform motion of a [body] around a point by having [that body] move with a complex composite motion closer to that point or away from it. If God were to grant me success in that, I would append it at the appropriate place, if God wills it'.

36 See Ibn al-Shāṭir's attack against earlier astronomers who used eccentric deferents at the beginning of his *Nihāyat al-Sūl*, ch. 2.

37 Ibn al-Shāṭir speaks of two spheres having identical radii, one carried at the cincture of the other. This could only mean that he was thinking of a Ṭūsī Couple, and not of two intersecting circles, for in medieval terminology such spheres would have had to intersect with each other which was never allowed.

4

Astronomy and Islamic society: Qibla, gnomonics and timekeeping

DAVID A. KING

(a) Qibla: The sacred direction

INTRODUCTION

In the Qur'ān. Muslims are enjoined to face the sacred precincts in Mecca during their prayers. The relevant verse (2.144) translates: 'turn your face towards the Sacred Mosque; wherever you may be, turn your face towards it ...'. The physical focus of Muslim worship is actually the Kaʿba, the cube-shaped edifice in the heart of Mecca. This formerly pagan shrine of uncertain historical origin became the physical focus of the new religion of Islam, a pointer to the presence of God.

Thus Muslims face the Kaʿba in their prayers, and their mosques are oriented towards the Kaʿba. The *miḥrāb*, or prayer-niche, in the mosque indicates the qibla, or local direction of Mecca. In medieval times the dead were buried on their sides facing the qibla; nowadays burial is in the direction of the qibla. Islamic tradition further prescribes that a person performing certain acts, such as the recitation of the Qur'ān, announcing the call to prayer, and the ritual slaughter of animals for food, should stand in the direction of the qibla. On the other hand, bodily functions should be performed perpendicular to the qibla. Thus in their daily lives Muslims have been spiritually and physically oriented with respect to the Kaʿba and the holy city of Mecca for close to fourteen centuries.

Muslim astronomers devised methods to compute the qibla for any locality from the available geographical data, treating the determination of the qibla as a problem of mathematical geography, as the Muslim authorities do nowadays. However, mathematical methods were not available to the Muslims before the late eighth or early ninth century. Furthermore, even in later centuries the qiblas found by computation were not generally used anyway. This is immediately clear from an examination of the orientations of medieval mosques, which are aligned towards Mecca, but not always according to the scientific definition of the qibla. The methods commonly used to find the qibla were derived from folk astronomy. Cardinal directions sanctioned by religious tradition and astronomical risings and settings were favoured. Thus the Muslims adopted different notions of a sacred direction different from those of the Jews and the Christians, who generally favoured praying toward the east. There was a most compelling reason for this independent development.

THE ORIENTATION OF THE KAʿBA

The Kaʿba itself is astronomically aligned, i.e. its rectangular base is oriented in astronomically significant directions. The earliest recorded state-

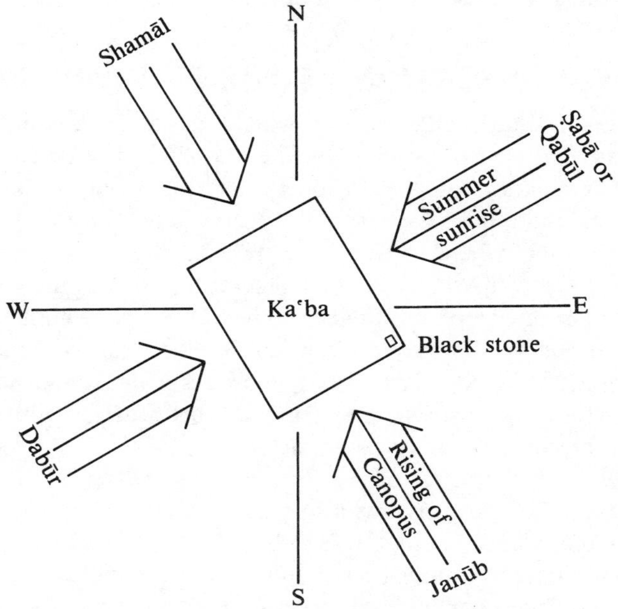

Figure 4.1 The astronomical orientation of the Kaʿba, mentioned in several medieval Arabic texts and confirmed by modern investigations. The associated wind-scheme shown here is also described in the medieval sources

ments about the astronomical alignment of the Kaʿba date from the seventh century, being attributed to Companions of the Prophet. The texts imply that the major axis points towards the rising of the star Canopus, the brightest star in the southern sky, and that the minor axis points towards midsummer sunrise. These two directions are roughly perpendicular to the latitude of Mecca (Figure 4.1). Modern plans of the Kaʿba and the surrounding mountains based on aerial photography essentially confirm the information provided by the medieval texts.

From these texts it is clear that the first generations of Muslims knew that the Kaʿba was astronomically aligned, so this was why they used astronomical alignments in order to face the Kaʿba when they were far away from it. In fact, they often used the same astronomical alignments to face the appropriate section of the Kaʿba as they would if they had been standing directly in front of that particular section of the edifice. One of several popular wind-schemes associated the four cardinal winds with the four walls of the Kaʿba (Figure 4.1).

For these reasons alignments with astronomical horizon phenomena and wind directions were used for qibla determinations for over 1,000 years.

THE ORIENTATION OF THE FIRST MOSQUES

The Prophet Muḥammad had said when he was in Medina: 'What is between east and west is a qibla', and he himself had prayed due south to Mecca. In emulation of the Prophet, and interpreting his remark as implying that the qibla was due south everywhere, certain Muslims used south for the qibla wherever they were. When mosques were erected from Andalusia to Central Asia by the first generation of Muslims known as the Companions of the Prophet (*ṣaḥāba*), some of these were built facing south even though this was scarcely appropriate in places far to the east or west of the meridian of Mecca. Certain early mosques from Andalusia to Central Asia bear witness to this. One may compare this situation with the eastern orientation of churches and synagogues.

Not only did the practice of the Prophet inspire later Muslims, but the practice of his Companions was also emulated. The Prophet himself had said: 'My Companions are like stars to be guided by; whoever follows their example will be rightly guided'. For this reason the qiblas adopted by the Companions of the Prophet in different parts of the new Islamic commonwealth remained popular in later centuries. In Syria and Palestine they adopted due south for the qibla, which was the generally accepted qibla in both regions thereafter. This qibla direction had the double advantage of having been used by the Prophet and by his Companions. In other parts of the Islamic commonwealth the first generation of Muslims adopted directions other than due south, for reasons which will become apparent below.

Some of the first mosques established outside the Arabian Peninsula were erected on the sites of previously existing religious edifices or were adapted from such edifices. Thus, for example, in Jerusalem the *Aqṣā* Mosque was built in the year 715 on the rectangular Temple area. Its *miḥrāb* was aligned with the major axis of the complex to face roughly due south. This direction was favoured as the qibla in Jerusalem in later centuries even though the astronomers had calculated that, according to the available geographical data, the qibla in Jerusalem was about 45° east of south.

Again, about the year 715, the Byzantine cathedral in Damascus, itself built on the site of a pagan temple, was converted into a mosque. The site was aligned in the cardinal directions, as was the orthogonal grid of the street-plan of the Greco-Roman city. The *miḥrāb* of the new mosque was placed in the southern wall. In Damascus the qibla of due south was favoured over the centuries in spite of the fact that the astronomers had calculated the qibla there at about 30° east of south. For this reason most medieval mosques in Damascus face south.

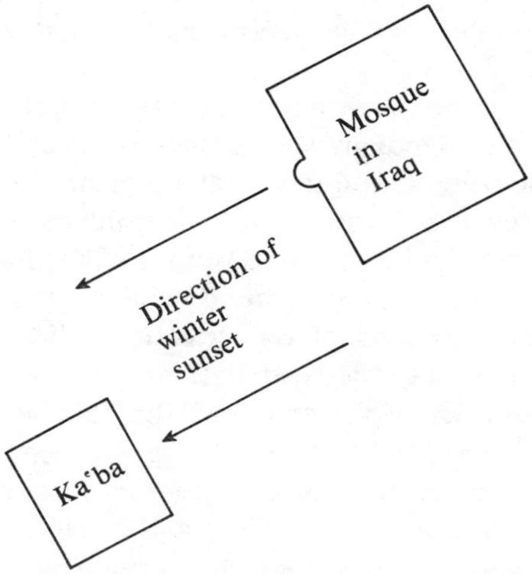

Figure 4.2 The qibla in Iraq was taken by certain authorities as the direction of winter sunset. One of the reasons for this was that the northeastern wall of the Kaʿba was associated with Iraq, and that if one stands in front of this wall one is indeed facing winter sunset

The first mosque to be built in Egypt was built facing winter sunrise, and it was this direction which remained the most popular throughout the medieval period amongst the religious authorities. Likewise some of the earliest mosques in Iraq were built facing winter sunset. These orientations were chosen so that the mosques would be 'facing' specific walls of the Kaʿba (Figure 4.2). Throughout the medieval period, winter sunrise and sunset were favoured in Egypt and Iraq respectively as the *qiblat al-ṣaḥāba*.

FINDING THE QIBLA BY NON-MATHEMATICAL METHODS

Simple practical means for finding the qibla by the sun, moon and stars, and even by the winds, are outlined in a wide variety of medieval texts. The methods advocated in these sources are adapted from the notions underlying the folk–scientific tradition which was widely disseminated in the Muslim world throughout the medieval period. This popular tradition of astronomy and meteorology was ultimately derived from pre-Islamic Arabia, but had been embellished by the indigenous as well as the Hellenistic traditions of folk science which had been practised in the areas overrun by the Muslims in the seventh century. It was quite distinct from the scientific tradition of the Muslim astronomers, but was far more widely known and practised.

Documented for the first time in the early centuries of the Islamic era, this astronomical lore was eventually applied on a popular level to the practical problems of organizing the agricultural calendar, regulating the lunar calendar and the religious festivals, reckoning the time of day by shadow lengths and the time of night by the positions of the lunar mansions and, what concerns us here, finding the direction of the qibla by non-mathematical means. Aspects of this scientific folklore are practised in agricultural communities in the Near East to this day.

Unlike the 'astronomy of the ancients', the popular scientific tradition relied solely upon observation of natural phenomena such as the sun, moon, stars and winds. As the Qur'ān states that these celestial bodies and natural phenomena were created by God, and specifically that men should be guided by the stars, folk astronomy, unlike mathematical astronomy and astrology, was not criticized by the legal scholars.

In the texts mentioned above, the qibla in individual localities is defined in terms of an astronomical horizon phenomenon, such as the rising or setting of a prominent star or of the sun at the equinoxes or solstices. Qibla directions are also given in terms of wind directions. These sources were not compiled by astronomers, but rather were texts dealing with the legal obligation of facing the qibla in prayer or texts dealing with folk astronomy.

Such non-mathematical methods for finding the qibla are occasionally cited in treatises on geography or history. The astronomers themselves are generally silent on these non-mathematical procedures.

The stars rise and set at fixed points on the horizon for a particular locality. At the equinoxes, sunrise and sunset define east and west, and the positions of sunrise and sunset at the solstices are some 30° north of these in midsummer and some 30° south of these in midwinter. The sources state that, for example, the qibla in Northwest Africa is towards the rising of the sun at the equinoxes (due east); that the qibla in the Yemen is towards the direction from which the north wind blows or is towards the Pole Star (which does not rise or set, but whose position defines north); that the qibla in Syria is towards the rising of the star Canopus; that the qibla in Iraq is towards the setting of the sun at midwinter; or that the qibla in India is towards the setting of the sun at the equinoxes (due west).

However, the situation was not quite so simple as this because different authorities proposed different means for finding the qibla in each region.

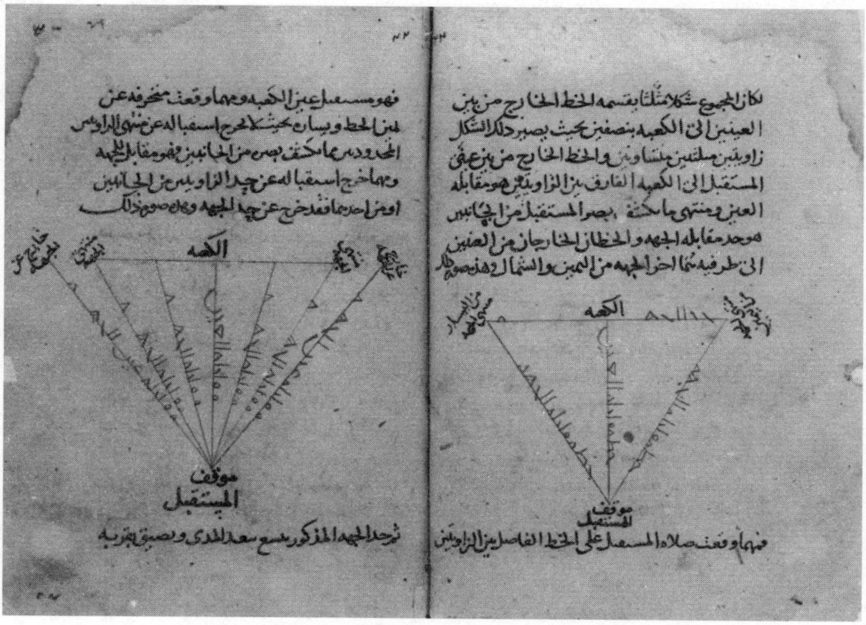

Plate 4.1 The two different general procedures for finding the qibla advocated by legal scholars. Taken from MS Oxford Bodleian Marsh 592, fols. 23ᵛ–24ʳ, of a twelfth-century Egyptian legal text on the qibla, with kind permission of the Keeper of Oriental Manuscripts, Bodleian Library, Oxford

In fact, sometimes different legal schools advocated radically divergent qiblas. In Central Asia, for example, one legal school favoured due west, which was the direction in which the road to Mecca left the region, and the rival legal school favoured due south because of the Prophetic dictum cited above. Others favoured the qibla used by the Companions who built the first mosques in the region, i.e. toward winter sunset. Yet others, of course, favoured the qibla computed by the astronomers.

In an attempt to resolve such problems, some legal scholars proposed that, whilst standing so that no one was actually facing the Kaʿba (in such a way that if one could actually see it, one's line of vision would be along a side of the edifice) was to be favoured, it was also permissible to pray in any direction which would be within one's field of vision in that optimal position (see Plate 4.1). The Arabic phrases *jihat al-Kaʿba* and *ʿayn al-Kaʿba* used to describe these two situations translate roughly as 'standing so as to face the Kaʿba head-on' and 'standing so as to face the general direction of the Kaʿba'. Since one's field of vision is slightly more than one quadrant of the horizon, both due west and due south were, at least to some, legally acceptable qibla directions for Central Asia. Likewise, due

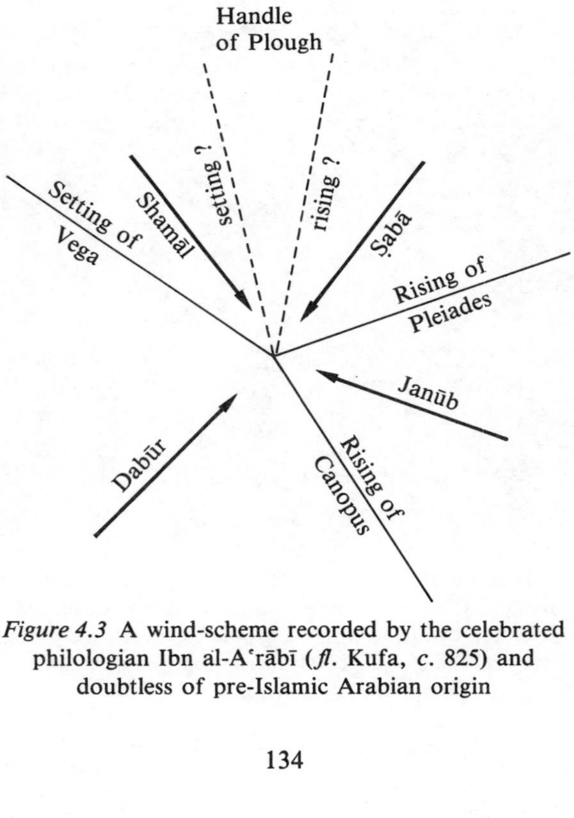

Figure 4.3 A wind-scheme recorded by the celebrated philologian Ibn al-Aʿrābī (*fl.* Kufa, *c.* 825) and doubtless of pre-Islamic Arabian origin

east and due south were both accepted by those Andalusian legal scholars who held the opinion that the entire southeastern quadrant constituted the qibla.

As noted above, we sometimes find qibla directions expressed in terms of wind directions, instead of astronomical horizon phenomena. Here we should bear in mind that several wind-schemes, defined in terms of solar or stellar risings and settings, were part of the folk astronomy and meteorology of the Arabian Peninsula before the advent of Islam. The limits of the winds in these schemes, which are recorded in various early Islamic sources, were defined either in terms of the rising or setting of such stars or star-groups as Canopus, the Pleiades and the stars of the handle of the Plough (which in tropical latitudes do rise and set), or in terms of the cardinal directions or sunrise and sunset at the solstices (Figure 4.3). One of the most popular wind-schemes was the one associating the four winds with the walls of the Kaʿba (see above and Figure 4.1). Thus when a wind direction is mentioned for the qibla, it is assumed that one knows the astronomically defined limits from between which the wind blows.

THE SACRED GEOGRAPHY OF ISLAM

The notion of a sacred geography, with a world divided into sectors about the Kaʿba and each facing a particular part of the Kaʿba, was widely accepted in the Muslim world in medieval times. The Islamic notion of a world oriented about the Kaʿba has its parallels in the medieval Jewish and Christian traditions of a world centred on Jerusalem, but is considerably more sophisticated than either.

One example of an Islamic scheme in this tradition is displayed in Plate 4.2, which is taken from an eighteenth-century Egyptian manuscript, although the scheme itself is much earlier, dating back at least to the twelfth century. The world is divided into eight sectors about the Kaʿba, and the *miḥrāb*s or prayer-niches in each sector face a specific segment of the perimeter of the edifice. The twelfth-century Egyptian legal scholar al-Dimyāṭī described the notion in the following terms:

> The Kaʿba with respect to the inhabited parts of the world is like the centre of a circle with respect to the circle. All regions face the Kaʿba surrounding it as a circle surrounds its centre, and each region faces a particular part of the Kaʿba.

The Kaʿba itself has various features which lend themselves to particular schemes. Since the edifice has four sides and four corners, a division of the world into four or eight sectors around it would be natural, and such four- and eight-sector schemes were indeed proposed. However, in other schemes

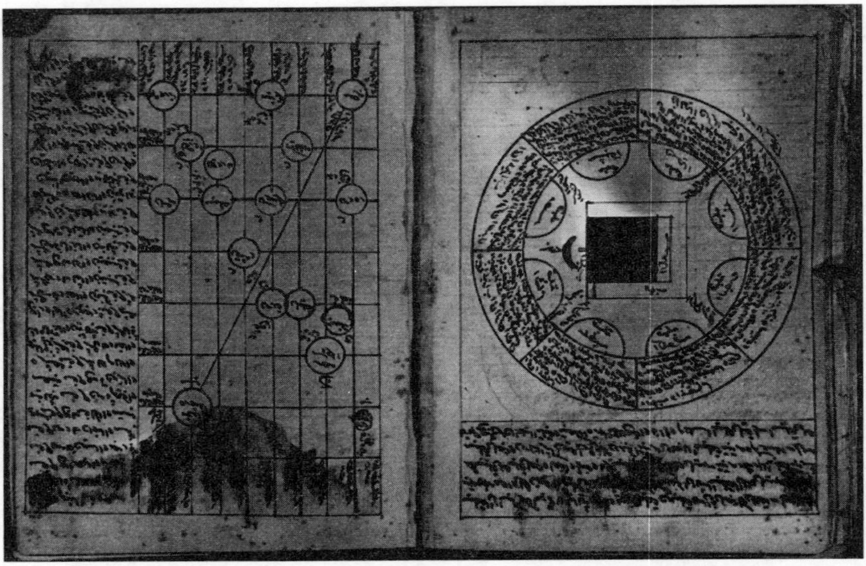

Plate 4.2 Two diagrams in an Ottoman Egyptian treatise on magic, mysticism and folk astronomy. On the right-hand side is an early eight-sector scheme of sacred geography. On the left-hand side is a latitude–longitude grid marked with the Kaʿba and various localities: an approximate value for the qibla can be found by measuring the inclination of the meridian of the line joining the locality to the Kaʿba. Taken from MS Cairo Ṭalʿat *majāmīʿ* 811,7, fols 60^{v}–61^{r}, with kind permission of the Director of the Egyptian National Library

the sectors were associated with segments of the perimeter of the Kaʿba, the walls being divided by such features as the waterspout on the northwestern wall and the door on the northeastern wall. In the scheme illustrated in Plate 4.2, the direction which one should face in each sector of the world is defined either in terms of the rising or the setting of a prominent star or star-group or in terms of a wind direction. In other such schemes the qibla is defined in terms of the cardinal directions or the rising or setting of the sun at the solstices. The directions of sunrise and sunset at midsummer, midwinter and the equinoxes, together with the north and south points, define eight (unequal) sectors on the horizon, and together with the directions perpendicular to the solstitial directions they define twelve (roughly equal) sectors. Each of these eight- and twelve-sector schemes was used in the sacred geography of Islam.

The sources for our knowledge of this tradition of sacred geography are treatises on folk astronomy; treatises on mathematical astronomy (especially almanacs of the kind produced annually); treatises on geography; treatises on cosmography; encyclopedias; historical texts; and, last but by no means least, texts dealing with the sacred law. Sometimes the schemes are described in words, sometimes with the aid of diagrams. Altogether more than thirty different sources compiled between the ninth and the eighteenth centuries have been found attesting to this tradition. Of these, only five are published; the remainder exist in unpublished manuscript form. We can be confident that more such works dealing with the subject were compiled but have not survived in the available manuscript sources.

The earliest Kaʿba-centred geographical scheme that is known is a simple four-sector scheme recorded in the published text of the geography of the ninth-century Baghdad scholar Ibn Khurradādhbih (Figure 4.4). One manuscript of the geography of the tenth-century Jerusalem-born geographer al-Muqaddasī contains a crude eight-sector scheme which has been much corrupted by copyists' errors. The scheme may not be original to al-Muqaddasī; it is probably by an even earlier writer.

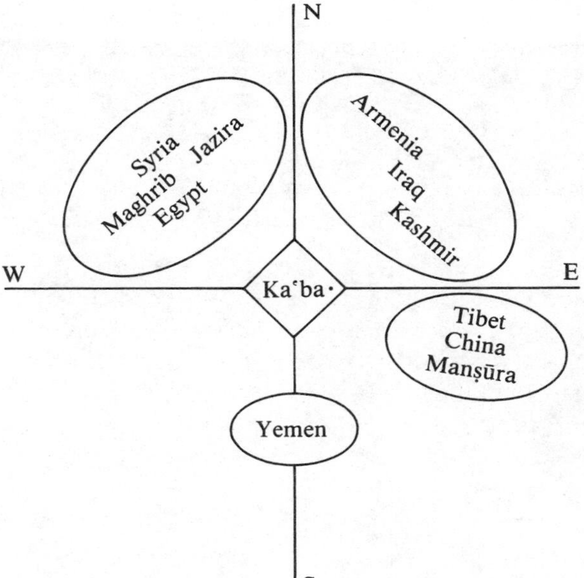

Figure 4.4 A simple scheme of sacred geography associated with Ibn Khurradādhbih

A more developed system of sacred geography was formulated by the tenth-century legal scholar Ibn Surāqa, a native of the Yemen who studied in Iraq. He produced three different schemes with eight, eleven and twelve sectors about the Kaʿba. His works on this subject have not survived in their original form, but his schemes were incorporated into various later treatises. His prescriptions for finding the qibla in each of the various regions about the Kaʿba are outlined in detail without any diagram. For each region he explains how people should stand with respect to the risings or settings of some four stars and the four winds. Thus, for example, the inhabitants of Iraq and Iran should stand in such a way that the stars of the Great Bear rise and set behind the right ear; a group of stars in Gemini rises directly behind the back; the east wind blows at the left shoulder and the west wind blows at the right cheek, and so on. In fact, the stars of the Great Bear do not rise or set in places as far north as Iraq and Iran – there they appear circumpolar. This feature of the instructions indicates that they were actually formulated in Mecca. When one stands there in the position described by Ibn Surāqa one is actually facing winter sunset, although this is not explicitly stated. The ultimate object of the exercise is to face the northeast wall of the Kaʿba.

Plate 4.3 Two different twelve-sector schemes of sacred geography with full instructions for finding the qibla by astronomical horizon phenomena, found in a thirteenth-century Yemeni treatise on folk astronomy. Taken from MS Milan Ambrosiana X73 sup., unfoliated, with kind permission of the Director of the Biblioteca Ambrosiana

In the eight-sector scheme illustrated in Plate 4.2, the qiblas are defined in terms of the stars which rise or set behind one's back when one is standing in the qibla, and in terms of the Pole Star. One would be thus facing these stars if one were standing directly in front of the appropriate section of the Kaʿba with one's back to the edifice. Various twelfth- and thirteenth-century Egyptian and Yemeni astronomical and legal texts contain two different twelve-sector schemes, one adopted from that of Ibn Surāqa. One such Yemeni treatise on folk astronomy presents both schemes – the diagrams are illustrated in Plate 4.3. Several medieval authors whose works were widely read in different parts of the Muslim world, such as the geographer Yāqūt and the cosmographers al-Qazwīnī and Ibn al-Wardī, copied these twelve-sector schemes but omitted the associated instructions for finding the qibla (Figure 4.5).

Yet another scheme occurs in the navigational atlas of the sixteenth-century Tunisian scholar al-Ṣafāqusī. It is distinguished from all others by the fact that there are forty *miḥrāb*s around the Kaʿba, and the scheme is

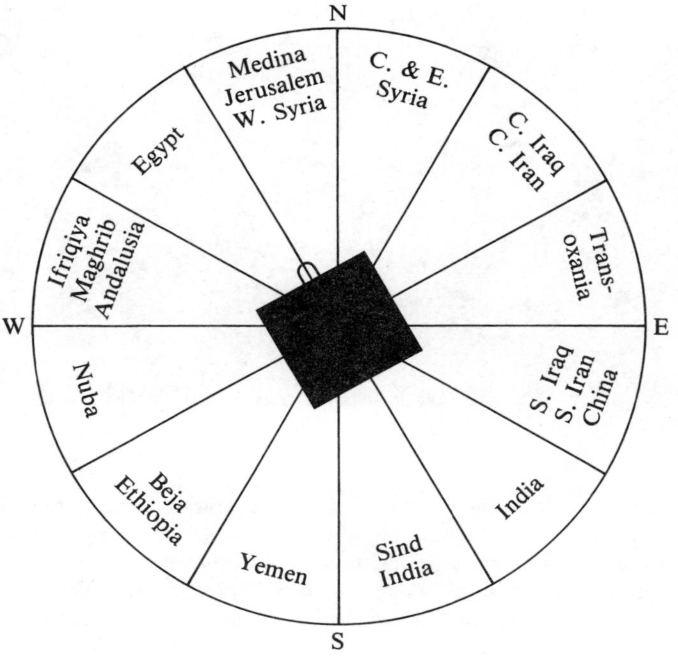

Figure 4.5 A simplified version of Ibn Surāqa's twelve-sector scheme of sacred geography as represented by various late-medieval cosmographers

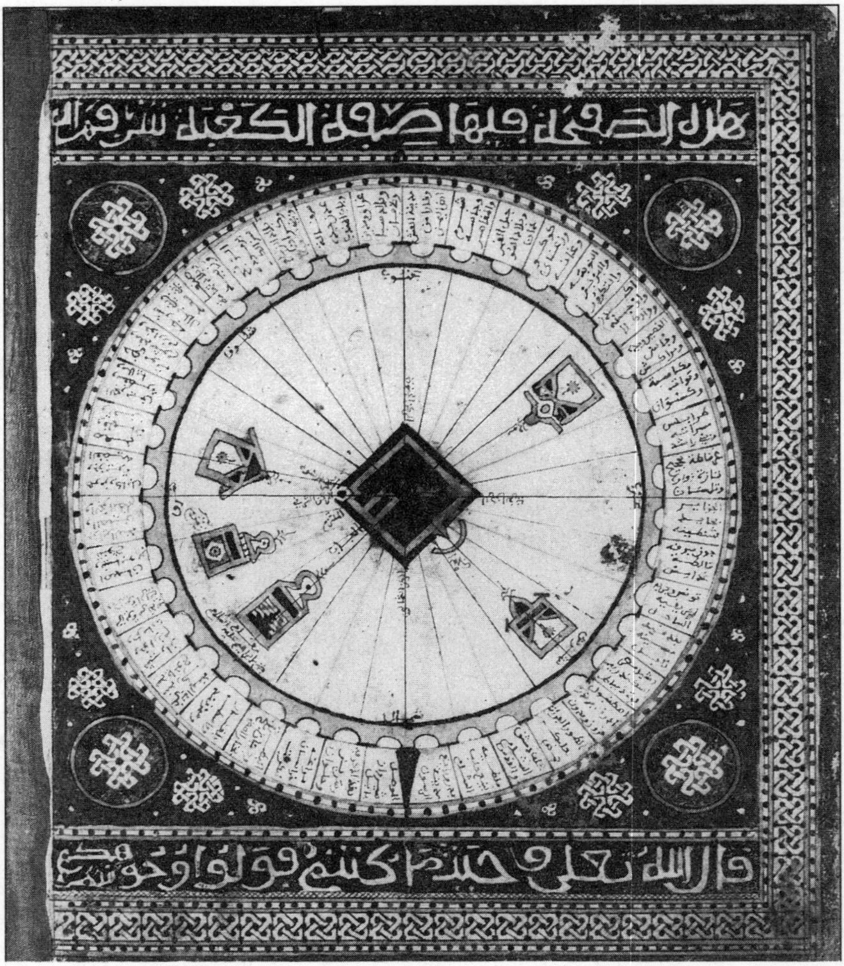

Plate 4.4 A forty-sector scheme of sacred geography in the *Atlas* of the sixteenth-century Tunisian scholar al-Ṣafāqusī. This scheme is superimposed on a thirty-two-sector wind-rose, a device used by Arab navigators for orientation by stellar risings and settings. Taken from MS Paris B.N. ar. 2273, with kind permission of the Director of the Bibliothèque Nationale

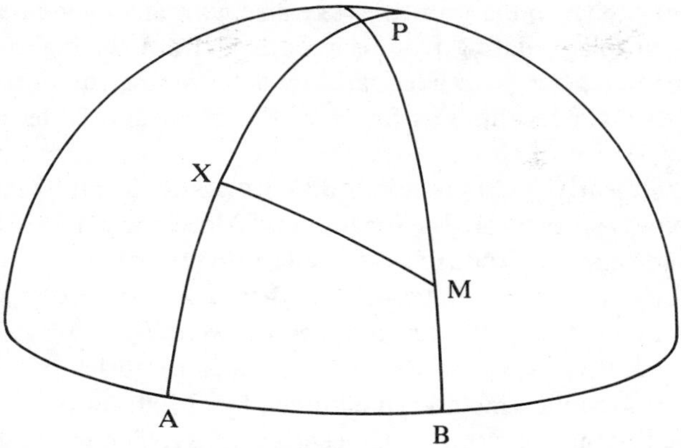

Figure 4.6 The qibla problem on the terrestrial sphere. A given locality and Mecca are represented by X and M, the North Pole by P and the equator by AB. The latitudes of X and M are $XA = \phi$ and $MB = \phi_M$, and their longitude difference $AB = \Delta L$. The angle AXM defines the qibla q

superimposed upon a thirty-two-division wind-rose, a device used by Arab sailors to find directions by the risings and settings of stars (see Plate 4.4).

No new schemes of sacred geography are attested in any known works compiled after the sixteenth century.

FINDING THE QIBLA BY MATHEMATICAL METHODS

The Muslim astronomers defined the qibla as the direction of the great circle joining the locality to Mecca, measured as an angle to the local meridian (Figure 4.6). From the ninth century onwards, they computed the direction of Mecca for various localities. Such calculations required a knowledge of latitudes and longitudes, originally adopted from Ptolemy's *Geography*, and they also involved the application of complicated trigonometric formulae or geometrical constructions, which the Muslims developed from a combination of Greek and Indian methods. The achievements of the Muslim astronomers in this field of endeavour are now fairly well documented in the modern literature, in so far as the methods of several medieval astronomers have been studied and analysed for their mathematical content.

Most Islamic astronomical handbooks with tables (known as *zīj*es and modelled after Ptolemy's *Almagest* and *Handy Tables*) contain a chapter on the determination of the qibla by such procedures. Independent treatises dealing only with the qibla problem were also compiled. The first solutions to the qibla problem, dating from the ninth – if not the eighth – century were approximate, but were adequate for determining the qibla to within a degree or two for localities as far from the meridian of Mecca as Egypt and Iran.

One of these early qibla methods, which owes its inspiration to cartography, involves representing the locality and Mecca on a plane orthogonal grid of latitude and longitude lines and measuring the orientation of the segment joining them (see Plate 4.2). Other approximate mathematical methods and a complicated accurate method were derived by solid geometry, but none of these was widely used in later centuries.

Another approximate method, mentioned by al-Battānī, was widely used and remained popular until the nineteenth century. The method could not be simpler. First draw a circle on a horizontal plane and mark the cardinal directions (Figure 4.7). Then draw a line parallel to the north–south line

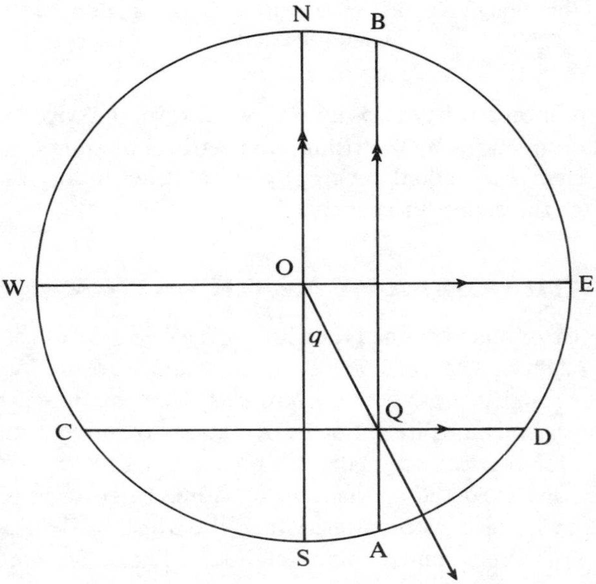

Figure 4.7 al-Battānī's approximate solution to the qibla problem. On the horizontal circle NESW the longitude difference ΔL is marked as SA and the latitude difference $\Delta\phi$ is marked as ED. Segments AB and CD are drawn parallel to NS and EW, respectively, to intersect at Q; then OQ defines the qibla

and at an angular distance – measured on the circle – equal to the longitude difference between Mecca and the locality ΔL, and another line parallel to the east–west line at an angular distance equal to the latitude difference $\Delta\phi = \phi - \phi_M$. Then the line joining the centre of the circle to the intersection of these two lines defines the qibla q. This procedure is equivalent to an application of the simple formula

$$\tan q = \frac{\sin \Delta L}{\sin \Delta\phi}$$

In the ninth and tenth centuries, more sophisticated accurate procedures were derived by plane or solid geometry or by spherical trigonometry. Most medieval scientists dealt with the qibla as a problem of spherical astronomy, in which it is required to determine the azimuth of the zenith of Mecca on the local horizon (Figure 4.8). In their procedures the altitude of the zenith of Mecca must be determined first; then the determination of its azimuth

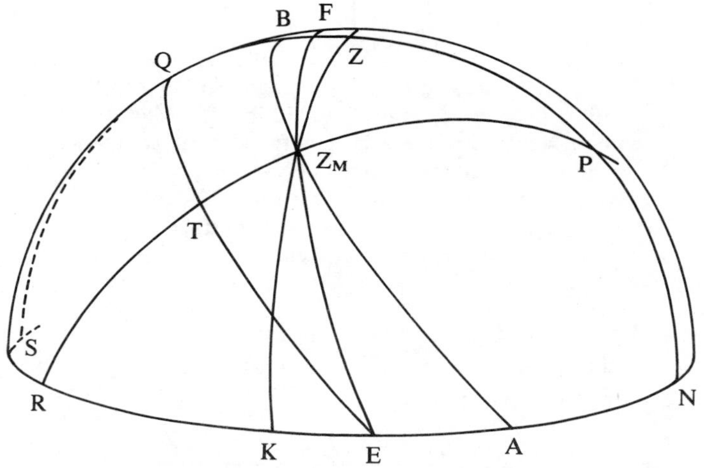

Figure 4.8 The qibla problem transferred to the celestial sphere (see Figure 4.6). It is required to find the azimuth of the zenith of Mecca Z_M. The problem is mathematically equivalent to finding the azimuth a of the sun with declination δ when the hour-angle is t: we have for latitude ϕ, $\delta = \phi_M$, $t = \Delta L$ and $a = q$. To solve this problem by medieval methods involved first finding the altitude of Z_M, namely h, and then deriving the corresponding azimuth a, which is the qibla. For the method of al-Nayrīzī we produce PZ_M to cut the equator at T and the horizon at R. For the method of the *zīj*es we draw the quadrant EZ_MF

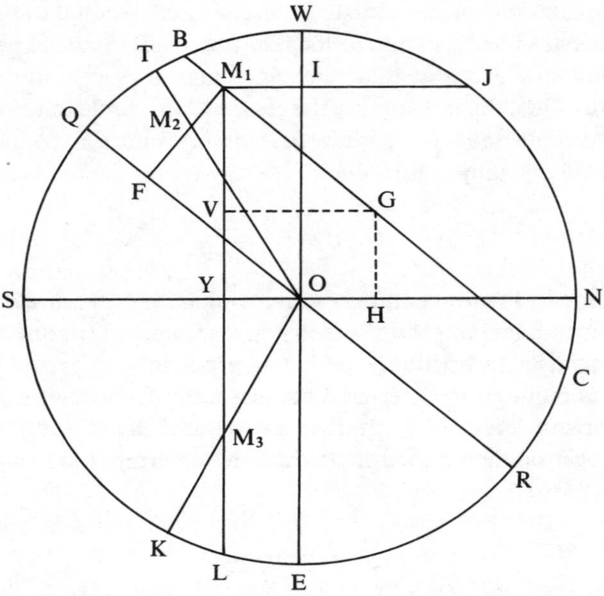

Figure 4.9 A diagram representing the solution to the qibla problem by Ḥabash al-Ḥāsib. This kind of solution, adopted by the Muslims from Greek sources, is known as an analemma. The various planes of operation, namely those of the meridian, celestial equator and horizon, are represented together in a single working plane

is a standard problem of spherical astronomy. These methods are all ultimately equivalent to an application of the modern co-tangent formula for spherical trigonometry, which yields

$$\cot q = \frac{\sin \phi \cos \Delta L - \cos \phi \tan \phi_M}{\sin \Delta L}$$

In order to illustrate the elegance of classical and medieval projection methods, we reproduce the geometric procedure outlined by Ḥabash al-Ḥāsib (*fl.* Baghdad and Damascus, *c.* 850), from which this formula follows immediately. His instructions refer to Figure 4.9 (the notation has been to some extent standardized). On a circle centre O mark the cardinal directions NESW, and then mark arc $WQ = \phi$, arc $QB = \phi_M$ and arc $QT = \Delta L$. Draw the diameter QOR and the parallel chord with mid-point G. Mark the point M_2 on OT such that $OM_2 = GC$ and draw the perpendicular M_2M_1 onto BC. Next draw M_1L parallel to WE and M_1IJ parallel to

SN to cut WE in I and the circle in J. Finally, construct the point M_3 on M_1L such that $OM_3 = IJ$ and produce OM_3 to cut the circle at K. Then OK defines the qibla.

This construction may be justified as follows. First, QOR and BGC represent the projections of the celestial equator and the day-circle of the zenith of Mecca in the meridian plane. Second, M_2 represents the projection of the zenith of Mecca in the equatorial plane. If we then imagine the equatorial plane to be folded into the meridian plane, M_2 moves to M_1, which is thus the projection of the zenith of Mecca in the meridian plane. Furthermore, M_1IJ is the projection in this plane of the almucantar (circle with fixed altitude) through the zenith of Mecca, whose radius is IJ. Also M_1I and IJ measure the distances from the zenith of Mecca to the prime vertical and to the line joining the local zenith to O, respectively. Finally, we consider the working plane to represent the horizon: by virtue of the construction, M_3 is the projection of the zenith of Mecca in this plane so that OM_3 produced indeed defines the qibla.

Alternatively the qibla problem could be solved by spherical trigonometry (see vol. II, Chapter 15). Al-Nayrīzī (*fl.* Baghdad, *c.* 900) proposed the following solution using four applications of the cumbersome Theorem of Menelaus. In Figure 4.8, we find successively the arcs TR, SR, MK and KS. First, find TR by considering SRE as the transversal of triangle TQP, thus:

$$\frac{\sin(PS)}{\sin(SQ)} = \frac{\sin(PR)}{\sin(RT)} \frac{\sin(TE)}{\sin(EQ)},$$

i.e.

$$\frac{\sin(180^\circ - \phi)}{\sin(90^\circ - \phi)} = \frac{\sin(90^\circ + TR)}{\sin(TR)} \frac{\sin(90^\circ - \Delta L)}{\sin 90^\circ}.$$

Second, find SR by considering QTE as the transversal of triangle RSP, thus:

$$\frac{\sin(PQ)}{\sin(QS)} = \frac{\sin(PT)}{\sin(TR)} \frac{\sin(ER)}{\sin(ES)},$$

i.e.

$$\frac{\sin 90^\circ}{\sin(90^\circ - \phi)} = \frac{\sin 90^\circ}{\sin(TR)} \frac{\sin(ER)}{\sin 90^\circ},$$

whence ER and SR ($= 90^\circ - ER$).

Third, find MK ($= h$) by considering SRK as the transversal of triangle Z_MZP, thus:

$$\frac{\sin(SP)}{\sin(SZ)} = \frac{\sin(PR)}{\sin(RZ_M)} \frac{\sin(Z_MK)}{\sin(KZ)},$$

i.e.

$$\frac{\sin(180^\circ - \phi)}{\sin 90^\circ} = \frac{\sin(90^\circ + \mathrm{TR})}{\sin(\mathrm{TR} + \phi_{\mathrm{M}})} \frac{\sin(\mathrm{Z_M K})}{\sin 90^\circ} .$$

Finally, find KS $(= q)$ by considering SZP as the transveral of triangle $\mathrm{Z_M}$RK, thus:

$$\frac{\sin(\mathrm{KS})}{\sin(\mathrm{SR})} = \frac{\sin(\mathrm{KZ})}{\sin(\mathrm{ZZ_M})} \frac{\sin(\mathrm{Z_M P})}{\sin(\mathrm{PR})} ,$$

i.e.

$$\frac{\sin q}{\sin(\mathrm{SR})} = \frac{\sin 90^\circ}{\sin(90^\circ - h)} \frac{\sin(90^\circ - \phi_{\mathrm{M}})}{\sin(90^\circ + \mathrm{TR})} .$$

Later Muslim astronomers also used the sine rule and the tangent rule to solve the problem in essentially the same way. The most popular procedure involving spherical trigonometry was known as the 'method of the *zīj*es'. It is recorded in several works from the ninth to the fifteenth century and simply involves finding the azimuth of the zenith of Mecca on the meridian and then on the local horizon. In Figure 4.8 we draw $\mathrm{EZ_M F}$ perpendicular to the meridian and then determine $\mathrm{Z_M F} = \Delta L'$ and $\mathrm{QF} = \phi'$, called the modified longitude difference and the modified latitude, respectively. These two quantities are found by two successive applications of the sine rule, as follows. From right triangles $\mathrm{Z_M FP}$ and TQP we have

$$\frac{\sin(\mathrm{Z_M F})}{\sin(\mathrm{TQ})} = \frac{\sin(\mathrm{Z_M P})}{\sin(\mathrm{TP})} ,$$

i.e.

$$\frac{\sin \Delta L'}{\sin \Delta L} = \frac{\sin(90^\circ - \phi_{\mathrm{M}})}{\sin 90^\circ} ,$$

and from right triangles FQE and $\mathrm{Z_M TE}$ we have

$$\frac{\sin(\mathrm{FQ})}{\sin(\mathrm{Z_M T})} = \frac{\sin(\mathrm{FE})}{\sin(\mathrm{Z_M E})} ,$$

i.e.

$$\frac{\sin \phi'}{\sin(90^\circ - \phi_{\mathrm{M}})} = \frac{\sin 90^\circ}{\sin(90^\circ - \Delta L')} .$$

Then we determine $\mathrm{FZ} = \Delta\phi' = \phi - \phi'$, called the modified latitude difference. Note that $\mathrm{Z_M F}$ and FZ are the coordinates of $\mathrm{Z_M}$ with respect to the zenith Z on the meridian. We now determine $\mathrm{Z_M K} = h$ and finally $\mathrm{KF} = q$, again by two applications of the same rule, as follows. From right triangles Z_MKE and FSE we have

$$\frac{\sin(Z_MK)}{\sin(FS)} = \frac{\sin(Z_ME)}{\sin(FE)},$$

i.e.

$$\frac{\sin(90^\circ - h)}{\sin(90^\circ - \Delta\phi')} = \frac{\sin(90^\circ - \Delta L')}{\sin 90^\circ},$$

and from right triangles KSZ and Z_MFZ we have

$$\frac{\sin(KS)}{\sin(Z_MF)} = \frac{\sin(KZ)}{\sin(Z_MZ)},$$

i.e.

$$\frac{\sin q}{\sin \Delta L'} = \frac{\sin 90^\circ}{\sin(90^\circ - h)}.$$

Some astronomers, such as Ibn Yūnus (*fl.* Cairo, *c.* 980), preferred solutions derived by projection methods. Others, such as Abū al-Wafā' (*fl.* Baghdad, *c.* 975), preferred solutions by spherical trigonometry. Ibn al-Haytham (*fl.* Cairo, *c.* 1025) wrote two treatises on the qibla, treating both kinds of solutions. His universal solution to the qibla problem by the 'method of *zīj*es', in which he considered sixteen possible cases, is of particular mathematical interest. Also al-Bīrūnī (*fl.* Central Asia, *c.* 1025) proposed solutions of both kinds.

Already in the early ninth century, simultaneous observations of a lunar eclipse were conducted at Baghdad and Mecca in order to measure the longitude difference between the two localities with the express purpose of finding the qibla at Baghdad. Al-Bīrūnī devoted an entire treatise to the determination of the qibla at Ghazna (now in Afghanistan). He used several different methods to measure the longitude difference between Mecca and Ghazna, took the average of the result and then calculated the qibla by several different accurate procedures. His treatise is a classic of mathematical geography and scientific method.

Muslim astronomers from the ninth century onwards also computed tables displaying the qibla as a function of terrestrial latitude and longitude, some based on approximate formulae and others based on the accurate formula. Some eight different tables are known from the manuscript sources, and one, by Ibn al-Haytham, has not been identified yet. An extract from one of the most remarkable of these tables, which was compiled by al-Khalīlī, a professional timekeeper (*muwaqqit*) at the Umayyad Mosque in Damascus in the fourteenth century, is displayed in Plate 4.5. Also, the

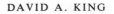

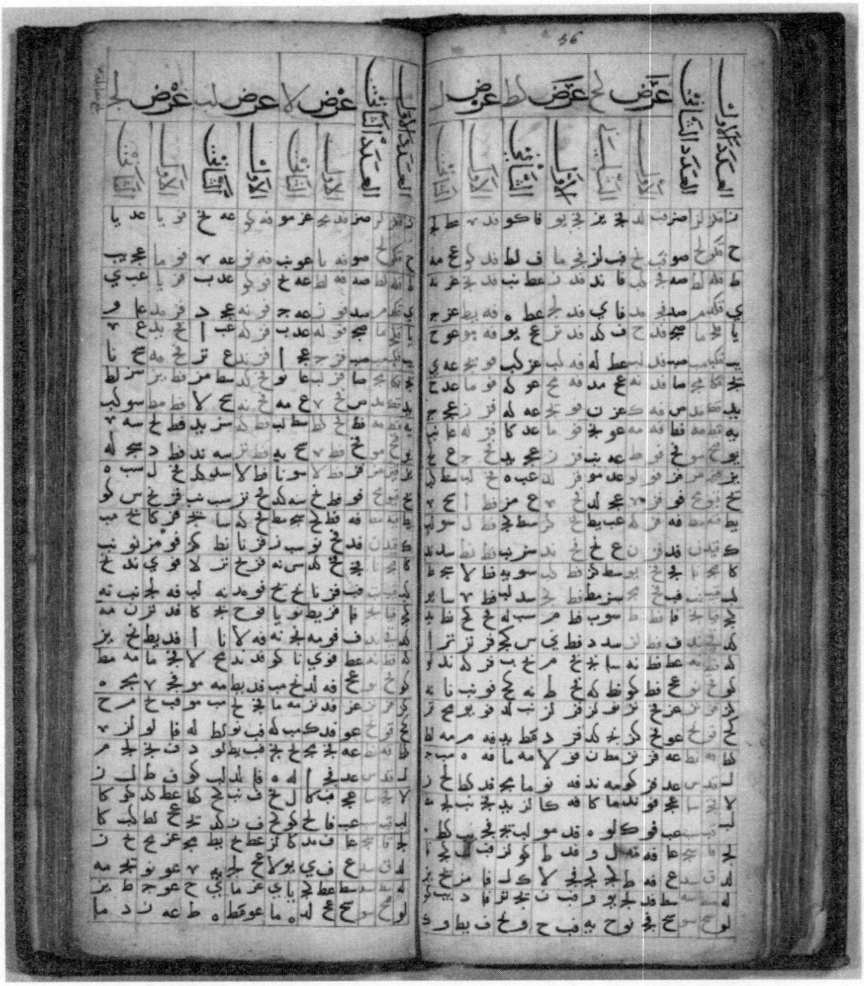

Plate 4.5 An abstract from the qibla-table of the fourteenth-century Damascus astronomer al-Khalīlī. This sub-table shows entries for latitudes 28°, 29°, ..., 33°, entered horizontally; the vertical arguments correspond to longitude differences ranging from 1° to 60°. Taken from MS Paris B.N. ar. 2558, fols 56ᵛ–57ʳ, with kind permission of the Director of the Bibliothèque Nationale

tables of geographical coordinates which were a feature of every Islamic astronomical handbook sometimes included qibla values for each locality.

Islamic treatises on the use of instruments such as the astrolabe and different varieties of quadrants usually included a chapter on finding the

Plate 4.6 An instrument for finding the qibla, from Iran (nineteenth century?). On the top half of the dial numerous localities are marked relative to Mecca at the centre; on the bottom half is a horizontal sundial for an unspecified latitude. Photograph courtesy of the Museum of the History of Science, Oxford

qibla by means of the instrument. From the fourteenth century onwards, compass boxes were available bearing lists of localities with their qibla directions or simple cartographic representations of the world about Mecca (Plate 4.6). Such devices have enjoyed a remarkable revival during the last few years: Saudia Airlines has recently purchased one million qibla boxes from a company in Switzerland for distribution to its passengers.

FINDING THE QIBLA FROM MECCA-CENTRED WORLD-MAPS

In 1989 a remarkable world-map centred on Mecca came to light (see Plate 4.7). It is engraved on a circular brass plate of diameter 22.5 cm, and was originally fitted with some kind of universal sundial. Some 150 localities between Andalusia and China are marked and named. The highly-sophisticated cartographical grid (see Figure 4.10) is conceived so that one can read the direction of any locality to Mecca from the circumferential scale and the distance to Mecca (in *farsakhs*) from the non-uniform scale on the diametral rule. From the calligraphy it is clear that the map dates from Isfahan, *c.* 1700, and the maker may be ʿAbd ʿAlī or his brother

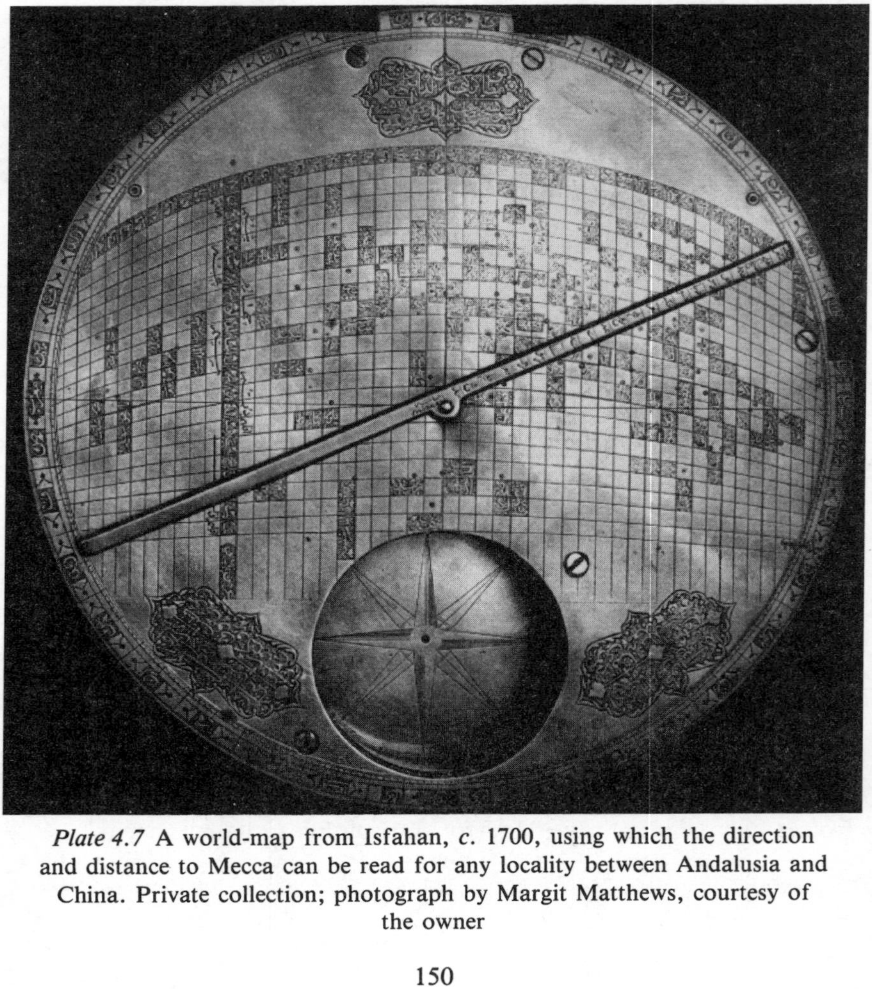

Plate 4.7 A world-map from Isfahan, *c.* 1700, using which the direction and distance to Mecca can be read for any locality between Andalusia and China. Private collection; photograph by Margit Matthews, courtesy of the owner

Muḥammad Bāqir, who together made the magnificent astrolabe presented to Shāh Ḥusayn in 1712, which is now in the British Museum. In 1995 another Mecca-centred world-map, like the first Isfahan one but somewhat later, came to light. It is fitted with a European-type universal, inclining sundial, and is signed by one Muḥammad Ḥusayn. The maker is probably to be identified as the son of the well-known Safavid mathematician Muḥammad Bāqir Yazdī, who in turn may be identical with the Muḥammad Bāqir mentioned above.

There is no parallel to these maps in Islamic cartography and no apparent trace of any European influence; indeed, they are without parallel in the history of cartography. Prior to the rediscovery of the first one in 1989 it was thought that the first person to construct a world-map centred on Mecca from which one could read off the qibla and the distance to Mecca was the German historian of Islamic science Carl Schoy, who published such a map *c.* 1920 (see Figure 4.11). The burning question remains: who designed the cartographical grid? And why would Muslim craftsmen attach to such grid a European-type sundial, a type quite useless for regulating any of the times of Muslim prayer other than midday?

In Safavid Iran no innovations of consequence were made in science, and when European notions were introduced – celestial maps on astrolabe plates, universal inclining sundials, and mechanical clocks – they are easily

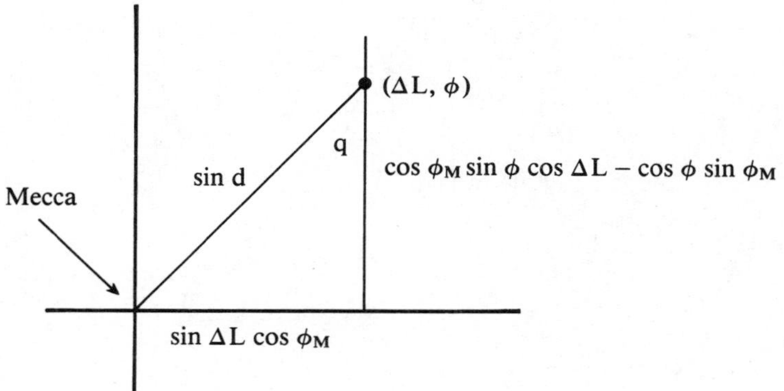

Figure 4.10 The mathematics underlying the theory of the grid on the Isfahan world-map, enabling the user to read the qibla on the circumferential scale and the distance on the diametral scale. An approximation has been used on the world-map so that the latitude curves are arcs of circles; this produces slight inaccuracies noticeable only on the edges of the map (that is, in Andalusia and China)

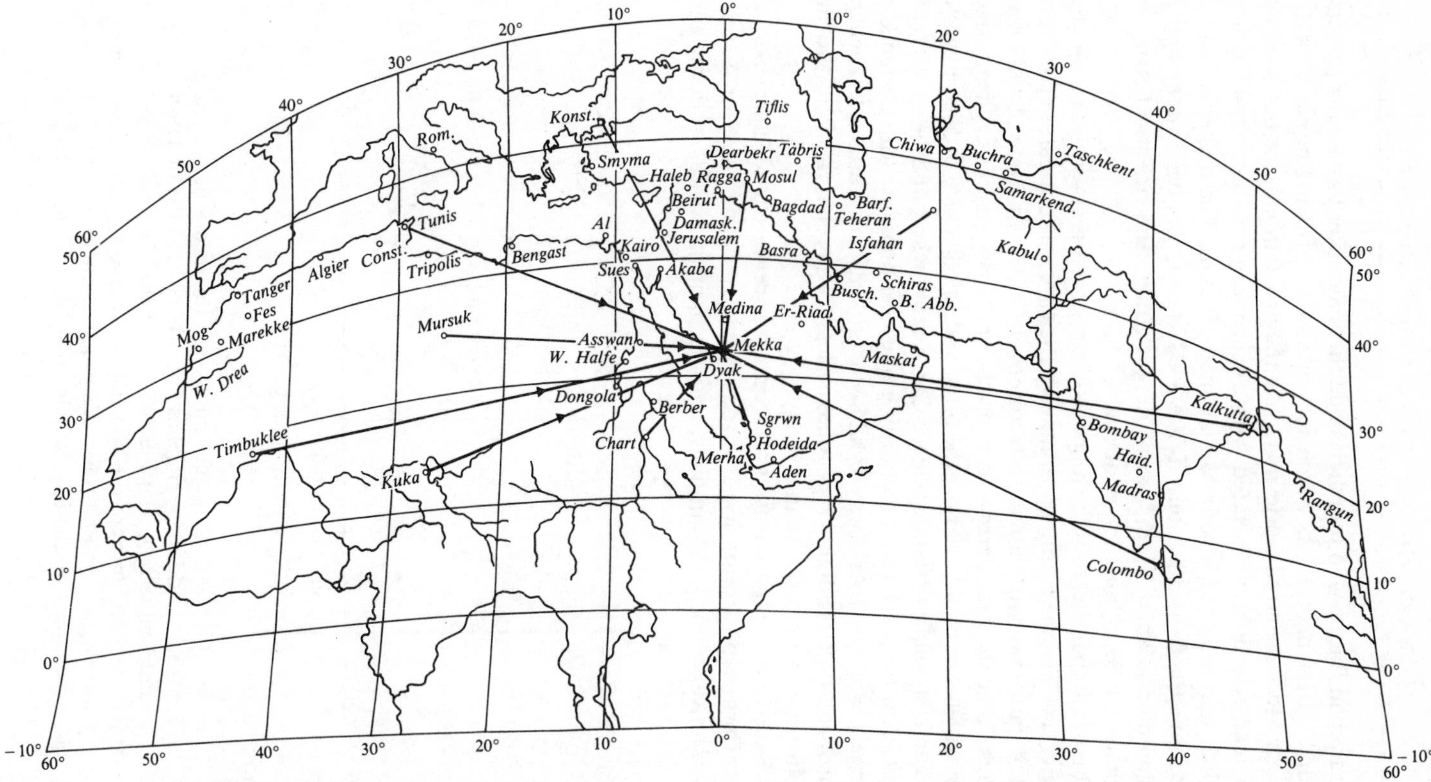

Figure 4.11 A world-map preserving direction and distance to Mecca, published by Carl Schoy, *c.* 1920

identifiable. Most Safavid treatises on qibla-determinations do not progress beyond the trivial, with extensive discussions of the quadrant in which one might expect to find the qibla and of the standard approximate method for calculating specific qibla-directions. One such is a treatise by Qāsim ʿAlī Qāyinī, a student of Muḥammad Ḥusayn ibn Muḥammad Bāqir Yazdī! Now, on the two maps the selection of localities is similar but not identical, and it is clear that the geographical data is taken from a more extensive list, and it seems likely that both are copies of a more detailed map based on the same principle. A fifteenth-century Persian geographical table, entirely within the Islamic tradition and known only from an early eighteenth-century copy, constitutes the common source, not only for the geographical data incorporated into both maps but also for the extensive geographical data engraved on various Safavid astrolabes (such as the one presented to Shāh Ḥusayn mentioned above). The compiler of this table not only listed longitudes and latitudes of some 250 localities but also accurately computed the qibla and distance from Mecca for each locality. Such competence in computation may well have been matched by the ingenuity needed to design a highly sophisticated grid based on the same principles.

Nevertheless at least this author would not be surprised if the idea behind the cartographical grid goes back far beyond the fifteenth century. Al-Bīrūnī (*fl.* Central Asia, *c.* 1025) wrote on a polar equi-azimuthal equidistant projection of the celestial sphere, but as yet no trace of a map by him centred on Mecca has been found (my assertions in 1994 that the geographical data from such a map had been located have been proven in 1995 to be misfounded). Al-Bīrūnī's surviving works on mathematical geography mark a high point between Antiquity and the Renaissance, and some ten other books by him on the subject, most focussing on the qibla-problem, are known to us only by title. Research on the development of this remarkable tradition of Mecca-centred world-maps is currently in progress.

ON THE ORIENTATION OF ISLAMIC RELIGIOUS ARCHITECTURE

Of course, the accuracy (judged by modern criteria) of a value of the qibla computed by a correct mathematical procedure for a particular locality depends on the accuracy of the available geographical data. Medieval latitude determinations were usually accurate to within a few minutes, but estimates of the difference in longitude between Mecca and various localities might be in error by several degrees. In Cairo, for example, the modern qibla is some 8° south of the qibla of the medieval astronomers, because they relied on a value for the longitude difference which was too small by 3°.

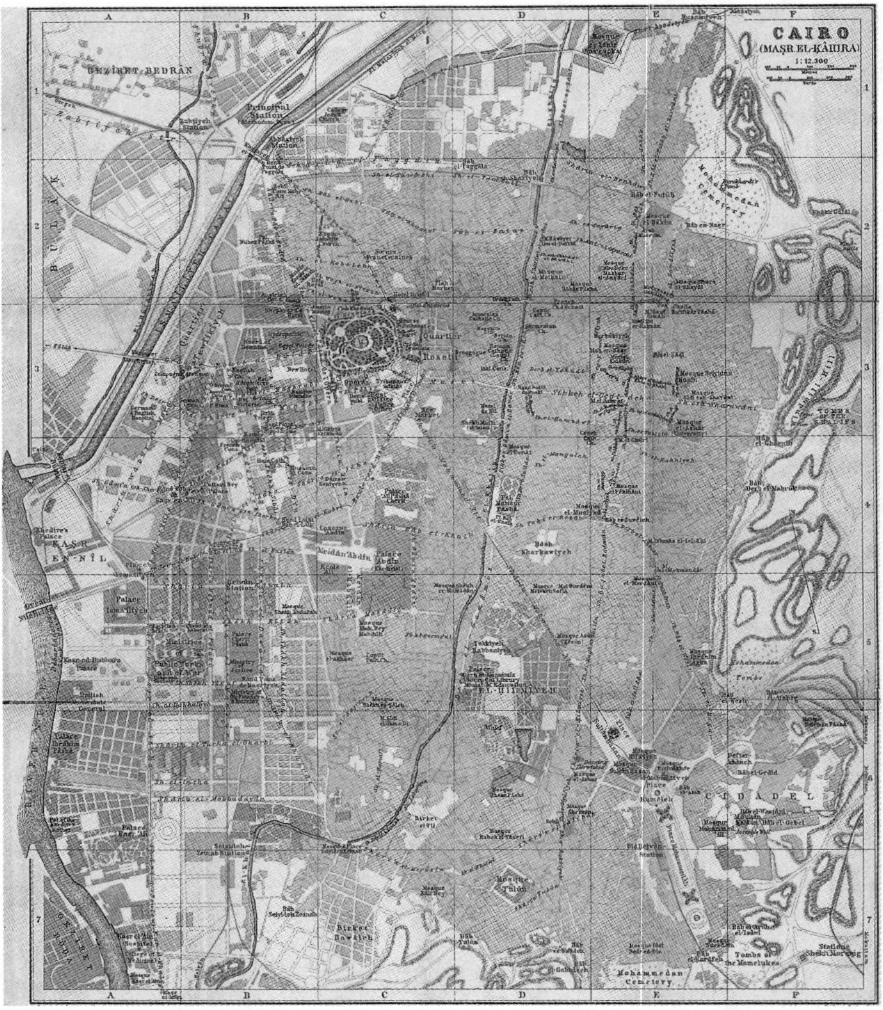

Plate 4.8 A plan of medieval Cairo showing the Mosque of al-Ḥākim and the Azhar Mosque inclined at about 10° to the street plan of the Fatimid city founded a few years earlier in the year 969. Both mosques were oriented in the qibla of the astronomers (*c.* 37° South of East) whereas the minor axis of the city is oriented towards the qibla of the Companions of the Prophet who conquered Egypt, i.e. towards winter sunrise (*c.* 27° South of East). The later Mamluk 'City of the Dead' was built entirely in the qibla of the astronomers. The modern qibla for Cairo is about 45° South of East but this is irrelevant to any discussion of the orientation of medieval mosques

Now it is quite apparent from the orientations of mosques erected between the seventh and the nineteenth centuries that the astronomers were not always consulted on the matter of the qibla. Some mosques, to be sure, are indeed oriented in the qiblas determined by the astronomers for the locality in question, but they constitute a minority. The different qiblas proposed in the various sources to some extent explain the diversity of mosque orientations in any given region of the Muslim world. For certain localities, yet more information on mosque orientations is available.

In Córdoba for example, as we know from a twelfth-century treatise on the astrolabe, some mosques were laid out towards winter sunrise because it was thought that this would make their qibla walls parallel to the northwest wall of the Kaʿba, itself thought by some authorities to be facing winter sunrise. The Grand Mosque faces a direction perpendicular to summer sunrise for the very same reason. Its axis is indeed parallel to the axis of the Kaʿba, a fact which explains why it faces the deserts of Algeria rather than the deserts of Arabia.

As noted already, the earliest mosque in Egypt, the Mosque of ʿAmr in Fustat, was laid out towards winter sunrise. The new city of al-Qāhira (Cairo) was laid out in the late tenth century a few miles to the north of Fustat with a more or less orthogonal street plan alongside the canal linking the Nile to the Red Sea. Now it happened quite fortuitously that the canal, first built by the ancient Egyptians and then restored once by the Romans and again by the Muslims, flowed past the new city in a direction perpendicular to the qibla of the Companions' mosque in Fustat. Thus the entire city lay in the qibla of the Companions (*c.* 27° South of East). But the Fatimids who built the city did not appreciate their good fortune, and besides, the Fatimid astronomer Ibn Yūnus computed the qibla mathematically as *c.* 37° South of East. So the first Fatimid mosques in Cairo, the Mosque of the Caliph al-Ḥākim and the Azhar Mosque, were erected at 10° skew to the street plan (Plate 4.8). In much of the later (thirteenth–sixteenth century) Mamluk religious architecture in the Old City, the exterior is aligned with the qibla of the Companions and the street plan and the interior is twisted so that the *miḥrāb* faces the qibla of the astronomers. In a suburb of al-Qāhira known as al-Qarāfa, the main urban axis and the various mosques along it have a southerly orientation, because that direction was preferred as the qibla. The entire 'City of the Dead', built by the Mamluks to the east of al-Qāhira, is laid out so that all the mausolea are facing the qibla of the astronomers, both internally and externally, and the roughly orthogonal street plan of the quarter is also aligned with this particular qibla.

In Samarkand, as we know from an eleventh-century legal treatise, the main mosque was oriented towards winter sunset in order that it should face the northeast wall of the Kaʿba. We have already noted that one legal school favoured due west for the qibla and another favoured due south; one would expect to find some of the religious edifices associated with the two schools reflecting this difference of opinion. Other religious architecture in the city was oriented in the qibla determined by the astronomers.

Only a preliminary survey has been made of mosque orientations, using over 1,000 plans available in the modern scholarly literature. Yet most of these plans are unreliable, so that no conclusions can be drawn from this data. Clearly, a proper survey of mosque orientations all over the Muslim world would be of extreme historical interest. Not only should all mosques, madrasas, mausolea and other religious edifices, as well as cemeteries, be carefully measured for their orientation, but also the local horizon conditions should be recorded in order to check for possible astronomical alignments. All measurements should be made with the accuracy achieved in the archaeoastronomical investigations that have been conducted in other parts of the world. This topic has yet to arouse the interest of historians of Islamic architecture – the latest general books on that topic and even regional studies of architecture ignore orientations altogether.

FURTHER READING

For an overview of the whole subject of the qibla see King (1985b). See also the articles '*Anwā*', '*Manāzil*', '*Maṭlaʿ*', '*Kaʿba*', '*Ḳibla*' and '*Makka* (as centre of the world)' in the *Encyclopaedia of Islam* (2nd edn, 8 vols to date. Leiden: E. J. Brill, 1960 to present) for various relevant topics. The 3rd, 5th and 6th are reprinted in King (1993).

On the popular methods of finding the qibla see Hawkins and King (1982) and King (1983a). On the notion of the world divided about the Kaʿba, see King 'The sacred geography of Islam', to appear.

On problems of orientation of religious architecture in Córdoba, Cairo and Samarkand, see King (1978b, 1983b, 1984). See also Barmore (1985) and Bonine (1990), for the only systematic surveys of mosque orientations in particular regions. See also King, 'The Orientation of Medieval Islamic Religious Architecture and Cities: Some Remarks on the Present State of Research and Tasks for the Future', *Journal for the History of Astronomy* (1995).

On the earliest mathematical procedures for finding the qibla, see King (1986a). Other studies on individual methods are Kennedy and Id (1974). Schoy (1921, 1922), Berggren (1980, 1981, 1985) and a study by Dallal (1995b) on Ibn al-Haytham's universal treatment of the qibla problem by spherical trigonometry.

Al-Bīrūnī's *Kitāb Taḥdīd nihāyāt al-amākin* was published by P. Bulgakov (Cairo, 1962) and translated by J. Ali as *The Determination of the Coordinates of Cities: al-Bīrūnī's Taḥdīd al-amākin* (Beirut: American University of Beirut Press, 1967). See further Kennedy (1973).

On medieval tables for finding the qibla, see, in addition to King (1986a), King (1975) and Lorch (1980).

On instruments for finding the qibla see Lorch (1982), Janin and King (1977) and King (1987b). For the sole surviving example of a compass-bowl in which the needle should float on water see S. Cluzan, E. Delpont and J. Mouliérac, eds., *Syrie, Mémoire et Civilisation*, Paris: Flammarion & Institut du Monde Arabe, 1993, pp. 440–1. (This compass is from 16th-century Syria, but the geographical information on it was carelessly copied from a much earlier instrument of the same kind.)

On world-maps centred on Mecca see King, 'Weltkarten zur Ermittlung der Richtung nach Mekka', in G. Bott, ed., *Focus Behaim-Globus*, 2 vols., Nuremberg: Germanisches Nationalmuseum, 1992, I, pp. 167–71, and II, pp. 686–91 and, more recently, *idem*, 'World-Maps for Finding the Direction and Distance of Mecca – a Brief Account of Recent Research', Symposium on Science and Technology in the Turkish and Islamic World, Istanbul, 3–5 June, 1994, and the article '*Samt*' in *Encyclopaedia of Islam*, 2nd ed. None of these is to be regarded as authoritative.

(b) Gnomonics: Sundial theory and construction

INTRODUCTION

One expression of the Muslim concern with timekeeping and regulating the times of prayer (see below) was an avid interest in gnomonics. Muslim astronomers made substantial contributions to both the theory and practice of the subject, and by the late medieval period there were sundials of one form or another in most of the major mosques in the Islamic world.

The Muslims came into contact with the sundial when they expanded into the Greco-Roman world in the seventh century. Already *c.* 700 the Caliph ʿUmar ibn ʿAbd al-ʿAzīz in Damascus was using a sundial for regulating the times of the daylight prayers in terms of the seasonal hours. This was probably a Greco-Roman sundial that had been found in the city. In antiquity the most common types were the hemispherical and the plane variety, and such sundials would have been known to the earliest Muslim scholars who dealt with mathematical astronomy. But at least al-Fazārī and Yaʿqūb ibn Ṭāriq, who worked in this field in the eighth century, are not known to have written on sundials.

EARLY TEXTS ON GNOMONICS

The earliest surviving Arabic treatise on sundials deals with their construction and was rediscovered only about ten years ago. Its author is stated to be al-Khwārizmī, the celebrated astronomer of Baghdad in the early ninth century. The work consists mainly of a set of tables of coordinates for constructing horizontal sundials for various latitudes (including the equator).

The basic mathematics is relatively straightforward although the precise means by which the tables were computed remains to be explained. With values of the solar altitude and azimuth (h, a) computed for the required ranges of solar longitude and time intervals, the radial coordinates of the points of intersection of the hour lines with the shadow traces are simply $(n \cot h, a)$ where n is the length of the gnomon (Figure 4.12). Each of al-Khwārizmī's sub-tables for a specific latitude displays for both of the solstices the solar altitude, the shadow of a standard gnomon (12 units) and the solar azimuth, i.e. triplets (h, s, a) for each seasonal hour of day (Plate 4.9). With these radial coordinates already tabulated, construction of the

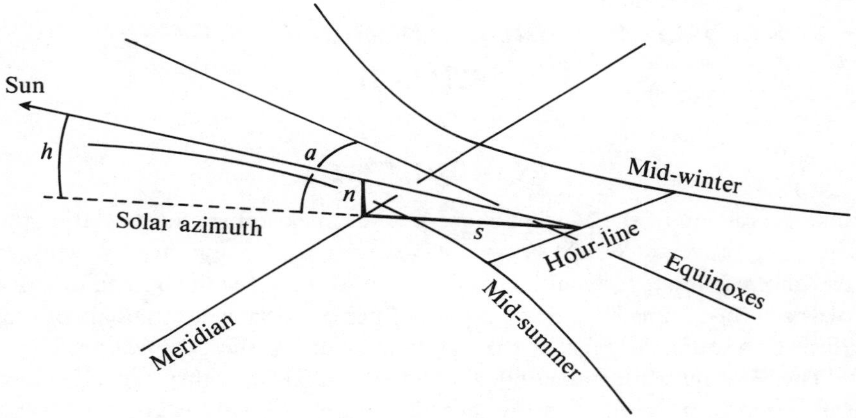

Figure 4.12 The basic theory underlying the construction of a horizontal sundial marked for the seasonal hours

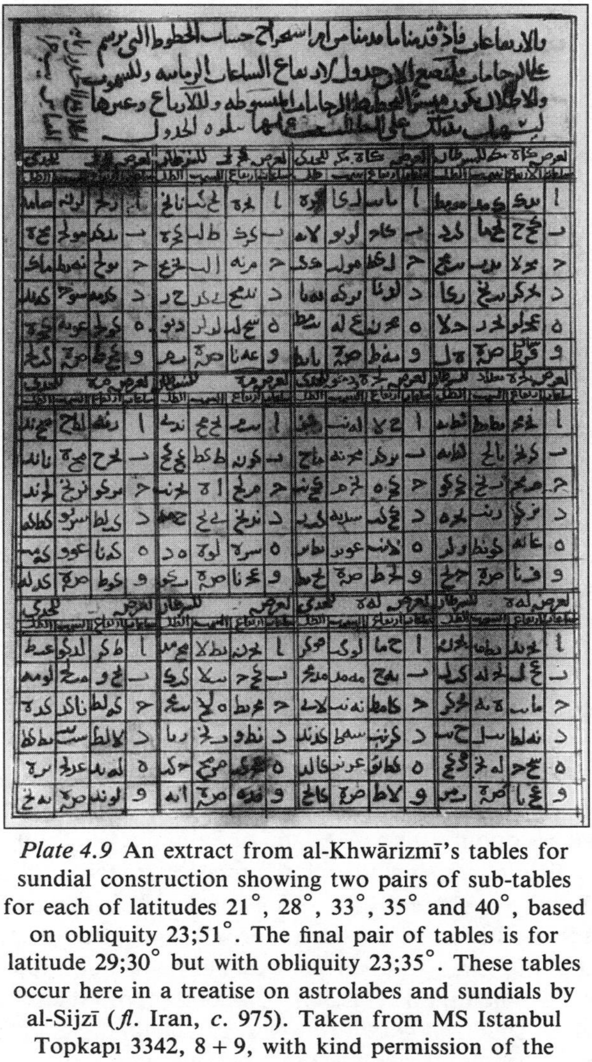

Plate 4.9 An extract from al-Khwārizmī's tables for sundial construction showing two pairs of sub-tables for each of latitudes 21°, 28°, 33°, 35° and 40°, based on obliquity 23;51°. The final pair of tables is for latitude 29;30° but with obliquity 23;35°. These tables occur here in a treatise on astrolabes and sundials by al-Sijzī (*fl.* Iran, *c.* 975). Taken from MS Istanbul Topkapı 3342, 8 + 9, with kind permission of the Director of the Topkapı Library

sundial would have been almost routine. We may presume that sundials were actually constructed using these tables, but none survives from this early period and no descriptions are known from contemporary historical sources.

The celebrated astronomer and mathematician Thābit ibn Qurra (*fl.* Baghdad, *c.* 900) wrote a comprehensive work on sundial theory which has survived in a unique manuscript. It is a masterpiece of mathematical writing, but has attracted remarkably little attention from historians of science since it was published in the 1930s. Thābit's treatise deals with the transformation of coordinates between different orthogonal systems based on three planes: (1) the horizon, (2) the celestial equator and (3) the plane of the sundial. The last may be the plane of (a) the horizon, (b) the meridian or (c) the prime vertical; or it may be (d) perpendicular to (b) with an inclination to (c); (e) perpendicular to (c) with an inclination to (b); (f) perpendicular to (a) with an inclination to (b); or (g) perpendicular to (c) with an inclination to (b), i.e. skew to (a), (b) and (c).

Thābit presents formulae for the solar altitude as a function of the hour-angle, declination and terrestrial latitude which are clearly derived by projection methods, and other formulae for coordinate conversion which are more easily explained by means of spherical trigonometry. Unfortunately he gives no clues how he derived the various formulae, and it is not known how he arrived at them. Even if Thābit had been familiar with Ptolemy's writings such as the *Analemma*, in which similar coordinate transformations are discussed, his own treatise is very much the result of a mature reworking of the material.

As far as we know, Thābit's main treatise on sundial theory is not referred to by any later astronomer. So it appears to have been of very limited influence in later Islamic gnomonics despite the fact that it is the most sophisticated Arabic account of the subject. Later Muslim astronomers were more interested in the practical side of gnomonics.

A unique fifteenth-century copy of a tenth-century treatise on the construction of vertical sundials has also survived. The work is by one of the two Baghdad astronomers Ibn al-Ādamī or Saʿīd ibn Khafīf al-Samarqandī: the copyist was not sure. Included in the treatise are tables of the functions $a(T, \lambda)$ and $z(T, \lambda)$ (where $z = 90° - h$ is the zenith distance of the sun) for each half seasonal hour of time since sunrise T and each 30° of solar longitude λ. Values are given to three sexagesimal digits and are computed for the latitude of Baghdad, taken as 33°. A second set of tables displays values of the functions $\sin \theta$ and $\cot \theta$ to three sexagesimal digits and each degree of argument. The base used for the sine function is 10, which is most unusual but simply means that the gnomon length was taken as 10. Two tables of the co-tangent function are presented, one to base 10 and another to base 1. The utility of these two sets of tables for generating pairs of orthogonal coordinates for drawing vertical sundials at any orientation to the meridian is obvious when we observe that for the sun at azimuth A from a vertical sundial with a perpendicular horizontal gnomon of length n – see Figure 4.13 – the orthogonal coordinates of the end of the gnomon shadow

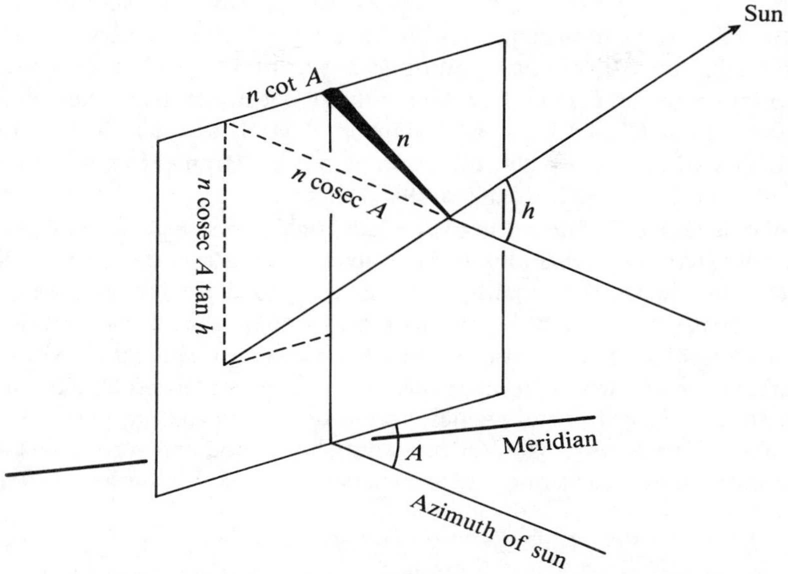

Figure 4.13 The basic theory underlying the construction of a vertical sundial inclined at an angle to the local meridian

measured with respect to the horizontal (x) and vertical (y) axes through the base of the gnomon are ($-n$ cot A, $-n$ cosec A tan h).

Even though several early works of consequence on gnomonics have been lost without trace, there is no shortage of other early material awaiting to be studied.

LATE TEXTS ON GNOMONICS

The major Islamic work on sundial theory from the later period of Islamic astronomy was a compendium of spherical astronomy and instrumentation appropriately entitled *Jāmiʿ al-mabādiʾ wa-l-ghāyāt fī ʿilm al-mīqāt* (*An A to Z of Astronomical Timekeeping*). It was compiled by Abū ʿAlī al-Marrākushī, an astronomer of Moroccan origin who worked in Cairo *c.* 1280. It is difficult to assess al-Marrākushī's own contribution to this enormous work (the Paris copy comprises 750 pages). The lengthy sections on sundial theory, with numerous tables mainly for Cairo, seem to be original, but we have no information on earlier Egyptian texts on sundial theory. In addition, the contemporary activity of al-Maqsī (see below) seems to be quite independent.

Al-Marrākushī's treatise was widely influential in later astronomical circles in Egypt, Syria and Turkey, and it survives in several copies. Although it is the most important single source of Islamic instrumentation, it has still not received the attention it deserves from historians. A French translation of the first half dealing with spherical astronomy and sundial theory was published by J.-J. Sédillot in 1834–5, and a rather confused summary of the second half dealing with other instruments was published by his son, L. A. P. Sédillot, in 1844.

Al-Marrākushī's discussion of sundials, richly illustrated with diagrams, concentrates on descriptions of the mode of construction; there is little underlying theory and usually no clue given as to how the numerous tables were constructed. The text deals with horizontal, vertical, cylindrical and conical sundials. There is also a discussion of 'winged' sundials in which the markings cover two adjacent plane surfaces, with a common axis in the horizon or vertical planes. A description of a compendium of scales and graphs for measuring shadows, converting horizontal and vertical shadows and calculating ascensions is also included. This device, known as *mīzān*

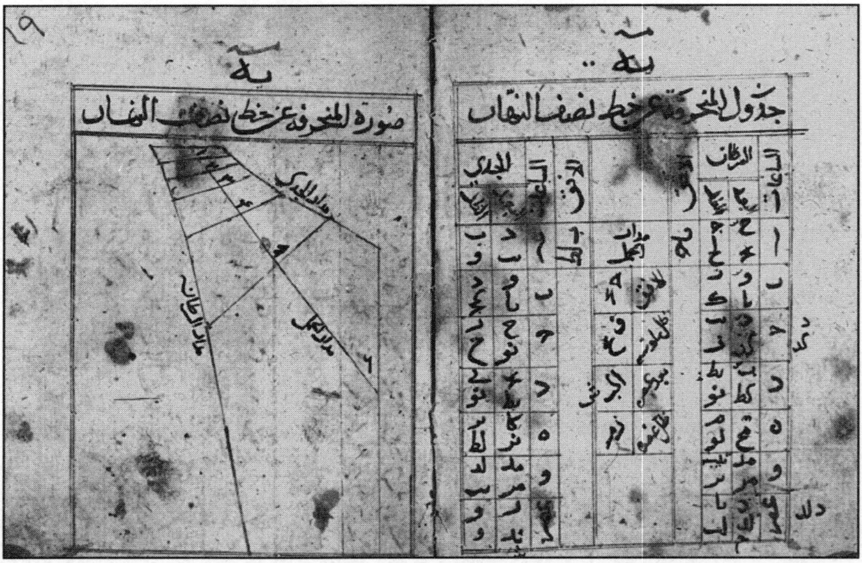

Plate 4.10 An extract from al-Maqsī's tables for constructing vertical sundials for the latitude of Cairo. This particular sub-table serves an inclination of 15° to the meridian. Taken from MS Cairo Dār al-Kutub *mīqāt* 103, fols 68ᵛ–69ʳ, with kind permission of the Director of the Egyptian National Library

al-Fazārī ('the balance of al-Fazārī') seems to be related to the eighth-century astronomer of that name.

A contemporary of al-Marrākushī, the Cairene astronomer al-Maqsī, compiled a set of tables for sundial construction which was also rather popular amongst later Egyptian astronomers. Al-Maqsī prepared tables for horizontal sundials for various latitudes, but the bulk of his treatise consists of tables for marking vertical sundials for the latitude of Cairo. For each degree of inclination to the local meridian he tabulated the coordinates of the points of intersection of the lines for the seasonal hours and the *ʿaṣr* with the shadow traces at the equinoxes and the solstices (Plate 4.10). Several later astronomers compiled extensive tables for sundial construction for specific latitudes, notably Cairo, Damascus and Istanbul; these still await study.

SUNDIALS

Only a few sundials survive from the medieval period. Hundreds or even thousands must have been constructed from the ninth century onwards, but the vast majority have disappeared without trace. Most, but not all, of the surviving sundials constructed before *c.* 1400 have been published; however, no inventory of Islamic sundials has been prepared yet.

Most Islamic sundials bear markings for the hours (seasonal or equinoctial) and for the midday (*ẓuhr*) and afternoon (*ʿaṣr*) prayers. Since the definitions of the beginnings of these two prayers were in terms of shadow lengths, the determination of the prayer times with a sundial was singularly appropriate.

HORIZONTAL SUNDIALS

The oldest surviving Islamic sundial (Plate 4.11) was made by Ibn al-Ṣaffār, an astronomer of some renown who worked in Córdoba about the year 1000. Only one half of the instrument survives but the remains are adequate to establish that gnomonics was not the maker's forte. The sundial is of the horizontal variety and there are lines for each of the seasonal hours, some with a kink at the shadow trace for the equinox, which is itself not straight. There are markings for the *ẓuhr* prayer and there would also have been markings for the *ʿaṣr*. The gnomon is now missing, but its length is indicated as the radius of a circle engraved on the sundial. Several other later Andalusian sundials which survive are singularly poor testimonials to their makers' abilities; most are marred by serious mistakes and one is from a practical point of view quite useless. Yet proper sundials must have existed in medieval Andalusia.

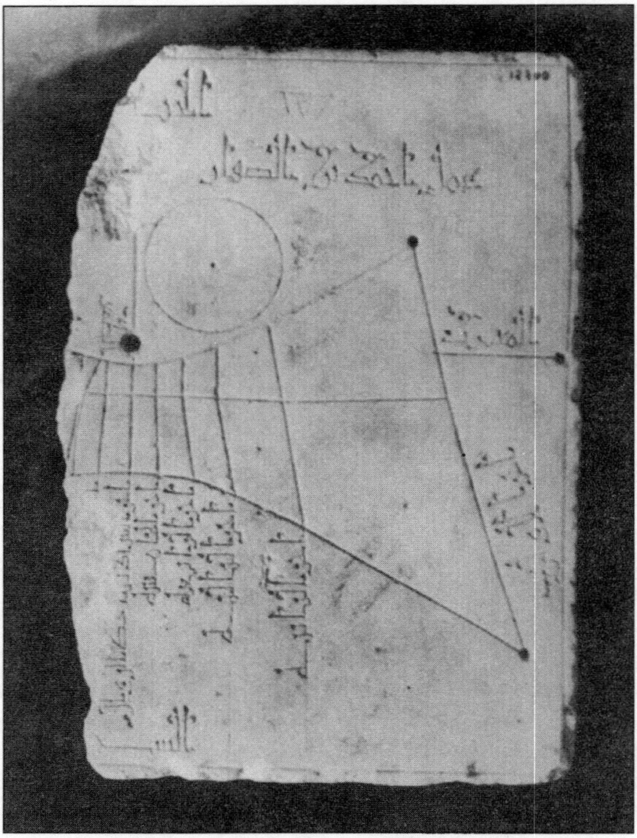

Plate 4.11 The oldest surviving Islamic sundial, made about the year 1000 in Córdoba by Ibn al-Ṣaffār. The curve for the *ẓuhr* is just visible on this fragment, and presumably there were curves for the beginning and end of the *ʿaṣr* as well. Photograph courtesy of the Museo Arqueológico Provincial de Córdoba

The Tunisian sundial shown in Plate 4.12 is a much neater production than the Andalusian sundials mentioned above. It was made in 1345/6 by Abū al-Qāsim ibn al-Shaddād and is of considerable historical interest because its markings display only the times of day with religious significance rather than the seasonal hours. For the afternoon (right-hand side) the curves for the *ẓuhr* and *ʿaṣr* are marked according to the standard Andalusian/Maghribī definitions. For the forenoon there is a curve for the *ḍuḥā*, symmetrical with the *ʿaṣr* curve with respect to the meridian, and a

Plate 4.12 A fourteenth-century Tunisian sundial indicating four times of day with religious significance. Property of the National Museum of Carthage; photograph courtesy of the late M. Alain Brieux, Paris

line for the times of the institution of *ta'hīb* one equinoctial hour before midday, associated with the communal worship on Friday. It was the symmetry of the *ḍuḥā* and *ʿaṣr* curves on this sundial which first led to an understanding of the definitions of the times of the daylight prayers in Islam. Close inspection of the markings on the sundial reveals that the solstitial curves are drawn as arcs of circles rather than hyperbolae. This sundial constitutes, therefore, a rather respectable example of a tradition of marking the solstitial traces in this way, which must have been widely known in medieval Andalusia and the Maghrib.

The astronomer Ibn al-Shāṭir, chief *muwaqqit* of the Umayyad Mosque in Damascus in the mid-fourteenth century, constructed in the year 1371/2 a magnificent horizontal sundial, some 2 m × 1 m in size (Plate 4.13). This was erected on a platform on the southern side of the main minaret of the Mosque, and fragments of it are on display in the garden of the National Museum in Damascus. An exact replica of the original made by the

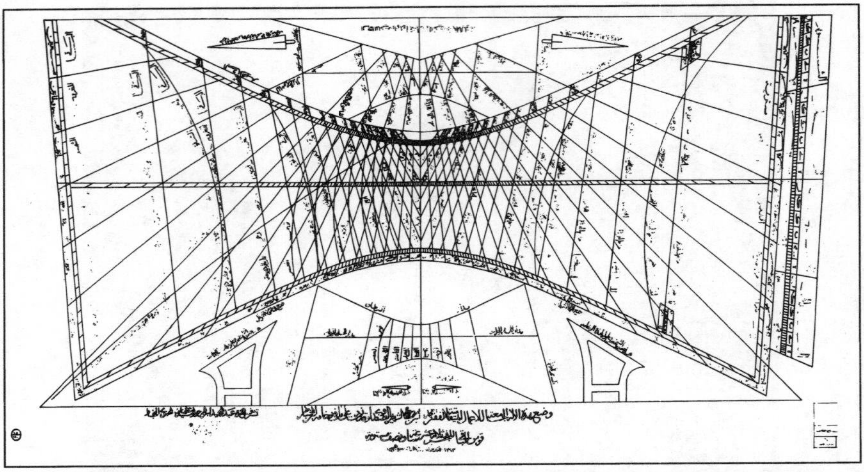

Plate 4.13 The markings on the sundial of Ibn al-Shāṭir which once graced the main minaret of the Umayyad Mosque in Damascus. The original sundial is in fragments, preserved in the garden of the Archaeological Museum in Damascus. This copy is made from an exact replica made by the nineteenth-century *muwaqqit* al-Ṭanṭāwī which is still *in situ* on the minaret. Courtesy of the Syrian Department of Antiquities and the late M. Alain Brieux, Paris

muwaqqit al-Ṭanṭāwī in 1876 is still *in situ* on the minaret. A long line of *muwaqqits* worked in the Mosque from the fourteenth to the nineteenth century, and presumably they used Ibn al-Shāṭir's sundial for regulating the prayer-times, along with the tables and other instruments that were also available to them.

Ibn al-Shāṭir's sundial has three main sets of markings. Indeed, there are actually three sundials inscribed on the marble slab. The small northern sundial with its own gnomon has markings for the seasonal hours and the *ʿaṣr* prayer. The small southern sundial has markings for the equatorial hours before midday and after midday, as well as after sunrise and before sunset. Its gnomon, parallel to the celestial axis, is ingeniously aligned with the larger gnomon of the third and main sundial. The latter bears markings for each twenty minutes before midday and after midday, as well as for each twenty equatorial minutes after sunrise up to midday and for each twenty minutes before sunset starting at midday. There are also curves for each twenty minutes up to the *ʿaṣr* prayer starting two hours before the prayer, as well as curves for the times three and four hours after daybreak and before nightfall. Finally, there is a curve for the time $13\frac{1}{2}$ hours before daybreak the next day, which al-Ṭanṭāwī says he himself added to Ibn al-Shāṭir's sundial.

Thus the sundial can be used to measure time after sunrise in the morning and time before sunset in the afternoon, and time before and after midday. It measures time relative to the *ẓuhr* and maghrib prayers, and the *ʿaṣr* curves enable measurement of time relative to the *ʿaṣr* prayer as well. The curves associated with nightfall and daybreak are for measuring time with respect to the *ʿishāʾ* and *fajr* prayers: when the shadow fell on these lines, the *muwaqqit* would know, for example, that the *ʿishāʾ* would begin in four hours or three hours and could see what the celestial configuration would be at nightfall from his astrolabe or quadrant. It is not clear why the *muwaqqit* would be interested in the times three or four hours after the *fajr* prayer, but when the shadow fell on al-Ṭanṭāwī's curve for $13\frac{1}{2}$ hours before daybreak, he could check with another instrument what the celestial configuration would be at daybreak the next day. The time $13\frac{1}{2}$ hours before daybreak was chosen because this was the latest time that could be shown on the sundial. A masterpiece of ingenuity and design, and an example of outstanding technical skill in stonemasonry, Ibn al-Shāṭir's sundial was first described in the scholarly literature in 1972. It is undoubtedly the most splendid sundial of the Middle Ages.

VERTICAL SUNDIALS

No vertical sundials survive from the first few centuries of Islamic astronomy, but we know they were made because of the treatises on their use which were compiled from the ninth century onwards.

The earliest surviving sundial from Muslim Egypt and Syria, made in 1159/60, is a simple vertical hand-dial. It serves for measuring the seasonal hours and bears two sets of markings on either side, one for latitude 33° (Damascus) and the other for latitude 36° (Aleppo). The instrument is known from texts such as the treatise of al-Marrākushī, where it is called *sāq al-jarrāda*, 'the locust's leg'. It is to be held in a plane perpendicular to that of the sun, with the gnomon attached to one of six holes at the top (which correspond to each pair of zodiacal signs between the solstices). The shadow of the tip of the gnomon will then fall on the markings and the time of seasonal hours can be measured. The inscription, which includes a dedication to the Ayyubid Sultan Nūr al-Dīn al-Zanjī, states that the markings are for determining the seasonal hours and the times of the prayers, from which one can conclude that the times of the *ẓuhr* and *ʿaṣr* prayers were regulated at particular seasonal hours.

The most common kind of vertical sundial was known from the ninth century onwards as *munḥarifa*, meaning simply 'vertical and inclined to the meridian'. There were usually markings for each seasonal hour and the *ʿaṣr* prayer bounded by two hyperbolic shadow-traces for the solstices. Tables such as those of al-Maqsī (see above and Plate 4.10) were particularly useful for constructing such sundials for the walls of mosques.

ASTRONOMICAL COMPENDIA

A compendium, or multi-purpose astronomical instrument, was devised by the fourteenth-century Syrian astronomer Ibn al-Shāṭir. The various movable parts of the instrument all fit into a shallow box with square base, which is covered by a hinged lid. On the outside of the lid there was fitted an alidade which could be rotated over a series of markings with which the user could compute oblique ascensions for Damascus and latitudes 30°, 40° and 50°. The lid could be opened so that it would lie parallel to the celestial equator for a series of six localities in Syria, Egypt and the Hejaz. Two sights could be erected at the end of the alidade and perpendicular to it, so that the alidade could be aligned equatorially with the sun or any northern star and the hour-angle could be read off a circular scale on the lid. A polar sundial, whose markings were engraved on the movable plate, could be set up so that it rested, somewhat precariously, on the sighting devices attached to the alidade now held horizontally. Using the polar sundial, supported in this way, one could read the equatorial hours before or after midday and also see when the time of the *ʿaṣr* had arrived. (However, Ibn al-Shāṭir erred in thinking that an *ʿaṣr* curve on a sundial for latitude zero could be used universally in this way.)

The fifteenth-century Egyptian astronomer al-Wafāʾī developed another compendium which he called *dāʾirat al-muʿaddil*, literally 'the equatorial circle'. The instrument consists of a semi-circular frame attached at the ends of its diameter to a horizontal base; this frame can be aligned in the celestial equator of any latitude. A special sight is attached radially to this frame so that the hour-angle of any celestial body with northern declination less than the obliquity of the ecliptic can be measured (Plate 4.14). The base of the instrument bore markings indicating the qiblas of various localities and occasionally also a horizontal sundial for a specific latitude.

The question of the influence of these Islamic compendia on the compendia which were so popular in Renaissance Europe has yet to be investigated. Otherwise the only Islamic treatise on sundials known in Europe seems to have been that incorporated in the thirteenth-century *Libros del saber*, which lacks, however, any sophisticated theory and tables, features which characterized most Islamic treatises on the subject.

Plate 4.14 A compendium of the variety known as *dā'irat al-muʿaddil*, particularly useful for measuring the hour-angle of the sun or any star at any latitude. Courtesy of the Director of the Museum of History of Science, Kandilli Observatory, Istanbul

FURTHER READING

For an overview see the article '*Mizwala*' in *The Encyclopaedia of Islam* (2nd edn, 8 vols to date, Leiden: E. J. Brill, 1960 to present), reprinted in King (1993). On Islamic sundial theory in general see Schoy (1923, 1924). On tables for constructing sundials see my forthcoming 'Survey of Islamic tables for sundial construction'.

On al-Khwārizmī's sundial tables see Rosenfeld (1983: 221–34) and also King (1983d: esp. 17–22). On Thābit's treatise see Garbers (1936) and Luckey (1937–8).

On al-Marrākushī's treatise see Sédillot J.-J. (1834–5) and Sédillot, L. A. (1844).

On Andalusian sundials see King (1978a). The Tunisian sundial is discussed in King (1977b). On Ibn al-Shāṭir's sundial see Janin (1971).

Other medieval sundials are described in Casanova (1923), Janin and King (1978) 'L'astronomie en Syrie à l'époque islamique,' in S. Cluzan, E. Delpont and J. Mouliérac, eds., *Syrie, Mémoire et Civilisation*, Paris: Flammarion & Institut du Monde Arabe, 1993, pp. 436–9, Bel (1905), Janin (1977) and Michel and Ben-Eli (1965).

The compendium of Ibn al-Shāṭir is discussed in Janin and King (1977), Brice *et al.* (1976) and Dizer (1977).

(c) ʿIlm al-mīqāt: Astronomical timekeeping

INTRODUCTION

The expression *ʿilm al-mīqāt* refers to the science of astronomical timekeeping by the sun and stars in general, and the determination of the times (*mawāqīt*) of the five prayers in particular. Since the limits of permitted intervals for the prayer are defined in terms of the apparent position of the sun in the sky relative to the local horizon, their times vary throughout the year and are dependent upon the terrestrial latitude. When reckoned in terms of a meridian other than the local meridian, the times of prayer are also dependent upon terrestrial longitude.

THE TIMES OF THE PRAYERS IN ISLAM

The definitions of the times of prayer outlined in the Qur'ān and *ḥadīth* were standardized in the eighth century and have been used ever since (Figures 4.14 and 4.15). According to these standard definitions, the Islamic day and the interval for the *maghrib* prayer begins when the disc of the sun has set over the horizon. The intervals for the *ʿishā'* and *fajr* prayers begin at nightfall and daybreak. The permitted time for the *ẓuhr* usually begins when the sun has crossed the meridian, i.e. when the shadow of any object has been observed to increase. In medieval Andalusian and Maghribi practice, it began when the shadow of any vertical gnomon had increased over its midday minimum by one-quarter of the length of the object. The interval for the *ʿaṣr* begins when the shadow increase equals the length of the gnomon and ends either when the shadow increase is twice the length of the gnomon or at sunset. In some circles, an additional prayer, the *ḍuḥā*, was

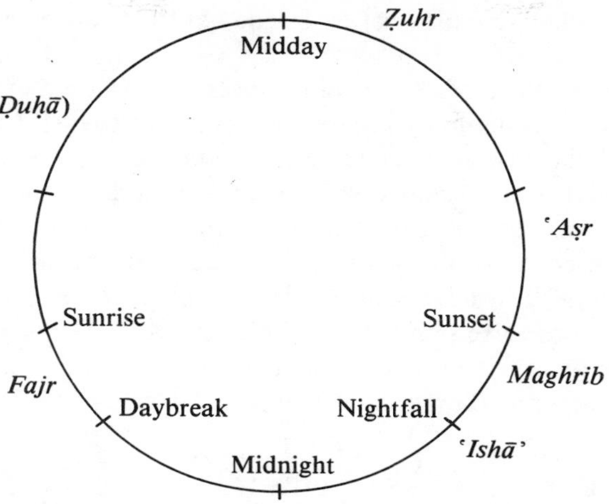

Figure 4.14 The Islamic day begins at sunset because the calendar is lunar and the months begin with the sighting of the crescent shortly after sunset. There are five canonical prayers; the times of the daylight prayers are defined in terms of shadow lengths, and the night prayers are determined in terms of horizon and twilight phenomena. A sixth prayer, the *ḍuḥā*, was practised at mid-morning in certain communities – see, for example, Plate 4.12 (Tunis) and Plate 4.18 (Istanbul)

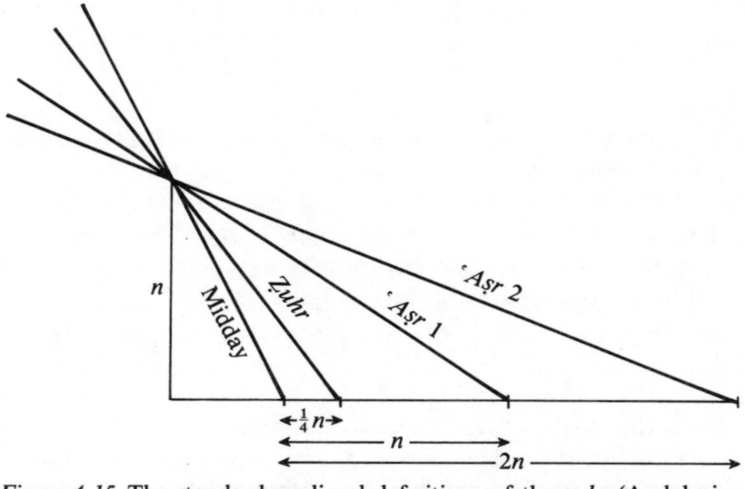

Figure 4.15 The standard medieval definitions of the *ẓuhr* (Andalusia and the Maghrib) and the *ʿaṣr* prayers in terms of shadow increases

performed at the same time before midday as the *ʿaṣr* was performed after midday.

The names of the daylight prayers appear to have been derived from the names of the corresponding seasonal hours in pre-Islamic classical Arabic, the seasonal hours (*al-sāʿāt al-zamānīya*) being one-twelfth divisions of the period between sunrise and sunset. The definitions of the times of these prayers in terms of shadow increases (as opposed to shadow lengths in the *ḥadīth*) represent a practical means of regulating the prayers in terms of the seasonal hours. The definitions of the *ḍuḥā*, *ẓuhr* and *ʿaṣr* in terms of shadow increases correspond to the third, sixth and ninth seasonal hours of daylight, the links being provided by an appropriate Indian formula relating shadow increases to the seasonal hours (see later).

SIMPLE ARITHMETICAL SHADOW SCHEMES FOR TIMEKEEPING

Before considering the activities of the Muslim astronomers in *ʿilm al-mīqāt*, it is important to note that, in popular practice, tables and instruments were not widely used. Instead, as we see from treatises on folk astronomy and on the sacred law, the daytime prayers were regulated by simple arithmetical shadow schemes of the kind also attested in earlier Hellenistic and Byzantine folk astronomy. Some twenty different schemes have been located in the Arabic sources. In most cases they are not the result of any careful observations, and the majority are marred by copyists' mistakes. Usually a single one-digit value for the midday shadow of a man 7 *qadams* ('feet') tall is given for each month of the year. One such scheme, attested in several sources, is (starting with value for January)

9 7 5 3 2 1 2 4 5 or 6 8 10.

The corresponding values for the shadow length at the beginning of the *ʿaṣr* prayer are 7 units more for each month.

Other arithmetical schemes presented in order to find the shadow length at each seasonal hour of day. The most popular formula advocated in order to find the increase (Δs) of the shadow over its midday minimum at $T\,(<6)$ seasonal hours after sunrise or before sunset is

$$T = \frac{6n}{\Delta s + n}$$

where n is the length of the gnomon. This is the formula which was first used to establish the values $\Delta s = n$ for the third and the ninth seasonal hours of daylight (the beginnings of the *ʿaṣr* and *ḍuḥā*) and $\Delta s = 2n$ for the tenth hour (sometimes taken as the end of the *ʿaṣr*).

Other simple kinds of time-regulation for irrigation purposes are practised in various rural areas of the Muslim world.

THE EARLIEST TABLE FOR TIMEKEEPING

It was al-Khwārizmī in Baghdad in the early ninth century who prepared the first known tables for regulating the times of the daylight prayers. Computed for the latitude of Baghdad, his tables display the shadow lengths for a gnomon of length 12 at the *ẓuhr* and at the beginning and end of the *ʿaṣr*, with values to one digit for each 6° of solar longitude (corresponding roughly to each six days of the year) (Plate 4.15). He also compiled some simple tables displaying the time of day in seasonal hours in terms of the observed solar altitude based upon an approximate formula.

The ninth-century astronomer ʿAlī ibn Āmājūr compiled a more exten-

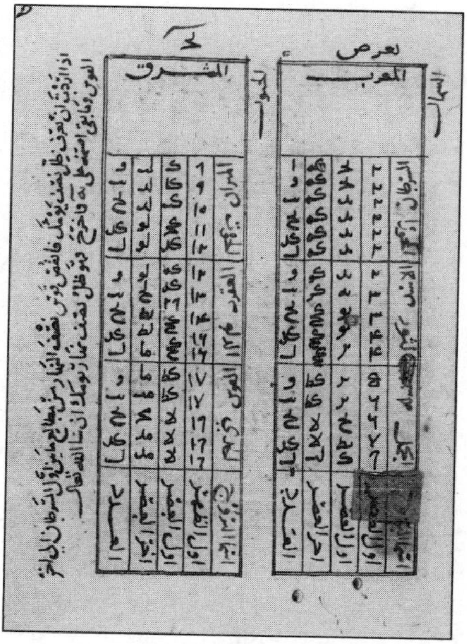

Plate 4.15 The earliest known Islamic table for regulating the times of the daylight prayers, associated with al-Khwārizmī. Taken from MS Berlin Ahlwardt 5793, fol. 94ʳ, with kind permission of the Director of the Deutsche Staatsbibliothek, Preussischer Kulturbesitz

sive table for timekeeping based on a simple approximate formula which serves all latitudes. The underlying formula is

$$T = \tfrac{1}{15} \arcsin\left(\frac{\sin h}{\sin H}\right)$$

where h is the observed altitude, H is the meridian altitude and $T(\leqslant 6)$ is the time in seasonal hours elapsed since sunrise or remaining until sunset. (Note that $T = 0$ when $h = 0$, and that $T = 6$ when $h = H$, as required for the cases when the sun is on the horizon and the meridian; in fact this formula is accurate only when the sun is at the equinoxes.) Ibn Āmājūr simply tabulated $T(h, H)$ for each degree of both arguments ($h < H$).

In astronomical handbooks from the ninth century onwards we find descriptions of an accurate method for finding the time elapsed since sunrise in equatorial degrees T or the corresponding hour-angle t from a pair of values h and H, or from h, ϕ and δ, where ϕ is the local latitude and δ is the declination (note that $H = 90° - \phi + \delta$). These involve the semi-diurnal arc D and the use of the versed sine function (vers $\theta = 1 - \cos\theta$) (see vol. II, chapter 15). The standard medieval formula, adopted by the Muslims from

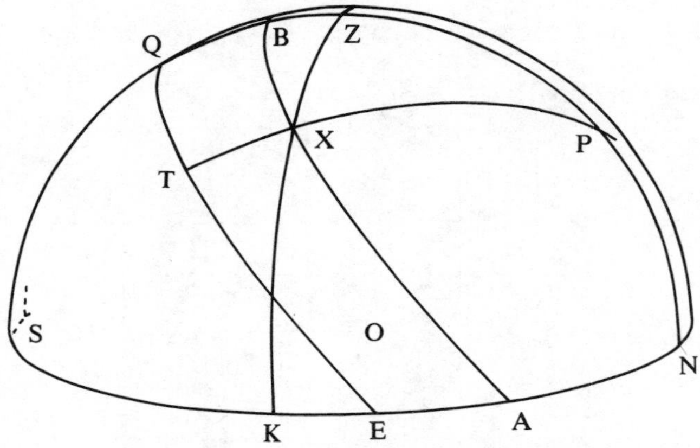

Figure 4.16 The celestial sphere about the observer at O. The horizon with the cardinal points in NES(W). The celestial equator is EQ(W), the celestial axis OP and the zenith Z. A celestial body rises at A, culminates on the meridian at B and sets at C. An instantaneous position is X with altitude measured by the arc XK; the hour-angle at that moment is then measured by the arc TQ of the celestial equator (or by angle TPQ) and the azimuth by arc EK of the horizon

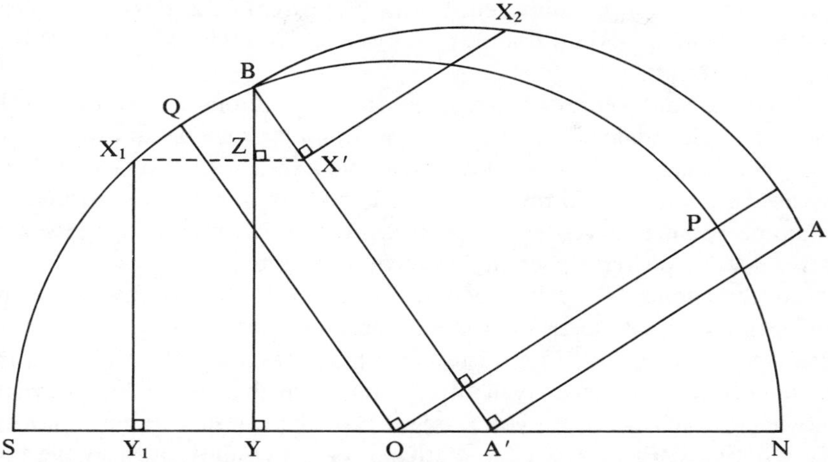

Figure 4.17 An analemma construction for finding the hour-angle from the observed altitude of the sun or any fixed star. The three-dimensional celestial sphere of unit radius is first projected orthogonally into the plane of the meridian SQBN. Then OQ, A′B and SN represent the celestial equator, the day-circle and the horizon, and X′ is the projection of X. The altitude circle (arc ZXK in Figure 4.16) is folded about its radius ZO into the plane of the meridian: X moves to X_1 such that $X_1S = h$, so that $X_1Y_1 = \sin h$. Note also that $BY = \sin H$. The day-circle is then folded about its diameter through B into the plane of the meridian (yielding an arc of a circle radius $\cos \delta$). X (in Figure 4.16) moves perpendicular to BA′ to X_2 (in this figure) and BX_2 measures the hour-angle t and BA the half arc of daylight D. Note that triangles BZX′ and BYA′ are similar. Therefore,

$$\frac{BX'}{BA'} = \frac{\text{vers } t}{\text{vers } D} = \frac{BZ}{BY} = \frac{\sin H - \sin h}{\sin H},$$

whence the standard medieval formula for $t(h,H)$

Indian sources, is (in modern notation)

$$\text{vers } t = \text{vers}(D - T) = \text{vers } D - \frac{\sin h \text{ vers } D}{\sin H}$$

This could be derived with facility by reducing the three-dimensional problem on the celestial sphere to two dimensions (Figures 4.16 and 4.17). The equivalent modern formula for the hour-angle t can also be derived by such procedures. It is

$$\cos t = \frac{\sin h - \sin \delta \sin \phi}{\cos \delta \cos \phi}$$

and is used in a form equivalent to this by later Muslim astronomers (see al-Khalīlī below). Several Islamic tables were universal in the sense that they served all terrestrial latitudes.

From the ninth century onwards we find descriptions of how to find the time of day or night using an analogue computer such as an astrolabe or using a calculating device such as a sine quadrant. In the first case there is no need to know the formula; in the second case one uses the formula to compute specific examples. Likewise numerous Islamic instruments were devised to be universal, serving all terrestrial latitudes.

ʿAlī ibn Āmājūr also compiled a table of $T(h, H)$ for Baghdad based on an accurate trigonometric formula. Some anonymous prayer-tables for Baghdad are preserved in a thirteenth-century Iraqi *zīj*; these display, for example, the duration of twilight in addition to the times of the daylight prayers for each day of the year, and are probably another Abbasid production, dating perhaps from the tenth century. Certainly quantitative estimates of the angle of depression of the sun at nightfall and daybreak occur in the *zīj* of the ninth-century astronomer Ḥabash al-Ḥāsib. Isolated tables displaying the altitudes of the sun at the *ẓuhr* and *ʿaṣr* prayers and the duration of morning and evening twilight occur in several other early medieval Islamic astronomical works, usually of the genre known as *zīj*.

Several examples of extensive tables for reckoning time by day for the solar altitude, or for reckoning time of night from altitudes of certain prominent fixed stars, have come to light. All of these tables were computed for a specific locality, and display either $T(h, H)$ or $T(h, \lambda)$, where λ is the solar longitude. To use any of them, one needed an instrument, such as an astrolabe, to measure celestial altitudes or the passage of time. There is no evidence that these early tables were widely used.

Of particular interest was the development in the ninth and tenth centuries of auxiliary trigonometric tables for facilitating the solution of problems of spherical astronomy, though not especially those of timekeeping. The auxiliary tables of Ḥabash (see above) and Abū Naṣr (*fl.* Central Asia, *c.* 1000) are the most impressive of these from a mathematical point of view, and al-Khalīlī's universal tables for timekeeping (see below) should be considered in the light of these earlier developments.

THE INSTITUTION OF THE *MUWAQQIT*

In practice, at least before the thirteenth century, the regulation of the prayer-times was the duty of the muezzin (Arabic, *mu'adhdhin*). These individuals were appointed for the excellence of their voices and their character, and they needed to be proficient in the rudiments of folk astronomy. They needed to know the shadows at the *ẓuhr* and the *ʿaṣr* for each month, and which lunar mansion was rising at daybreak and setting at nightfall, information which was conveniently expressed in the form of mnemonics; they did not need astronomical tables or instruments. The necessary techniques are outlined in the chapters on prayer in the books of sacred law and the qualifications of the muezzin are sometimes detailed in works on public order (*ḥisba* or *iḥtisāb*).

In the thirteenth century there occurred a new development, the origins of which are obscure. In Egypt at that time we find the first mention of the *muwaqqit*, a professional astronomer associated with a religious institution, whose primary responsibility was the regulation of the times of prayer. Simultaneously, there appeared astronomers with the epithet *mīqātī* who specialized in spherical astronomy and astronomical timekeeping, but who were not necessarily associated with any religious institution.

TIMEKEEPING IN MAMLUK EGYPT

In Cairo in the late thirteenth century, a *mīqātī* named Abū ʿAlī al-Marrākushī compiled a compendium of spherical astronomy and instruments from earlier sources which was to set the tone of *ʿilm al-mīqāt* for several centuries. His treatise, appropriately entitled *Jāmiʿ al-mabādiʾ wa-l-ghāyāt fī ʿilm al-mīqāt*, (*An A to Z of Astronomical Timekeeping*), was first studied by the Sédillot *père et fils* in the nineteenth century.

Al-Marrākushī's contemporary, Shihāb al-Dīn al-Maqsī, compiled a set of tables displaying the time since sunrise as a function of solar altitude and longitude for the latitude of Cairo (apparently based on an earlier, perhaps less extensive, set by the tenth-century astronomer Ibn Yūnus). In the fourteenth century these tables were expanded and developed into a corpus covering some 200 manuscript folios and containing over 30,000 entries. The Cairo corpus of tables for timekeeping was used for several centuries and survives in numerous copies, no two of which contain the same tables. Besides tables displaying the time since sunrise, the hour-angle (time remaining until midday) and the solar azimuth for each degree of solar longitude, which with about 30,000 entries make up the bulk of the corpus (Plate 4.16), there are others displaying the solar altitude and hour-angle at the *ʿaṣr*, the solar altitude and hour-angle when the sun is in the direction of the qibla (see section (a)), and the duration of morning and evening twilight.

In some late copies of the Cairo corpus there are tables for regulating the time when the lamps on minarets during Ramadan should be extinguished and when the muezzin should pronounce a blessing on the Prophet Muḥammad. In some copies, early and late, there is a table for orienting

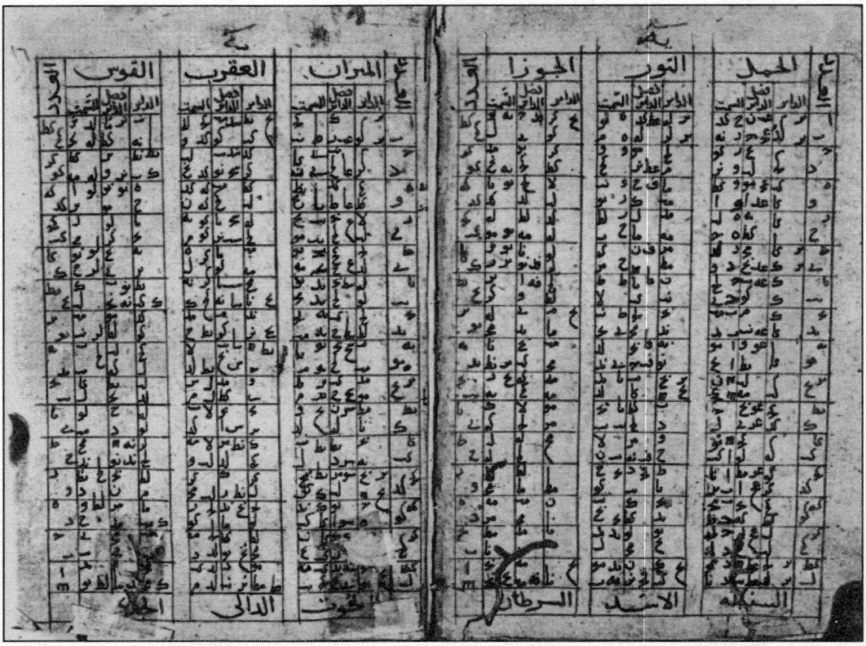

Plate 4.16 An extract from the Cairo corpus of tables for timekeeping. This particular sub-table displays values of three functions – the hour-angle, time since sunrise and azimuth – for each degree of solar longitude when the sun has altitude above the horizon 15°. Taken from MS Cairo Dār al-Kutub *mīqāt* 690, fols 15ᵛ–16ʳ, with kind permission of the Director of the Egyptian National Library

the large ventilators which throughout the medieval period were a prominent feature of the Cairo skyline. These were aligned with the roughly orthogonal street plan of the medieval city, itself astronomically aligned towards winter sunrise.

Al-Maqsī also compiled an extensive treatise on sundial theory, including tables of co-ordinates for making the curves on horizontal sundials for different latitudes and vertical sundials at any inclination to the local meridian for the latitude of Cairo. The latter were particularly useful for constructing sundials on the walls of mosques in Cairo, and the special curves for the *ẓuhr* and *ʿaṣr* enabled the faithful to see how much time remained until the muezzin would announce the call to prayer.

A contemporary of al-Marrākushī and al-Maqsī, the Cairo astronomer Najm al-Dīn, compiled a table of timekeeping which would work for any latitude and which could be used for the sun by day and the stars by night.

The function tabulated is $T(h, H, D)$, where D was the half arc of visibility of the celestial body in question over the horizon, and the entries number over a quarter of a million. This table was not widely used (if at all) and is known from a unique copy, perhaps in the hand of the compiler.

Another region of the Islamic world in which the writings of al-Marrākushī and the output of the early Cairo *muwaqqit*s were influential was the Yemen. Under the Rasulids mathematical astronomy was patronized and practised. In particular the Sultan al-Ashraf (*reg.* 1295–6) compiled a treatise on instrumentation inspired by that of al-Marrākushī. The Yemeni astronomer Abū al-ʿUqūl, who worked for the Sultan al-Mu'ayyad in Taiz, compiled a corpus of tables for timekeeping by day and night which is the largest such corpus compiled by any Muslim astronomer, containing over 100,000 entries.

In Cairo in the fourteenth century there were several *muwaqqit*s producing works of scientific merit, but the major scene of *ʿilm al-mīqāt* during this century was Syria.

TIMEKEEPING IN FOURTEENTH-CENTURY SYRIA

The Aleppo astronomer Ibn al-Sarrāj, who is known to have visited Egypt, devised a series of universal astrolabes and special quadrants and trigonometric grids, all for the purpose of timekeeping: his works represent the culmination of the Islamic achievement in astronomical instrumentation. Two other major Syrian astronomers, al-Mizzī and Ibn al-Shāṭir, studied astronomy in Egypt. Al-Mizzī returned to Syria and compiled a set of hour-angle tables and prayer-tables for Damascus modelled after the Cairo corpus. Ibn al-Shāṭir compiled some prayer-tables for an unspecified locality, probably the new Mamluk city of Tripoli. Al-Mizzī also compiled various treatises on instruments, but Ibn al-Shāṭir turned his attention to theoretical astronomy and planetary models. This notwithstanding, he also devised the most splendid sundial known from the Islamic Middle Ages.

It was a colleague of al-Mizzī and Ibn al-Shāṭir named Shams al-Dīn al-Khalīlī who made the most significant advances in *ʿilm al-mīqāt*. Al-Khalīlī recomputed the tables of al-Mizzī for the new parameters (local latitude and obliquity of the ecliptic) derived by Ibn al-Shāṭir (Plate 4.17). His corpus of tables for timekeeping by the sun and regulating the times of prayer for Damascus was used there until the nineteenth century. He tabulated the following functions for each degree of solar longitude λ: the solar meridian altitude; half the diurnal arc; the number of hours of daylight; the solar altitude at the beginning of the *ʿaṣr*; the hour-angle at the beginning of the *ʿaṣr*; the time between the beginning of the *ʿaṣr* and sunset; the time between midday and the end of the *ʿaṣr*; the duration of night; the duration of

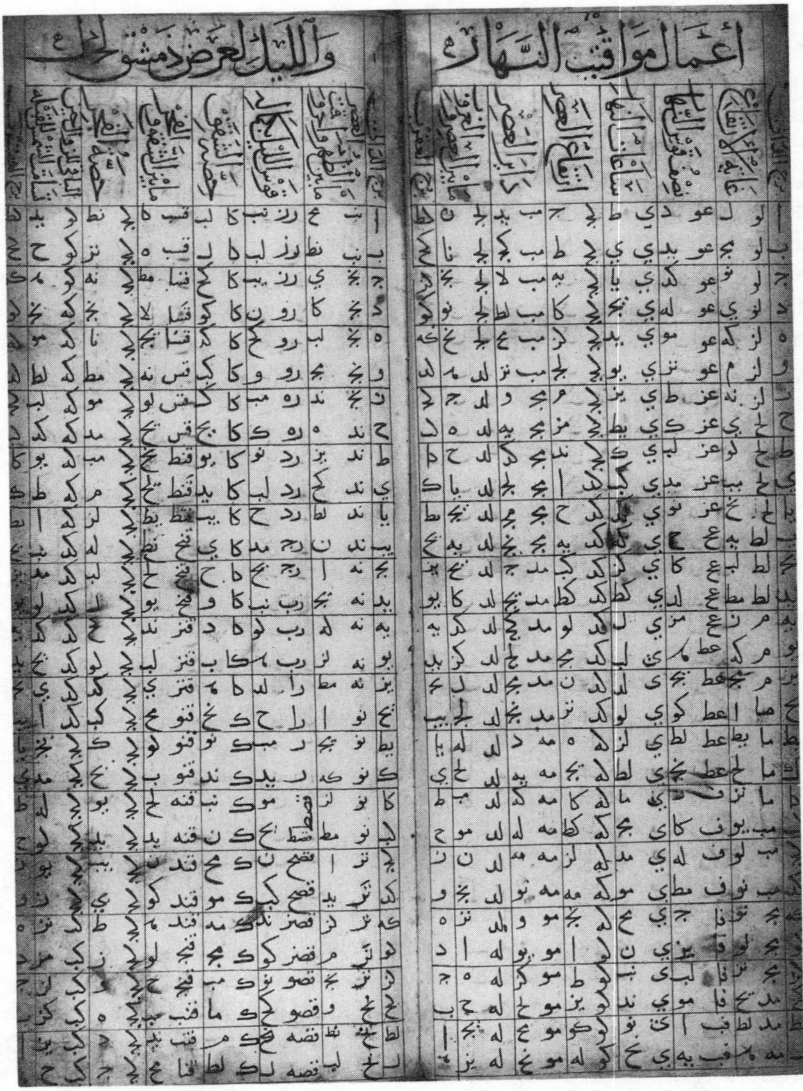

Plate 4.17 An extract from the prayer-tables for Damascus prepared by al-Khalīlī. This particular sub-table serves solar longitudes in Aquarius and Scorpio, and the twelve functions are tabulated for each degree of longitude across the double-page. Taken from MS Paris B.N. ar. 2558, fols 10ᵛ–11ʳ, with kind permission of the Director of the Bibliothèque Nationale

evening twilight; the duration of darkness (from nightfall to daybreak); the duration of morning twilight; and the time remaining until midday from the moment when the sun is in the same direction as Mecca. Entries for all but the third function are in equatorial degrees and minutes (where $1°$ corresponds to 4 minutes of time). These tables contain 2,160 entries. Al-Khalīlī also tabulated the hour-angle t as a function of solar altitude h and solar longitude λ for the latitude of Damascus. His tables of $t(h, \lambda)$ contain about 10,000 entries.

In addition, al-Khalīlī compiled some tables of auxiliary trigonometric functions for any latitude considerably more useful than the earlier tables of this kind by Ḥabash (see above). The functions tabulated are

$$f(\phi, \theta) = \frac{R \text{ Sin } \theta}{\text{Cos } \phi}$$

$$g(\phi, \theta) = \frac{\text{Sin } \theta \text{ Tan } \phi}{R}$$

and

$$K(x, y) = \text{arcCos}\left(\frac{Rx}{y}\right)$$

where the trigonometric functions are to base $R = 60$. The total number of entries in these auxiliary tables exceeds 13,000. Values are given to two sexagesimal digits and are invariably accurately computed. With these tables the hour-angle can be found with a minimum of calculation. Al-Khalīlī presents the procedure

$$t(h, \delta, \phi) = K([f(\phi, h) - g(\phi, \delta)], \delta)$$

which is equivalent to the modern formula. Likewise the corresponding azimuth a (measured from the meridian) is given by

$$a(h, \delta, \phi) = K([g(\phi, h) - f(\phi, \delta)], h).$$

These tables serve to solve numerically any problem which can, in modern terms, be solved by means of the spherical cosine formula.

Al-Khalīlī also compiled a table displaying the qibla or local direction of Mecca as a function of terrestrial longitude and latitude. He appears to have used the universal auxiliary tables to compile this qibla table.

Some of the activities of the Damascus school became known in Tunis in the fourteenth and fifteenth centuries. Extensive auxiliary tables and prayer tables for the latitude of Tunis were compiled there by astronomers whose names are not known to us. Prayer-tables were also prepared for various latitudes in the Maghrib.

TIMEKEEPING IN OTTOMAN TURKEY

More significant was the influence of the Cairo and Damascus schools on the development of *ʿilm al-mīqāt* in Ottoman Turkey. The Damascus astronomers of the fourteenth century had already prepared a set of prayer-tables for the latitude of Istanbul, but several new sets of tables were prepared by Ottoman astronomers for Istanbul and elsewhere in Turkey after

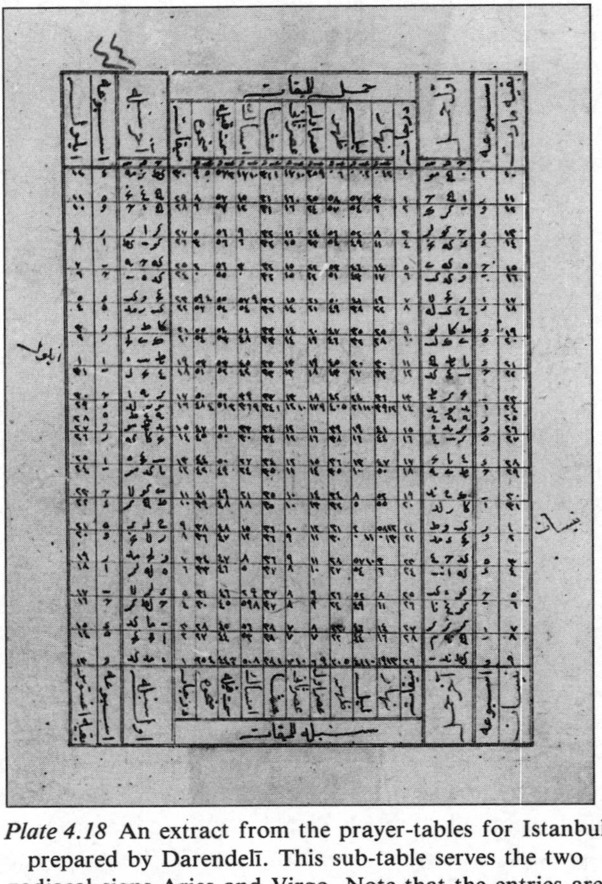

Plate 4.18 An extract from the prayer-tables for Istanbul prepared by Darendelī. This sub-table serves the two zodiacal signs Aries and Virgo. Note that the entries are written in Indian numerals, rather than the alpha-numerical (*abjad*) notation which was more usual for astronomical tables even under the Ottomans. Taken from MS Cairo Ṭalʿat *mīqāt turkī* 29, fol. 44ʳ, with kind permission of the Director of the Egyptian National Library

the model of the corpuses for Cairo and Damascus. Prayer-tables for Istanbul are contained in the very popular almanac of the fifteenth-century Sufi Shaykh Vefā and in the less widely distributed almanac of the sixteenth-century scholar Darendelī (Plate 4.18). The latter displays the lengths of daylight and night, as well as the *times* (expressed in the Turkish convention, see below) of midday, the first and second *ʿaṣr*, nightfall and daybreak, the moment when the sun is in the qibla and a morning institution called the *ẓaḥve* (related to the *ḍuḥā*, see above). These two sets of tables remained in use until the nineteenth century.

Large sets of tables for timekeeping by the sun and/or stars were prepared for Istanbul and for Edirne. One set for the sun was compiled by Taqī al-Dīn ibn Maʿrūf, director of the short-lived Istanbul Observatory in the late sixteenth century. In the eighteenth century the architect Ṣāliḥ Efendī produced an enormous corpus of tables for timekeeping which was also very popular amongst the *muwaqqit*s of Istanbul.

A feature distinguishing some of these Ottoman tables from the earlier Egyptian and Syrian tables is that values of the time of day are based on the convention that sunset is 12 o'clock. This convention, inspired by the fact that the Islamic day begins at sunset (because the calendar is lunar and the months begin with the sighting of the crescent shortly after sunset), has the disadvantage that clocks registering 'Turkish' time need to be adjusted by a few minutes every few days. Prayer-tables based on this convention were compiled all over the Ottoman Empire and beyond: examples have been found in the manuscript sources for localities as far apart as Algiers and Yarkand and Crete and Sanaa. In the late Mamluk and Ottoman periods the *muwaqqit*s compiled numerous treatises on the formulae for timekeeping and the procedures for computing the time of day or night, or the prayer-times, using either an almucantar quadrant (modified from the astrolabe) or a sine quadrant.

MODERN TABLES FOR THE PRAYER TIMES

In the nineteenth and twentieth centuries, the times of prayer have been or still are tabulated in annual almanacs, wall-calendars and pocket-diaries, and the times for each day are listed in newspapers. In Ramadan, special sets for the whole month are distributed. These are called *imsākīya*s, and indicate in addition to the times of prayer, the time of the early morning meal called the *suḥūr* and the time shortly before daybreak when the feast should begin, called the *imsāk*. Modern tables are prepared either by the local surveying department or observatory or by some other agency enjoying the approval of the religious authorities; usually they display the times of the five prayers and sunrise. Recently, electronic clocks and watches have appeared on the market which are programmed to beep at the prayer-times for different localities, and to pronounce a recorded prayer-call.

FURTHER READING

On the prayers in Islam see the article '*Ṣalāt*' in the *Encyclopaedia of Islam* (2nd edn, Leiden, 1960 onwards). For an overview of Islamic timekeeping see the article '*Mīḳāt*' in the *Encyclopaedia of Islam*, reprinted in King (1993).

On the definitions of the times of prayer as they appear in the astronomical sources, see Wiedemann and Frank (1926). For al-Bīrūnī's discussion, see Kennedy *et al.* (1983: 299–310). On the origin of these definitions see King, 'On the Times of Prayer in Islam', to appear.

On the procedures advocated by the legal scholars and in treatises on folk astronomy see King (1987a).

On the formulae for timekeeping used by the Muslim astronomers see the papers by Davidian, Nadir and Goldstein reprinted in Kennedy *et al.* (1983: 274–96) and the studies listed below.

On solutions (i.e. tables and instruments) serving all latitudes see King (1987c, 1988, 1993).

On the earliest known tables for regulating the prayer times and reckoning time of day from solar altitude, see King (1983d: esp. 7–11). On al-Marrākushī and his treatise see the section 'Gnomonics' in this chapter and also King (1983c: esp. 539–40 and 534–5). On the institution of the professional mosque timekeepers see King, 'On the role of the Muezzin and the Muwaqqit in medieval Islamic society', to appear in S. Livesey and J. F. Ragep, eds., Proceedings of the Conference 'Science and Cultural Exchange in the Premodern World' in Honor of A. I. Sabra, University of Oklahoma, Norman, Ok., Feb. 25–27, 1993, Leiden: E. J. Brill, 1995.

On the corpuses of tables for Cairo, Taiz, Damascus and Jerusalem, Tunis and Istanbul, see respectively, King (1973a; 1979: esp. 63; 1976), King and Kennedy (1982: esp. 8–9) and King (1977a). Each of these papers is reprinted in King (1987b).

On the auxiliary tables of Ḥabash, Abū Naṣr and al-Khalīlī see, respectively, Irani (1956), Jensen (1971) and King (1973b).

For an analysis of all available tables see King, *Studies in Astronomical Timekeeping in Islam, I: A Survey of Tables for Reckoning Time by the Sun and Stars*, and *II: A Survey of Tables for Regulating the Times of Prayer* (forthcoming).

On the Ottoman convention of reckoning sunset as 12 o'clock, see Würschmidt (1917). On the *muvakkithanes*, the buildings adjacent to the major Ottoman mosques which were used by the *muwaqqits*, see Ünver (1975).

5

Mathematical geography

EDWARD S. KENNEDY

INTRODUCTION

The historian of the Islamic exact sciences is frequently confronted with an *embarras de richesse* – hundreds of manuscript sources which have never been studied in modern times. For descriptive geography the situation may well be the same. The reader will find indications to this effect in the surveys of S. Maqbul Ahmad (1965a,b). But for those parts of the subject which employ mathematics, frustration arises from a paucity rather than a plethora of sources. For instance, it is reliably reported (Shawkat 1962: 12) that the astronomer Ibn Yūnus (*fl.* 1000) made a world map for the Fatimid caliph al-ʿAzīz. But precise information as to the projection method is not available, much less the map itself.

What information is available can be regarded as involving either geodesy or cartography, and the presentation below is organized under these two main topics. Under the first, the subject of latitude determinations leads to that of geodesy proper, thence to the fixing of longitudes and the zero meridians upon which they were based. The section concludes with an indication of the end-products of these operations – lists of place names with co-ordinates.

In the cartographical section which follows, the lack of precise information alluded to above severely hampers an assessment of the degree to which Hellenistic geography penetrated the Muslim world. The situation of al-Bīrūnī will be seen to be the reverse of al-Idrīsī's. For the former, projections are adequately described, but no applications to actual maps can be exhibited until the Renaissance or later. For the latter, many copies of the map survive, but the projection methods are largely a matter of conjecture. The maps of other scientists are described, but no attempt is made to cover Muslim navigation or sea charts.

GEODESY

Determination of latitudes

Since the latitude ϕ of a locality equals the altitude of the celestial pole at that place, this quantity is easily determined by astronomical methods (Figure 5.1). For instance, the observer may note h, the meridian altitude of the sun on a particular day, and calculate δ, its declination at the time of the observation. Then, for localities in the northern hemisphere,

$$\phi = 90^\circ - (h - \delta)$$

since the elevation of the culminating point on the celestial equator is the complement of the polar altitude. Or the meridian altitude of a fixed star

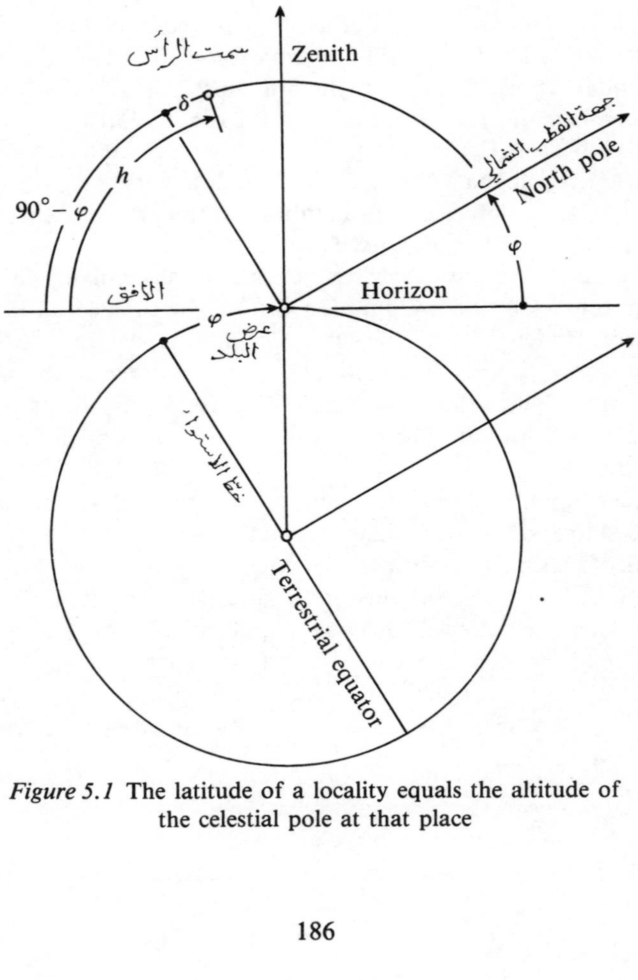

Figure 5.1 The latitude of a locality equals the altitude of the celestial pole at that place

of known declination may be observed at night, and the same expression can be applied.

Alternatively the observer may note the altitudes of a circumpolar star at its two meridian crossings. The ϕ is the mean of the two altitudes.

Worked examples of these methods from the records of his predecessors and contemporaries are given by al-Bīrūnī (*fl.* 1010) in his *Taḥdīd* (Kennedy 1973: 16–31).

Given the ease of latitude determinations, it would be expected that the values which have come down to us might be fairly accurate. Out of the 506 localities whose co-ordinates are reported by al-Kāshī (*fl.* 1400), modern values for the latitude have been found for 381. The mean of the differences between al-Kāshī's latitudes and the modern determinations is only four minutes of arc. However, the mean of absolute values of the same set of differences is $1^\circ 15'$, which is not very impressive. The results for al-Kāshī are typical of the fifty or so sources for which these statistics were calculated. In extenuation, it must be remembered that no author was in a position to verify personally any more than a very few latitudes. The rest he must accept on faith, and there must have been many cities which could boast no competent resident astronomer. The records show many latitudes which are precise to within a quarter of a degree.

The size of the Earth

It is appropriate to discuss this topic next, because the commonest medieval method of finding the length of a degree along a terrestrial meridian depended upon latitude determinations.

The caliph al-Ma'mūn (reigned 813–33) mounted one or more expeditions charged with this task. The sources vary as to details, but there is general agreement as to the method used (see Barani 1951). The idea was to choose a suitably flat expanse of the Syrian desert, and from some initial point observe ϕ. The observers then set out either due north or due south, measuring the distance traversed as they proceeded. This continued until the expedition arrived at a station at which ϕ was just one degree different from that of the first locality. Then the distance travelled is the length of a meridian degree.

It would have seemed vastly more practical to have travelled any distance, the farther the better, and then simply have divided the distance by the change in ϕ. For to demand that $\Delta\phi = 1^\circ$ implies a continued setting up of additional stations until the desired integer difference is observed. Perhaps in practice the observers adopted the reasonable procedure.

However it was arrived at, the value of $56\frac{2}{3}$ Arab miles per degree was obtained, and was consistently used by subsequent investigators, e.g.

al-Bīrūnī (1967, say) and al-Ṭūsī (Kennedy 1948: 115). Other results are cited by the sources, but they are all close to this canonical value. Its multiplication by $360/\pi$ gives the corresponding diameter of the earth.

To inquire concerning the method's precision is to pose the difficult and perhaps insolvable metrological problem of conversion between medieval and modern units. The question was exhaustively investigated by Nallino (1892–3). He concluded that the $56\frac{2}{3}$ Arab miles is equivalent to 111.8 km per degree, which is astonishingly close to the accurate value of 111.3. This is probably a coincidence. But he gives the results of investigations by nine other scholars, which range between 104.7 and 133.3. So the Ma'mūnic result is probably rather good.

Base meridians

All of the geographical lists described below, except two, may be divided into a pair of categories depending upon the zero meridian of the particular table. Ptolemy (*fl.* AD 150), the father of mathematical geography, measured longitudes eastward from the Fortunate Isles (*al-jazā'ir al-khālidāt*, the Canaries). About half of the Muslim sources followed him, and the group thus constituted is called for convenience the C class. The second group, designated by A, followed al-Khwārizmī (*fl.* AD 820) in using as prime meridian the 'western shore of the encompassing sea' (*sāḥil al-baḥr al-muḥīṭ al-gharbī*), it being agreed in the literature that the A meridian is 10° east of the C (e.g. al-Bīrūnī 1967: 121). It is not clear how this division originated. Nallino has shown (1944: 490) that it was not al-Khwārizmī's intention to change the zero point. For some reason, the astronomers of al-Ma'mūn decided that the longitude of the Abbasid capital, Baghdad, should be 70°. However, if Baghdad were reasonably plotted on a map based on Ptolemy's *Geography* it would have a longitude near 80°, and over half of the Muslim sources give this value. The notion reported below, that the 'Cupola of the Earth' as conceived by the 'Easterners' was $13\frac{1}{2}°$ east of Ptolemy's 'Cupola', was probably involved, for $13\frac{1}{2}°$ is not far from ten. Bīrūnī explicitly gives the displacement as 10° (al-Bīrūnī 1967: 120, 121). Al-Khwārizmī corrects by 10° Ptolemy's gross overestimate of the length of the Mediterranean, but this did not affect the base meridian.

However it came about, the existence of the A and C categories is a fact. Longitudes of the same city in tables of the two groups tend to differ by precisely 10°. Furthermore, for localities of known modern (Greenwich) longitude, calculations have been made of the mean difference between medieval and modern longitudes. There is considerable divergence between the means for individual sources, but those of the A class cluster about 24°; those from C are near 34° (cf. Kennedy and Regier 1985).

Longitudes measured from a third base meridian are reported by one source. Al-Hamdānī (d. 946; Müller 1884: 27,45) states that the 'Easterners' (*ahl al-mashriq*), the Indians and those who follow them, measure longitude west from the eastern edge of China. It was commonly agreed that the inhabited portion of the globe is the surface of a hemisphere bounded by a great circle through the poles. The apex of this, called the 'Cupola of the Earth' (*qubbat al-arḍ*), is the point on the equator which has the bounding circle as pole. Hamdānī goes on to say that the Easterners take the Cupola as being 90° west of their base meridian. Since he also mentions the Sindhind (from Sanskrit *siddhānta*), the Cupola is probably supposed to be on the meridian through Ujjain, the Greenwich of ancient Indian astronomy. In the Arabic literature the name was corrupted from Uzain (by the omission of a dot over one letter) to *arīn*, hence *qubbat arīn*. Hamdānī states further that Ptolemy's Cupola is, reasonably enough, 90° east of his base meridian, and that the two cupolas do not coincide, the Indian one being $13\frac{1}{2}°$ east of Ptolemy's. If longitudes measured from east and west are denoted by λ_E and λ_W respectively, then the Indian and Ptolemaic longitudes of a particular locality should satisfy the relation

$$\lambda_E + \lambda_W = 90° + 13\tfrac{1}{2}° + 90° = 193\tfrac{1}{2}°.$$

Hamdānī gives the Indian co-ordinates of twenty-two cities, most of them in the Arabian peninsula, but including also Jerusalem and Damascus. Of these, three towns are not found among the other Muslim lists of place names with co-ordinates. But of the remaining nineteen, the longitudes of nine conform to the above rule within a degree, for many sources of the C (Ptolemaic) category.

Honigmann (1929: 132–55) writes of a 'Persian system' in which longitudes are measured west of a prime meridian passing through the easternmost point of Asia. He is doubtless referring to the meridian of Hamdānī's 'Easterners', for the latter attributes some co-ordinates to al-Fazārī (*fl.* AH 760) and some to Ḥabash al-Ḥāsib (*fl.* 850), and both of these were influenced by the astronomy of Sasanian Iran as well as that of India.

Al-Bīrūnī implies (Kennedy 1973: 126) that in at least one set of tables, no longer extant, the base meridian was that of the Cupola itself.

One source, contained in Leiden MS Utr. Or. 23, is unique in that its longitudes are reckoned from Basra, presumably the anonymous compiler's station. However, since the column heading of the longitude entries is 'longitude difference', rather than the usual 'longitudes', the Basra meridian is not to be regarded as a base.

Longitude determinations

Once a prime meridian has been agreed upon, finding the longitude of a given locality resolves itself into the problem of determining the longitude difference between it and a place of known longitude. In theory this is even simpler than a latitude determination, for by virtue of the earth's rotation, in which twenty-four hours corresponds to 360°, the longitude difference equals the difference in mean local time between the two places. But in practice, what is needed is a time signal available simultaneously at both localities, and in medieval times, with no radio, this was far from simple.

A lunar eclipse is such a signal, for its phases appear the same from any point on the earth at which the eclipse is visible. A pair of observers, one at each locality, could observe the respective local times at which contact, and maximum immersion or totality, begin and end. Al-Bīrūnī (Kennedy 1973: 164) reports such a joint operation carried out between him, observing at Kāth (in Central Asia), and Abū al-Wafā' at Baghdad. A difficulty is that the phases of a lunar eclipse, unlike those of the solar variety, are not sharply defined events.

Al-Bīrūnī also exploited to the full, in his *Taḥdīd* (1967; Kennedy 1973), a geodetic method of finding longitude differences. Suppose the latitudes of the two localities are known, as well as the great circle distance between them. A meridian and a parallel of latitude pass through each of the two points. These four circles intersect in four points which constitute a determinate isosceles plane trapezoid. To this al-Bīrūnī applies a theorem of Ptolemy involving the sides and diagonals of cyclic trapezoids which gives him the equivalent of the following formidable expression (Kennedy 1973: 152):

$$\Delta\lambda = \text{arc crd} \sqrt{\left[\frac{\text{crd}^2(\widehat{AB}) - \text{crd}^2(\Delta\phi)}{\cos\phi_A \cos\phi_B}\right]}$$

where Δ indicates a difference, λ is terrestrial longitude, crd θ is the length of the chord of the unit circle subtended by a central angle θ and A and B are the localities in question.

Al-Bīrūnī approximated great circle distances by obtaining the length of caravan routes in leagues (*farsakhs*), multiplying by a coefficient which depended upon the directness and difficulty of the route, thence converting to miles and degrees. Seeking the $\Delta\lambda$ between Baghdad and Ghazna (in modern Afghanistan), his patron's capital, he made successive applications of his algorism for the stages through Rayy, Jurjāniya and Balkh. Being rightly suspicious of his result, he made additional calculations along a southern traverse through Shīrāz and Zaranj, trying also a branch through Bust. He accepted the arithmetic mean of the three findings thus obtained.

The final result is in error by about a third of a degree out of twenty-four, which, considering the crudeness of his original data, is very good.

No other geographers are known to have adopted al-Bīrūnī's method, and a geodetic solution explained by al-Kāshī (Kennedy 1985: 30) is astonishingly inaccurate. By and large, the longitudes appearing in the texts are much less reliable than the latitudes.

Geographical lists

Indicative of the amount and extent of geographical knowledge current in the world of medieval Islam is a collection of lists giving place names with latitudes and longitudes (published as Kennedy and Kennedy 1987). The sources may be divided into three categories: (1) astronomical handbooks (*zīj*es), unpublished manuscripts for the most part, containing geographical tables enabling the user to reduce observations made at one locality to consistency with those made at any other place in the table; (2) compilations made to form the basis for a map; and (3) more general geographical works which contain the co-ordinates of localities. To date, seventy-four sources have been entered on the magnetic tape on which the material is stored, and the number continues to rise. The sources vary in size from over 600 localities to as low as two. Most of the cities listed are in the Mediterranean basin, the Middle East and central Asia, but there are scatterings of localities in Europe north of Spain, central Africa, India and China.

It is possible to establish families of related sources, but no two are identical. On the other hand, no source is completely independent of the others.

CARTOGRAPHY

The Hellenistic heritage

The earliest cartographer whose work influenced the Muslims was Marinus of Tyre (*c.* AD 100). In Marinus' world map the co-ordinate net consisted of two families of mutually orthogonal parallel lines (Figure 5.2). Since the sphere is not applicable to the plane, any plane map of a portion of the earth's surface involves distortion. The cartographer may choose a mapping which is conformal (shape preserving), which is area preserving or in which some distances are preserved, but he cannot have everything. In Marinus' map, distances are preserved along all the meridians and along the parallel of latitude through Rhodes ($\phi = 36^\circ$; Neugebauer 1948: 1037–9). But since latitude circles decrease in size as ϕ increases, distances along parallels north of Rhodes in the Marinus map are stretched, and those south of Rhodes are compressed.

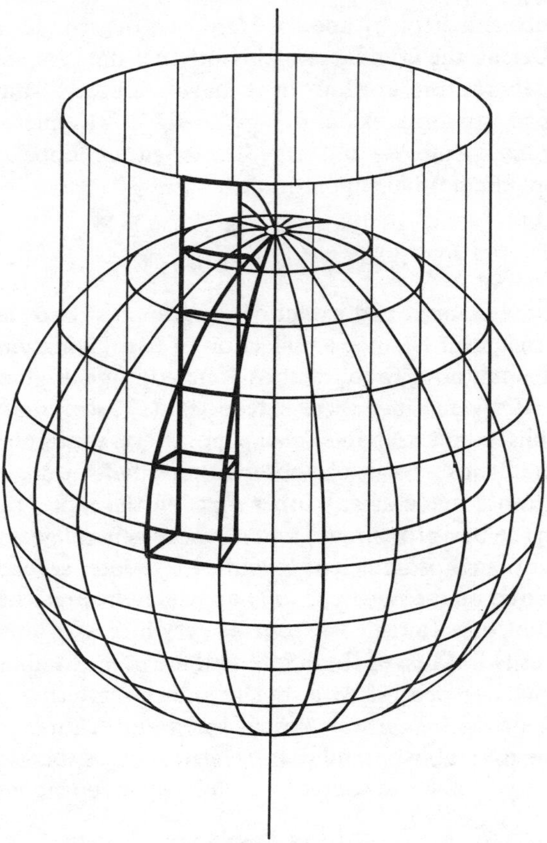

Figure 5.2 The principle of the Marinus mapping

Ptolemy developed two map projections, in both of which the meridians converge, in contrast to Marinus' cylindrical scheme.

In the first Ptolemaic map, distances are preserved along all the meridians, which make up a family of concurrent straight lines. Parallels of latitude map into concentric circles, orthogonal to the meridians, which therefore pass through the common centre. The latter point is so chosen that (1) distances are preserved along the latitude circle which passes through Rhodes, and (2) the ratio of distances is preserved along the parallels through Thule ($\phi = 63^\circ$) and the equator ($\phi = 0^\circ$).

Ptolemy's second scheme retained concentric circles as maps of the parallels of latitude, but now distances are preserved along three of them, for latitudes of 63°, $23;50^\circ$ and $-16;25^\circ$. As a consequence of this, the maps of the meridians can no longer be straight lines. They are now a family of circles, each one being determined by the three points having the same longitude on the three latitude circles named above. The effect of this is to damage slightly the preservation of distances along the meridians.

Note the progression in these three maps. In the first, the co-ordinate net is rectilinear and orthogonal; in the second, one set of co-ordinate curves is made up of circles; in the third, both sets are circles.

It is well-nigh certain that, in some form or other, Ptolemy's world map was available to the geographers of the Abbasid Empire. Al-Masʿūdī (*Murūj al-Dhahab*, vol. 1, p. 183; and *Kitāb al-Tanbīh wa-l-Ishrāf*, p. 33) claims that he had seen one or more examples, and that they had been surpassed in excellence by al-Ma'mūn's map (*al-ṣūrat al-ma'mūniya*). But no versions from Abbasid times are known to have survived. The earliest extant copies of the *Geography* were made in thirteenth and fourteenth century Constantinople. From these, Arabic translations were made *c.* 1465 by order of the Ottoman sultan Mehmet II. One of the translations is MS Aya Sofya (Istanbul) 2610, and the world map from it has been reproduced in facsimile in Fischer (1932) and in Maqbul Ahmad (1965b). The entire manuscript was published in facsimile in Egypt (Cairo?) in 1929 (Bagrow 1955: 27n), although the book has no indication of its provenance or date.

All this is much too late to have any bearing on the Abbasids, and indeed the nature of the Ptolemaic material that did reach them is a matter of dispute. Thus Mžik (1915) thinks it probable that they used a Syriac version of the *Geography*, perhaps with no world map at all. Ruska (1918), on the other hand, considers they may well have worked from the Greek directly.

Al-Ma'mūn's map

It is well known that during his reign (813–33) the caliph al-Ma'mūn attracted eminent savants to his 'House of Wisdom' (*bayt al-ḥikma*). One of the fruits of their collaboration was a picture of the known world which, in important respects, was an improvement upon Ptolemy's (Nallino 1944: 458–532). But of this, only the related geographical table by al-Khwārizmī (1926), together with three regional maps, has survived. No copy of the main map has turned up. Al-Masʿūdī (*Kitāb al-Tanbīh wa-l-Ishrāf*, p. 44) states that on it the climate boundaries, which are parallels of latitude, are rectilinear. This can be taken to imply that the projection was of the Marinus type.

The conjecture is made well-nigh certain by the geographical table of Suhrāb (*fl.* 930) which is closely related to that of al-Khwārizmī. In the introduction to Suhrāb's work (Mžik 1930) are careful directions as to how to lay out the co-ordinate net on which the localities are to be plotted. It is to consist of two families of mutually orthogonal parallel lines which form squares. Hence distances along the equator and the meridians are preserved. Because of greater east–west stretching in the temperate zone it is inferior to the Marinus map proper.

The 'Atlas of Islam'

In the tenth century a group of geographers including al-Balkhī, al-Iṣṭakhrī, al-Maqdisī and Ibn Ḥawqal composed works which have so many features in common that they have been given the appellation the 'Atlas of Islam' (Kramers 1931–2). Each one has a standard set of twenty maps, of which the first is a world map. However, these are so strongly schematized as to become, as Kramers puts it, cartographical caricatures.

Al-Bīrūnī's contributions

Fairly early in his career (*c.* 1005, cf. Richter-Bernburg 1982), the great polymath of Central Asia wrote a short work on mappings of the sphere. Berggren (1982) is a recent translation, together with a commentary and a bibliography of earlier translations and editions. To it a facsimile of the Leiden manuscript copy has been appended. Al-Bīrūnī discusses in this treatise eight varieties of map projection. Three of them are described below. Of these, the first and third seem to have been originated by him. The names given to them here conform to modern standard usage.

1 The *doubly equidistant* map is laid out as follows. Choose a pair of fixed points on the sphere, A and B. In the middle of the paper on which the map is to appear draw the straight segment A′B′, the length of which, to a suitable scale, shall equal the length of the great circle arc AB. Then the map of any point P on the sphere is the vertex P′ of the plane triangle A′B′P′ of which the sides A′P′ and B′P′ have the lengths of the great circle arcs AP and BP respectively, and are on the proper side of the base. This mapping has been discussed in modern times, but no modern applications are known (Deetz and Adams 1945: 176), much less medieval ones.

2 The *azimuthal equidistant* map is equally easy to describe. Choose a fixed point on the sphere, say A, and a zero direction through it. Then the point A′ at the centre of the map is the image of A, and a fixed ray through A′ determines directions. For any point P on the sphere, its map P′ is the endpoint of the straight segment A′P′ having as length the length of the great circle arc AP. The azimuth of A′P′ with respect to the fixed ray must equal the azimuth of AP on the sphere. Bīrūnī describes the process in mechanical terms as being a rolling of the sphere on top of the map from an initial

tangent position at A′ in the direction of P until P is the point of tangency, thus determining P′.

A primitive and presumably intuitive example of this system is the world map drawn by ʿAlī b. Aḥmad al-Sharafī of Sfax in 1571 (Brice 1981: vi; Nallino 1916). He was doubtless ignorant of the work of al-Bīrūnī, as was Postel, the first to apply this mapping in Europe, in 1581 (Deetz and Adams 1945: 175). The azimuthal equidistant projection is widely employed nowadays (Figure 5.3).

3 The *globular system* maps a hemisphere onto a circle (Figure 5.4). Consider a pair of diameters EW and NS, which intersect at O and which divide the circle into quadrants. EOW is the map of half the equator such that E has longitude $\lambda = 0^\circ$, O has $\lambda = 90^\circ$ and W has $\lambda = 180^\circ$. Graduate all four radii and all four quadrants into convenient equal divisions, say ninety; one per degree. Number the divisions upward and downward from E, O and W in such manner that N, the map of the north pole, has $\phi = 90^\circ$, and for S, the south pole, $\phi = -90^\circ$. The co-ordinate net is composed of two families of circular arcs. The map of the meridian having longitude λ is the unique circular arc passing through N, S and the point on EW determined by the given λ. The map of the parallel of latitude ϕ is the circular arc passing through the three points, on each of NES, NOS and NWS, for which ϕ has the given value.

Al-Bīrūnī was clearly pleased with his construction, for he derives expressions for calculating the radii and locating the centres of the co-ordinate curves. He had every right to be complacent; distortion is slight in the central portion of the map, and radial distances are very nearly preserved throughout. The region of greatest stretching is along the periphery. Since the map resembles the stereographic projection described below, it is almost conformal.

Conjectures have been made as to how al-Bīrūnī came to think of this mapping. Berggren (1982) suggests that, because of the co-ordinate net composed of equally divided circular arcs, it is an expansion of Ptolemy's second mapping to cover an entire hemisphere. It seems more probable, especially since al-Bīrūnī may have been ignorant of Ptolemy's maps, to think of it as a close approximation to the azimuthal equidistant system when the centre is a point on the equator, and only a hemisphere is mapped. For this special case the meridians map into smooth symmetrical curves, each one passing through the two poles and one of the equally spaced graduations on the rectilinear map of the equator. The parallels of latitude map into smooth curves, each one passing through the two points on the circumference and the one point on the vertical diameter for which ϕ has a particular value. These curves are not circles, but they are close to being

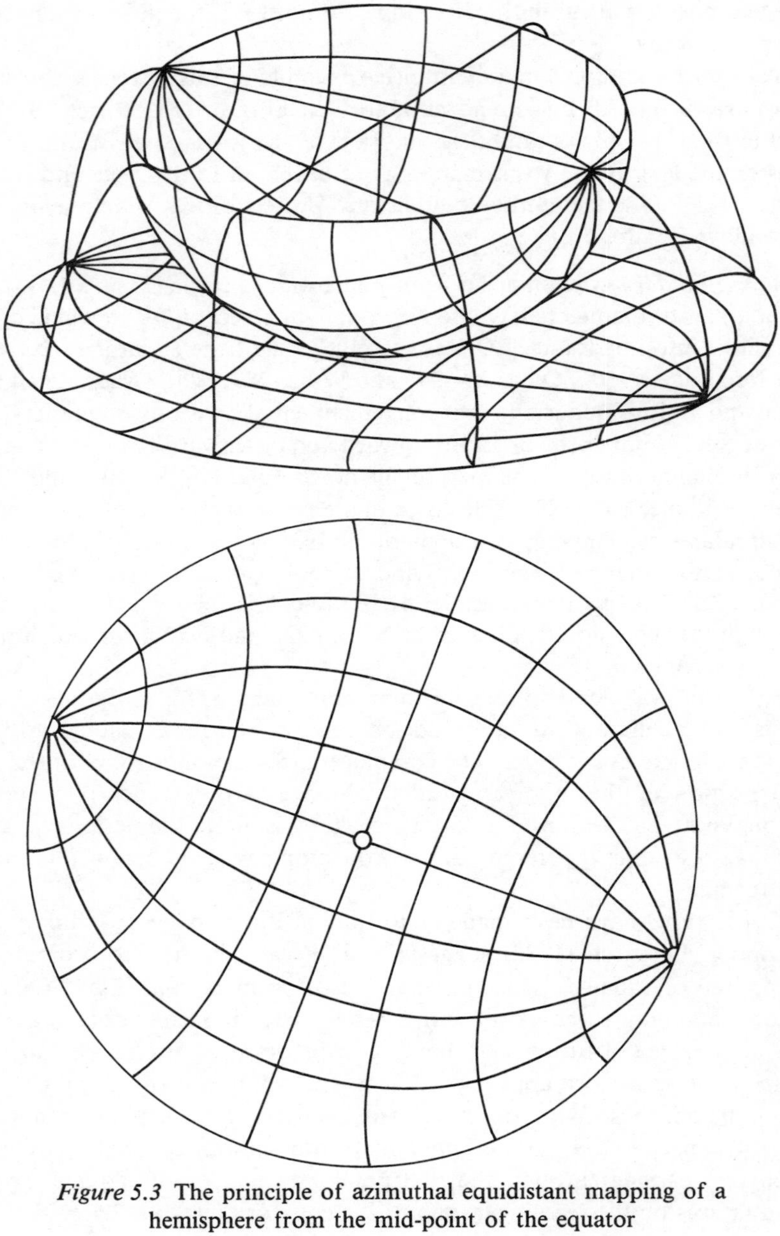

Figure 5.3 The principle of azimuthal equidistant mapping of a hemisphere from the mid-point of the equator

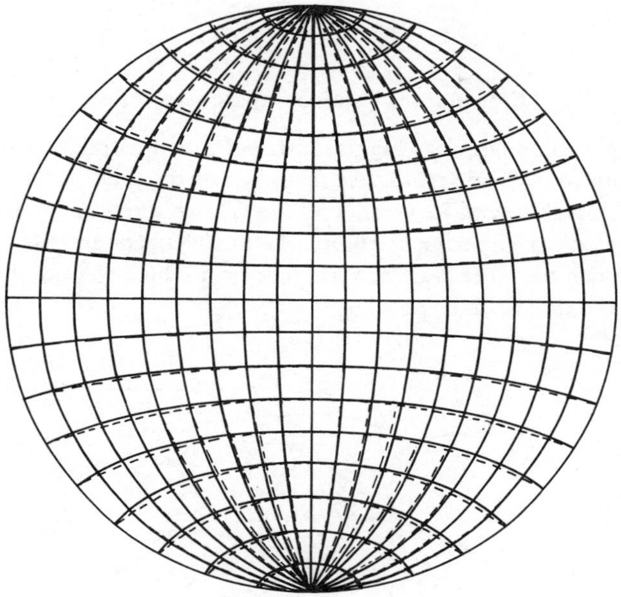

Figure 5.4 Superposed co-ordinate nets of a hemisphere mapped by the azimuthal equidistant (broken line) and global projections

circles, and al-Bīrūnī drew them as such. In Kennedy and Debarnot (1984) superposed co-ordinate nets of the azimuthal equidistant and globular maps are displayed, and they are seen to be very near to each other.

No oriental examples of the globular map are known. However, after a lapse of six centuries, it reappeared, independently of al-Bīrūnī, in Europe. In 1660 a Sicilian, Gianbattista Nicolosi, published two examples, one a representation of the eastern, the other of the western hemisphere (d'Avezac 1863: 342). Another application appeared, in Paris, in 1676, and others followed. In 1701 the French scientist, Philippe de la Hire, described a perspective mapping invented by him for which some of the co-ordinate curves are elliptical. However, the resulting net is very similar to that of the globular map.

The English cartographer, Aaron Arrowsmith, in 1794 published a world map. In the explanatory material accompanying it he says he has chosen de la Hire's projection as being the best. He then describes laying out the co-ordinate net with circular arcs in exactly the same manner as al-Bīrūnī (d'Avezac 1863: 359). There is no question of al-Bīrūnī's having influenced Arrowsmith, but it would be curious if both men, one in the eleventh century, the other in the eighteenth, had the same motive for choosing the simpler curve.

The equatorial stereographic projection

In a stereographic mapping (Figure 5.5), points on a sphere are projected onto the plane of a fixed great circle from one of the poles of the circle. The projection, together with its leading property, that circles map onto circles, was discovered early, perhaps around 150 BC (Neugebauer 1949). Its main application has been the standard astrolabe, in which the point of projection is the south celestial pole.

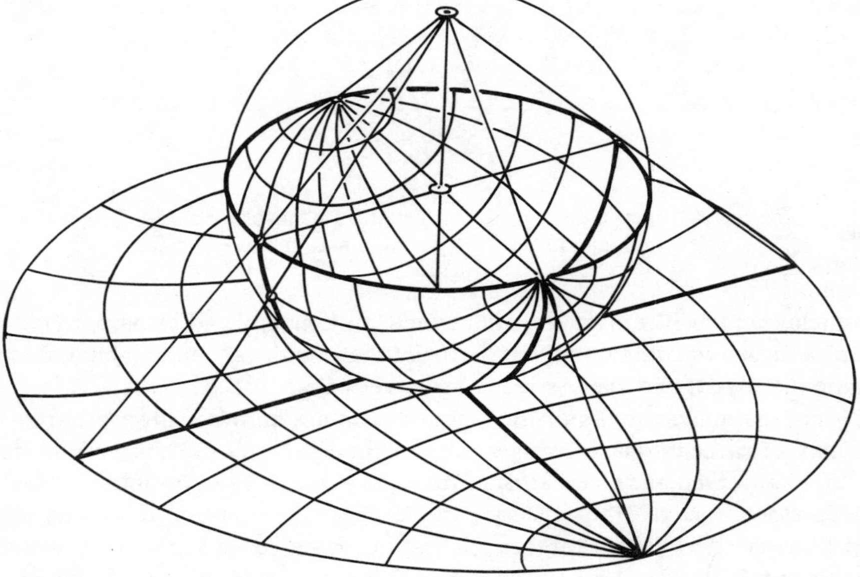

Figure 5.5 The principle of stereographic projection of a hemisphere from the mid-point of the equator

About 1050, however, the Spanish Arab al-Zarqāllu (Azarquiel) invented a form of astrolabe called *al-ṣafīḥa* (in the Latin West, *saphea*) which employs stereographic mapping from a point on the equator (Millás-Vallicrosa 1943–50). This instrument was popularized in Europe; its form of projection was adopted for terrestrial maps. By the end of the sixteenth century it had become the prevailing system for presenting world maps

(Keuning 1955: 7–9). These have been confused with the globular projections described above. The two types can be distinguished by noting that in the stereographic maps the distances between graduations on the equator increase slightly toward the edge of the map; in the globular case the distances are constant.

Al-Idrīsī's map

The Norman king Roger II of Sicily included geography among his many intellectual interests. He commissioned the Moroccan Abū ʿAbdallāh Muḥammad al-Sharīf al-Idrīsī to compile a comprehensive atlas of the world. Roger supported the project lavishly, sending travellers to distant places whose reports supplemented the written sources at al-Idrīsī's disposal. After fifteen years of work, in 1154, the job was complete. It comprised a circular world map (Miller 1926–31: vol. 5, p. 160), the much larger rectangular map described below and an accompanying Arabic text.

The large map (most recently published as Miller (1981)), is made up of seventy rectangular sheets, to be assembled in seven rows of ten sheets each, north being at the bottom, opposite to the modern convention. Many hundreds of geographical features and cities are shown, but the method by which they were plotted is not obvious. The upper and lower edges of each sheet coincide with the upper and lower boundaries of one of the seven 'climates' of classical antiquity (see Honigmann 1929; and Dallal 1984). The standard definition of these zones on the earth's surface is astronomical. In principle, the first climate begins at the parallel of latitude along which the length of maximum daylight is $12\frac{3}{4}$ hours. It ends, and the second climate commences, at the latitude enjoying a maximum daylight of $13\frac{1}{4}$ hours. Thence the successive climates advance northward, each boundary marking a half hour increase in maximum daylight length.

It is a consequence of this definition that the widths of the climates decrease as they proceed north. On the Idrīsī map, however, they tend to have a constant width of 6°, as can be seen from a partial scale of latitudes along the right edge of the map (cf. Miller 1926–31: vol. 5, p. 164).

All indications are that al-Idrīsī was mathematically unsophisticated and innocent of trigonometry, but that his rough and ready methods were well suited to reconciling the mass of frequently contradictory data available to him. The introduction to his text (al-Idrīsī 1970; Jaubert 1836–40) lists twelve sources, only one of which, Ptolemy's *Geography*, is known to be based upon co-ordinates. However, most Muslim geographers tended to present their material arranged by climates, so it would be natural for al-Idrīsī to plot localities judiciously within their proper climates, without bothering about the precise boundaries of the latter. The naïve investigation described in Kennedy (1986) demonstrates that in fact he did not err drastically.

As for longitudes, no trace of a horizontal scale appears on the map. It has been explained above why medieval longitude determinations were extremely unreliable, and al-Idrīsī's diffidence is understandable. If he assumed (as was then common) that the inhabited portion of the globe comprised 180° of longitude, it follows that each sheet covers 18°. Comparison of this with the climate widths demonstrates that the map is of the Marinus type, in the sense that a degree of longitude is about $\frac{6}{10}$° of latitude. Hence only in the sixth and seventh climates is distortion minimal. Everywhere else east–west distances appear shorter than they should in comparison with north–south distances.

In his introduction al-Idrīsī mentions a plotting board (*lawḥ al-tarsīm*) and an iron scale. The precise form and function of these objects is not clear. However, his sources frequently gave the distances between localities. A reasonable procedure would have been to commence by plotting widely separated cities whose positions seemed reliable, thence filling in intermediate points by successive triangulation on the plotting board for eventual transfer to the final map, originally engraved on sheets of silver.

Whatever method was used, the result was the *chef d'oeuvre* of Islamic cartography. A large body of literature has grown up about it, including studies of particular regions on the map, e.g. the British Isles in Beeston (1949), Scandinavia in Tuulio-Tallgren (1936) and Tuulio-Tallgren and Tallgren (1930), Germany in Hoernerbach (1938), Spain in Dozy and de Goeje (1866), Bulgaria in Nedkov (1960), Africa in Mžik (1921) and India in Maqbul Ahmad (1960).

Iranian rectangular co-ordinate maps

There exist several copies of a geographical work written *c*. 1340 by one Ḥamdallāh al-Musṭawfī al-Qazwīnī which contain a map published in facsimile in Miller (1926–31: vol. 5, plates 34, 35 and 86). This covers the region between Syria and Kashmir from west to east and from the Yemen through Khwārizm south to north. The field was broken into rectangles by families of orthogonal parallel lines drawn at 1° intervals. Some 170 cities were located by writing their names inside the appropriate rectangle determined by their respective latitudes and longitudes. Examination of a dozen or so cases demonstrates that their co-ordinates, to integer degrees, coincide with the geographical tables of the late Persian *zīj*es. Geographical features are lacking except for coast lines.

The map described above is a sensible if primitive example of a co-ordinate net, the only one extant from medieval Muslim cartography. It is an application of the directions in the introduction to Suhrāb's table mentioned above. A world map which also appears in al-Mustawfī's book is a less happy effort along the same lines. It is best discussed in conjunction with the world map of Ḥāfiẓ-i Abrū (d. 1430), published in Miller (1926–31: vol. 5, plates 72 and 82), for the later geographer seems clearly to have depended upon his predecessor, and the vagaries of copyists' errors make it easier to draw conclusions from as many manuscripts as possible. Two copies of al-Mustawfī's world-map appear in Miller (1926–31: vol. 5, plate 83).

The general idea was to lay down a square rectilinear co-ordinate net with longitudes ranging from 0° to 180° and latitudes (in modern terminology) from $-90°$ to 90°. For al-Mustawfī the interval between lines was 10°, for Ḥāfiẓ 5°. Inside the square a circle was inscribed representing the inhabited hemisphere. Inside this was the map proper, with the regions having co-ordinates falling within the excluded corners either ignored or fudged. Al-Mustawfī wisely refrained from plotting cities, confining himself to regions only. Ḥāfiẓ displays a good many cities, but they tend to be in the central portion of the map where distortion is less disastrous.

NOTE

The author is greatly beholden to Professor Fuat Sezgin for the hospitality of the Frankfurt Institute, and to Dr. Reinhard Wieber for pointing out errors and omissions in a first draft of the chapter.

6

Arabic nautical science

HENRI GROSSET-GRANGE

(in collaboration with HENRI ROUQUETTE)

INTRODUCTION

Nautical 'knowledge' is principally founded on the accumulated experience of navigators, but it is also a 'science' which stands at the cross-roads of different disciplines: in particular, astronomy, geography and meteorology – without forgetting the question of measuring and observational instruments.

It is difficult to retrace the history of Arabic nautical science, because the ancient texts are currently lacking. The only works that are available were composed at the end of the fifteenth and at the beginning of the sixteenth century, and describe exclusively the art of navigating in the Indian Ocean. This account is therefore limited by force of circumstance to the analysis of the nautical instructions of their two authors, Ibn Mājid and Sulaymān al-Mahrī, navigators who were, we can say, the inheritors of a tradition whose historical development we cannot rediscover with our present knowledge of the sources.

It is helpful to recall first of all the historical and geographical framework in which the work of these two mariners was undertaken, to note the 'routes' and the vessels which they used, and to discuss some basic facts of navigation both ancient and modern, together with a brief definition of some nautical terminology; all of which is necessary to enable the texts to be presented and analysed, and the importance of the Arabic nautical experience to be fully understood.

The geographical and historical setting

The experience of the two navigators Ibn Mājid and al-Mahrī is set within a very precise geographical framework, that of the Indian Ocean: the traditional route of contact between the Western (Roman and then Arabic) and the Chinese civilizations, it is the domain of regular and alternating winds, the monsoons, which have always favoured extremely active commercial exchanges between its different shores.

The era in question covers about a century (1450–1550), and is generally considered to be that of the transition between the Middle Ages and modern times; it was the era of 'great discoveries' which saw Portuguese mariners round Africa and penetrate the Indian Ocean, which had been the exclusive domain of Arab, Persian, Indian and Chinese navigators for more than half a millennium.

In this ocean, the Arabs of the time operated from two main areas: on one side, the east coast of Africa, in the fief of Oman, with its numerous ports (thirty-seven, it appears) of which the most important were Mogadishu and particularly Malindi (modern Kenya), Kilwa (Tanzania) and Sofala (Mozambique); on the other side, the Sultanate of Delhi (since 1206; in 1310 it controlled nearly all the Deccan). The mariners were thus required to navigate, with the aid of the south-western monsoon, between these two coasts, and even beyond, towards the straits. In about 1420 an Indian (or Arabic) vessel rounded the Cape and entered the Atlantic.

On these voyages, the navigators crossed the paths of Chinese mariners, who were pushing into the area. From 1402 a Korean map included the tip of Africa. In 1405 the great maritime expeditions of the Chinese Admiral Zheng He began; in the course of several attempts, he reached Indonesia and India, then passed them, headed for Africa in around 1417 and returned there in 1431–3.

Was the Indian Ocean therefore a Sino-Arabic condominium? It seems that the Arabs maintained a more permanent presence there, of an essentially commercial nature.

In the fifteenth century the closure of the overland silk route, due to the xenophobic and isolationist policies of the Mings, gave the Muslims a monopoly of east–west trade. But they profited from it only until the intervention of the Portuguese.

The latter progressively circumnavigated Africa. Bartholomeu Dias reached the Cape in 1488. Vasco da Gama sailed along the coast of Mozambique (where, at Quelimane, he met four Arabic vessels heavily laden with gold, jewels, diamonds and spices). In order to rival his counterpart in Mombasa, the sultan of Malindi secured for da Gama the services of the most skilful pilot of the Indian Ocean, Ibn Mājid, known since 1462 for his nautical treatises. In twenty-three days, Ibn Mājid led the Portuguese fleet to Calicut (south of Mahé, in present-day Kerala state).

Although this feat indicates an experienced pilot, the identification of that pilot as Mājid the author of the navigational treatises has not been formally demonstrated. At all events, an Arab mariner became the unwitting instrument if not of the ousting of the Arabs from the navigation of the Indian Ocean – since it continues to be active in our own times between east Africa, Somalia, the Arabian peninsula, the Indian sub-continent and Laccadives–Maldives – at least of the ending of a private hunting ground.

The routes and the vessels

The phenomenon of the monsoon favoured the establishment of regular routes, exploited by family shipowners.

Having set out from the bustling and competing ports of Africa, the Arab navigators put in at Goa or Calicut on the West coast of India, and pressed on as far as Malaysia. Their reaching China is more uncertain (there may have been a trading post at Canton). They transported ivory and gold, the raw materials for luxury goods, and also slaves from west to east. The return freight included cotton, silk, spices, ceramics and porcelain.

The monsoon then, was the major influence on the orientation of these routes: from November to March the movement of air from India (cool) towards Africa (hot) generates the monsoon from the northeast; from April, the sun reheats India, causing the monsoon to reverse direction and blow from the southwest. From June to September, it sweeps over the whole of the Arabian sea and the Bay of Bengal.

There were two main shipping routes. The first was the route serving Malacca; for various reasons this rounded Ceylon at a great distance (only the condensation covering its contours, or, at night, 'false lights', were visible), then it continued, with the aid of observations, towards Nicobar. The second route was the crossing from India to Oman, at the end of the eastern monsoon; heading first for Socotra, which was sometimes sighted before the first signs of the reverse monsoon were felt, then sailing back up by hugging the wind in the direction of Arabia, and then travelling along the Arabian coast; if the coast was not reached in time, it was necessary to return to India and wait there fore several months. At best this would take twice as long as the direct route.

Routes which basically followed a straight line, such as sailing up the Red Sea, were equally not without serious dangers.

There is, however, evidence of breaks in this web of maritime exchanges. The manuscripts hint at the existence of some kind of interdicts operating south-east of Sumatra, beyond Singapore, in the Bay of Bengal, in the Arab–Persian Gulf itself and even north of Jeddah. On the other hand, the accuracy of the latitude figures between the Sunda Islands, the Chagos Islands and Pemba suggests that there may have been recent direct contacts. As al-Mahrī wrote: 'the mariners of the Indian Sea and the Christians are in agreement on such a value . . . but the people of China, Java and beyond

. . .' It would appear that documents as yet unfound but indispensable to complete our knowledge should be sought in India and in Portugal.

Because of its meteorological characteristics, the Indian Ocean requires ships that make good speed into the wind (tacking close to the wind) and that perform especially well with following winds. In fact the dhows – still found today, made of teak boards assembled side by side, with a tall stem and a raised deck at the stern – the baghlas and the sambuks are all rigged with the 'Arabic' lateen sail, operated according to local custom. These are excellent seasonal ships, long and slim.

We know that the vessels of Ibn Mājid and al-Mahrī's time were capable of tacking close to the wind at the end of a season, and thus with gentle breezes, so as to reach home without being trapped in a foreign anchorage when the monsoon turned. On the other hand, we cannot describe with certainty the construction and rigging of these ships, which also varied. The drawings which probably most resemble them figure on certain Portuguese maps of the early sixteenth century. We can recognize in them a type of steering apparatus still sometimes used on sturdy smaller boats, the helmsman being practically at the foot of the rear mast (on a boat with two masts).

Nautical terminology

Altitude or elevation angle from the direction of a celestial body to the horizontal plane of the observation point (altitude + zenith distance = 90°).
Astrolabe ancient instrument that determines the moment when a star reaches a given altitude above the horizon.
Azimuth the angle (measured from the south towards the west) between the vertical plane of a star and the meridian plane of a given place.
Co-ordinates longitude and latitude of a star:

(a) *ecliptic co-ordinates* relative to the largest circle described in a year by the earth on the celestial sphere in its motion around the sun.
(b) *equatorial co-ordinates* relative to the great circle described on the celestial sphere by the plane of the earth's equator.

Dead reckoning a means of determining the ship's position on a sea chart by an estimate based on the preceding course, the speed, or even the wind and the current. This estimated 'position' must be verified as soon as the opportunity permits by the most precise observation possible of seamarks or celestial bodies.
Following wind wind from behind or thereabouts.

Free wind wind received at about 30° in relation to the rear of the boat (on the port or starboard side) (wind ... 60° ...)
Gnomon column or pin on an early sundial whose shadow indicates the time of day. Sometimes applied to the sundial itself.
Landing(s) the approach(es) to land.
Latitude angle formed at a given place by the vertical between that place and the equatorial plane (measured from the equator – positive northwards, negative southwards). To determine the longitude and the latitude of the ship is to 'take a bearing'.
Longitude dihedral angle formed at a given place between the meridian plane of that place and the meridian plane of an acknowledged standard place (now usually Greenwich observatory), measured westwards.
Mansion (or 'house') position of the sun on the celestial sphere with respect to well-known constellations (Sagittarius, Aquarius, etc.) on a particular day.
Meridian (a) plane defined by the vertical of a given point and the earth's rotational axis.
Meridian (b) measure of the highest apparent position of a celestial body (preferably the sun) from a given place, on a given day; it permits easy calculation of the vessel's latitude, useful for routes travelling approximately north to south.
Nautical ephemeris (pl. ephemerides) table(s) giving the values for certain variable astronomical measurements for each day of the year, in particular the co-ordinates of the planets, of the sun and of the moon.
Nautical instructions collection of useful navigational information relating to coasts, winds, currents, seamarks, lights and lighthouses.
Nautical mile unit of length, used only in sea or air navigation, corresponding to the distance between two points of the same longitude and whose latitude differs by one minute of arc (approximately 1,852 m).
Ocean navigation navigation of the high seas (out of sight of land and seamarks).
Precession very slow conical movement, made by the earth's rotational axis around a mean position corresponding to a direction perpendicular to the ecliptical plane.
Rhumb angle made by two of the thirty-two divisions of the compass ('areas of wind'): N, N1/4NE, NNE, etc. 1 rhumb = 11° 15′ (see Figure 6.1).
Seamark fixed and highly visible object, situated on the coast, enabling the navigator at sea to determine the ship's position.
Shore bed sea bed close to the coast which plunges vertically into the sea.

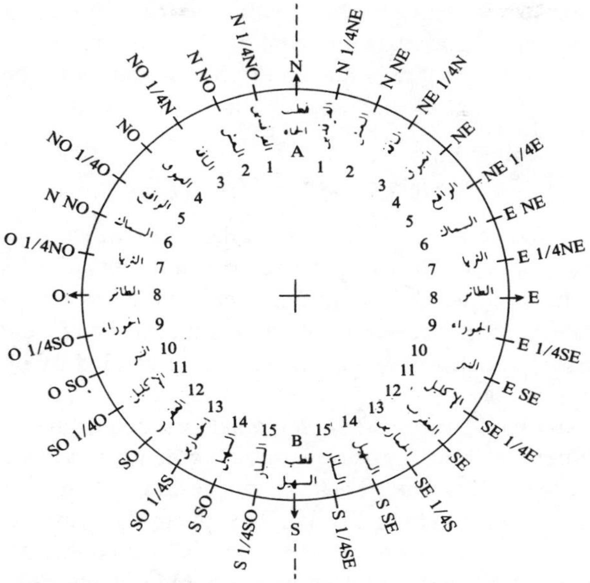

Figure 6.1 The map compass which is described in Arabic manuals of nautical instruction

A North Pole	5 Vega Lyrae	9 Orion	13 Argo Navis
1 Ursa Minor	6 Arcturus	10 Sirius	14 Canopus
2 Ursa Major	7 Pleiads	11 Scorpio	15 Southern Cross
3 Cassiopia	8 Altair	12 Antares	B South Pole
4 Aries			

Tack action of catching the wind alternately from the port and starboard sides, generally to sail into the wind.
Tacking point, gybing point inferior angles fore and aft of a sail.

Some basic facts of modern astronomical navigation

The reader with no particular nautical knowledge will probably best appreciate the technical skill of the contemporaries of Ibn Mājid by taking a brief look at the principal methods of determining a ship's position used by sailors in around 1950, just prior to the intensive, indeed exclusive, use of radio-electrical equipment for navigation.

Navigation within sight of land

In order to locate the ship correctly when in sight of land, a triangle was drawn by plotting, with the alidade of the compass, the azimuth of three seamarks (if possible) and marking them on the map. The triangle obtained by the connection of the three plots had to be as small, and thus as precise, as possible.

Navigation out of sight of land

In the case of fog, or at night (no coastal lights) or when on the high seas, the ship's route was drawn from the last definite plot taken, by means of dead reckoning: a combined estimate of the course, of the assumed speed (at the surface) and, if necessary, of the wind and the current. The result was, of course, only approximate and had to be verified by observation of seamarks or stars as soon as possible.

Astronomical navigation usually included two procedures.

For all routes, the ship's position was obtained by plotting three stars, considered as seamarks. The 'elevation' of the star, taken with the sextant, was converted with the help of the nautical ephemerides into a 'geometrical place', i.e. the site of those points where the star is seen under the same elevation at the same moment, plotted in a roughly straight line on the map. By plotting three stars simultaneously, situated if possible at 120° from one another, a triangle could be obtained, as with the seamarks, whose internal area, and therefore accuracy, depended on the measuring precision of the sextant. This in turn depended on the clarity of the star and of the horizon (at night and by day in 'poor visibility'), as well as on refraction, on the stability of the ship and the steadiness of the operator's arm, etc.; in short it could be haphazard.

For approximately north–south routes, where it was necessary to rectify the dead reckoning principally by latitude (except in the case of strong cross-currents), the faster meridian method would be used. The operator focused on a star at its diurnal apogee (given by the ephemerides when it passed the meridian of the supposed place, then measured the altitude and made a simple calculation to obtain the latitude of the observation point. Using the sun at precisely midday, this method was generally more accurate, at least for moderate altitudes (less than 45°).

It is easy, therefore, to understand the importance, for seafarers of all eras, of the observation of seamarks, of the visibility and altitude of the stars, and of the meridian.

It goes without saying that the contemporaries of Ibn Mājid and al-Mahrī, while basing themselves on the same elements, employed much more rudimentary methods. In the first place, there was no question of locating the ship's position on a map, and no possible comparison between that map (a 'portulan' sea chart) and modern-day route charts (large-scale ocean maps that permit the plotting of an approximate route, which is then transferred onto detailed small-scale charts). In sight of land as on the open sea, the navigators used their own reckoning (of speed, voyage time and drift) compared with texts, such as the poems of Ibn Mājid, that served as nautical instructions:

> to go from Aden to Goa, take such a course up to point x, where you will find such a wind regime at such a time of year. Then take such a course until you measure such a star at such an altitude which corresponds to the landing at Goa. Do this from the east, correcting your deviation from the route by using the altitude of the star each night. After such a voyage time, start to sound . . .

Thus we can see that the modern idea of bearings was not conceivable because of the lack of precise equipment, such as charts, measuring instruments and ephemerides. Nevertheless, Ibn Mājid led Vasco da Gama from Malindi to Calicut in twenty-three days.

SOURCES FOR THE STUDY OF ARABIC NAUTICAL KNOWLEDGE

As we have previously stated, this study is not a detailed review of Arabic nautical knowledge, but an analysis of the quintessential experience of two navigators covering the northern and western parts of the Indian Ocean – and far beyond in the case of Ibn Mājid – during the period 1450–1550. The relativity of Arabic nautical knowledge is indeed recognized by its principal possessor, Ibn Mājid, who, probably because of his collaboration with the Portuguese, advised his compatriots of the Indian Ocean to enlist 'in the school of the Francs [Westerners], whence nautical science and art now come'.

This experience of an essentially utilitarian and empirical technique is related in various manuscripts written between about 1460 and 1550. Extant copies of the originals are the source for most of the notes and comments which form the substance of this chapter.

Ibn Mājid and al-Mahrī were both navigators. If the former was at the height of his art in 1496 (date of Vasco da Gama's expedition, which he could have piloted) and thus experienced the incursion of the Portuguese in the 'Arab lake', then al-Mahrī is probably his junior. According to different hypotheses, he died between 1511 and 1554. The dating of his books is therefore difficult, all the more so because certain of his works contain the same material.

Manuscript sources

Three manuscripts were referred to in writing this chapter.

1 A copy of manuscript no. 992 by Ibn Mājid: fols 82ʳ–106ʳ, Oriental studies, Academy of Sciences, Leningrad.
2 Manuscript no. 2292 of the Arabic collection in the Bibliothèque Nationale, Paris. This contains some of the books of Ibn Mājid.
3 Manuscript no. 2559 of the Arabic collection in the Bibliothèque Nationale, Paris. This contains some of the books of Ibn Mājid and those of al-Mahrī.

These manuscripts are themselves only copies with variations (where comparison between two texts is possible). Through these copies we encounter the titles of other books so far unknown.

Other collections of Arabic knowledge

The Indian Ocean was a place of frequent meetings, even of collaboration and exchanges between mariners. Consequently the parameters of 'Arabic knowledge' cannot be drawn as neatly as one would like: do not important components of this knowledge come from Chinese seafarers? And does not the abundant Portuguese maritime literature of the sixteenth century rely in part on the legacies of Ibn Mājid and his contemporaries?

Therefore we can say that nautical knowledge transcends time and history; it is a common storehouse drawn from predecessors and rivals and enriched with each generation. However, the preponderance of Arab mariners in the Indian Ocean for some centuries gives weight to the part of that knowledge conveyed by Ibn Mājid and al-Mahrī.

Having said that, the authors of works published in Arabic in around the tenth century and later are mostly of foreign origin, and the Arabic nautical books themselves highlight the differences between Arabs, Ormuzians, Indians etc. Well before Marco Polo, books of astronomy called the *Sind* were known in Andalusia, and Marco Polo referred to the methods and documents of the mariners of the Far East. There were also some Chinese and Japanese charts.

We must thus expect to have to contrast the Arabic nautical books with many others of the same genre. In due course the Portuguese benefited from these former sources and enriched them with their own observations: 'for the single period from 1538 to 1552, more than 4700 documents, nearly all in Portuguese and nearly all unpublished' (Aubin 1972).

The study of the nautical instructions of Ibn Mājid and of al-Mahrī must thus depend on comparisons with a whole group of other texts from various periods.

Discussion of the sources

Before undertaking the interpretation and analysis of the authors, which often entails questions of authenticity, i.e. the faithfulness of the copies to the original, it is necessary to have overcome the obstacle of language.

The instructions are written in terminology that we find too vague – although more precise than some in modern vocabulary – despite the stability of Arabic over the centuries. Thus 'tacking point' and 'gybing point' were expressed then, as they are today, by specialized terms; but in certain cases, right and left are written identically. Similar examples abound.

But in what spirit should one approach the reading of Ibn Mājid and al-Mahrī? How much of a critical eye should the informed reader apply to their claims? A knowledge both of the personalities of the authors and of their work (we have about forty highly varied volumes) can be helpful; extensive analyses can be found in Ferrand (1921, 1924), Khoury (1970, 1972) and Tibbets (1971).

At first sight al-Mahrī's sober and lucid didacticism is appealing, whereas Ibn Mājid seems unmethodical and pretentious. However, the scientific verification of the authors' statements, and Ibn Mājid's greater familiarity with navigational practices, lead the reader to one conclusion: Mājid has sailed the seas far more than his emulator. We can see in al-Mahrī a wise man spurred by his curiosity for things of the sea, but a poor navigator, and in Mājid perhaps something of a 'Captain Marius'[1] but undoubtedly a fellow enthusiast of the sea.

Certainly his books, which were apparently written for apprentice navigators, cause the reader a great many difficulties: this is poetry, composed of evocation and allusion in which certain hints enabled the informed and perceptive individual to understand the rest.

Moreover, the techniques of critical analysis can refine the essential research needed to determine the authenticity of certain texts. Thus in the *Sufaliyya*, one of the three nautical texts in manuscript 992, certain passages appear apocryphal because of inadmissible nautical blunders on the part of Mājid, and would be difficult to attribute to lack of attention by the copyist. And this is not the only text which reveals the intention of 'playing Mājid'.

Finally we note that Mājid, the traditional practitioner, remains silent on the theory of latitude extracted from a meridian altitude (although he refers to declination tables), whereas al-Mahrī demonstrates this point masterfully, but betrays himself by omitting to adapt the formula to southern latitudes: thus he never crossed the equator, which explains certain of his results.

The study of Ibn Mājid and al-Mahrī leads us to ask where the line should be drawn between science and empiricism. An empirical and traditional mariner such as Mājid based himself on direct and lengthy experimentation. But should we regard either of these two navigators as men of science on that account? We can certainly grant al-Mahrī the status of a scholar, who was simply intrigued by the sea, and hail Ibn Mājid as a craftsman whose skill put him 'in a class of his own', despite the undoubted flaws that marred his personality.

THE MEANS OF ARABIC NAVIGATION

This discussion is not intended to be a didactic account of Arabic nautical knowledge, but is an attempt to make some progress, albeit often conjectural, in the understanding of a body of knowledge which is itself largely imperfect and lacking in overall coherence.

We should not picture an Arab mariner such as Ibn Mājid as being like an officer of the watch, observing the seamarks or the stars with the relative accuracy of his day, and plotting them in a triangle on a chart to correct an earlier position obtained by dead reckoning.

Using his own experience and that of his precursors, Ibn Mājid practised what we might call a 'refined dead reckoning' – an improved method of estimation. The charts were probably only used as a guide to the distances between lands or the general orientation of coasts and the locations of ports; they would scarcely have allowed any other purpose. The elevations of the stars, for their part, helped to locate the ship in a particular zone. And for the rest, the 'nautical instructions', the knowledge of the navigators and their intuition determined the reckoning. Although the Indian Ocean is a sea with stable winds, offering the advantages already noted, the regularity of the monsoons is not such as to make a good assessment of the force and direction of the winds and currents unnecessary.

Units of measurement

In a world that pre-dated the unifying and pro-scientific effects of the metric system, what means of measurement did the Arabs use? Essentially fingers, or digits, *zams* and *tirfas*. As in modern times, it was the measure of height which enabled the determination of distance: *zams* and *tirfas* were defined in relation to the finger. However, the concept of a constant unit of measurement was not yet entirely familiar to the minds of the time, thus constituting a major obstacle – made worse by the absence of sufficiently precise instruments – to the adoption of a truly scientific approach. Basically, however, invariance is not important provided that the order of magnitude of the variations is consistent with the degree of precision of the observations.

Fingers and the dubbān

Fingers were measured by means of the 'wood' (see pages 225–8). Consequently the maximum measurement was twelve fingers, or about 20°. Thus only the elevations of the lower stars could be measured.

Different human cultural groups have naturally based their measurements on the finger, or on the palm, the elbow, the foot, etc., but seeking to measure 'fingers', in the sense of very fine angles carved with a knife on small boards, would appear to be attempting the impossible. In fact accurate elevations could only be reached up to 20° and very probably less (cases of accuracy lower than 5° are too frequent to be attributed to chance).

Recourse was therefore had to the hand, which provided the *dubbān*, a term used to define the angle covered by 4 fingers – a crude though personalized standard. (The mariners of Mājid's time could, of course, have obtained the standard of 4 fingers by means of the rotation of the Pole star – if the diameter did not change with time. In any case the sky could provide an invariant means of reference, the angular distance between most of the stars remaining stable over the centuries.)

The term *dubbān* is used in connection with two stars, of which α Cocher was one of Mājid's favourites: 'α Cocher has a *dubbān* to its east (β) and to the south of the *dubbān* is a star of the same size (θ) which is called the *dubbān* of the *dubbān*, they have a distance between them of 4 fingers'.

In spite of this, Ibn Mājid never formally refers to this standard for the wood, in contrast to al-Mahrī: 'the wood of the measuring *dubbān* must correspond to the *dubbān* of α Cocher at the culminating point of Leo, and the rest of the woods will be correct by being divided according to this standard; it is an angular measure, and that is more exact than extending the arm.'

The angular distance is 7° 36′ between α and β and 7° 42′ between β and θ. Since there is no example of precise measurements being taken with the wood other than vertically, and that at the culminating point given, α and β are vertical, about 30° of altitude in al-Mahrī's land, the reference of four fingers for the *dubbān* on the wood is an accurate definition, at least according to him. In these conditions (Figure 6.2) measurement by the woods is 'out' by about 1°, whence an observed angle of 6° 40′ (compared

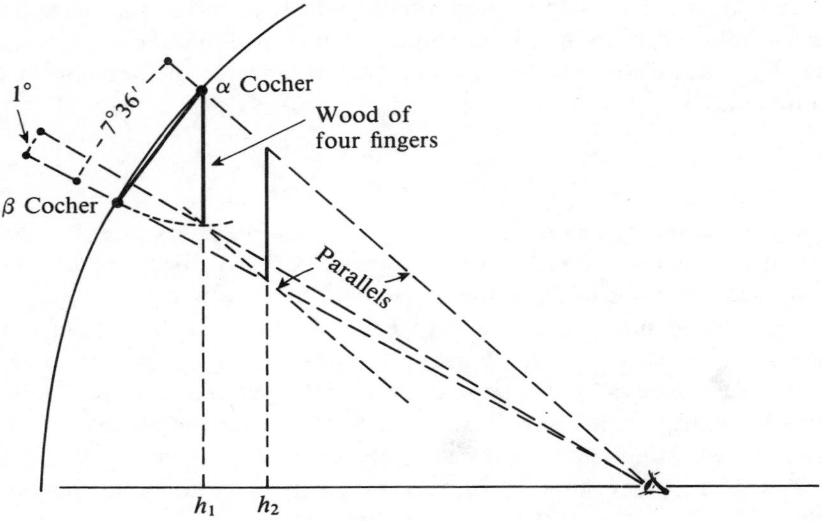

Figure 6.2 (Note: proportions greatly exaggerated)

with the true angle of 7° 36′). This implies that the arm must have contracted from h_1 to h_2.

To clarify these ideas and to achieve an equivalence between fingers (as corresponding to the meridians of stars noted by Ibn Mājid) we have incorporated several corrections of a 'modern' nature: refraction (bending of the rays by the atmospheric layers), true altitude (the height of the observation point above the sea influences the measured altitude of the celestial body), and the Pole star, which is not true north (the true altitude of the Pole star by meridian gives the latitude). The results of these calculations are given in Table 6.1. We have used the observations of the stars reported by Ibn Mājid, leaving aside al-Mahrī as too questionable despite his scientific qualities (except in the case of agreement with Mājid).

This table results from a very great number of comparisons between meridians of the α Southern Cross, α Eridian, and above all the Pole star, as well as several paired stars taken as quasi-meridians. The mean values between the second and the twelfth degree, being 1° 36′, corresponds to the figure given by the Portuguese. As for the huge first finger, we can explain it by the haziness of the nocturnal horizon, which necessitates an exaggerated elevation of the wood in order to distinguish the horizon clearly from the lower part of the wood. This hypothesis appears to be confirmed by the exaggerated measurements (sometimes of the order of a degree) observed in the pairing of large southern stars that are too high to be measured by

Table 6.1 Altitudes (or elevations) by fingers and latitudes

Fingers		*Differences*	*Refraction correction*	*True altitudes (observation altitude 5 m)*	*Pole star correction*	*Latitudes by Pole Star*
1st	2° 54′		20′	2° 34′	3° 31,8	6° 05,8
		1° 39′				
2nd	4° 33′		15′	4° 18	″	7° 49,8
		1° 44,5				
3rd	6° 17,5		12′6	6° 04,9	″	9° 36,7
		1° 37,5				
4th	7° 55		11′	7° 44	″	11° 15,8
		1° 30				
5th	9° 25		10,2	9° 14,8	″	12° 46,6
		1° 42,5				
6th	11° 07,5		9,3	10° 58,2	″	14° 30
		1° 42,3				
7th	12° 49,8		8,6	12° 41,2	″	16° 13
		1° 30,55				
8th	14° 20,3		8,1	14° 12,2	″	17° 44
		1° 25,6				
9th	15° 45,9		7,7	15° 38,2	″	19° 10
		1° 29,58				
10th	17° 15,7		7,5	17° 08,2	″	20° 40
		1° 44,56				
11th	19° 00,3		7,1	18° 53,2	″	22° 25
		1° 22,5				
12th	20° 22,8		6,7	20° 16,2	″	23° 48

the meridian method; indeed it is recorded that 'α Navire and β Centaur ... must be measured in the first northern climate .. by the light of the moon; this is the particularity of the southern stars ...'.

Certainly the clarity of the horizon by moonlight would have avoided an exaggerated elevation of the wood and consequently an inflated altitude figure.

The modern reader is surprised by the inequality of the fingers in the table, whereas the Arabs did not ask themselves whether the fingers differed in value. A close analysis of the texts, which would overburden the present study without enhancing it, would reduce a certain number of inaccuracies in the notations of altitude, but not all of them.

Zams

Estimated distances were calculated in terms of a unit known as the *zam*, duly defined by al-Mahrī: 'the *zam* is of two types, *definite (or customary) and technical. The former is the eighth of a distance such that a star increases or diminishes in height by one finger*, when going towards it or turning away from it, whether in theory or reality ...'.

Elsewhere, he also qualifies as *ḥaqqī*, 'true' or 'in its true sense', the *zam* obtained by measurement (though it may be obvious that the meridian method could be used, and al-Mahrī was probably aware of that, Ibn Mājid believed at first that the procedure was valid whatever the azimuth of the observed star, provided that it was situated in the axis of the ship, which is mathematically false). Al-Mahrī specifies that the definite *zam* implies 'a stable wind of average force'. On the other hand, he does not mention the 'general *zams*' (my interpretation of *Jumma*), which Ibn Mājid refers to a great deal, noting in particular: 'the exact value of technical 'general *zams*' exceeds the *zams* of routes and of distances (actually travelled ...)'. (This passage allows the questioning of certain estimated distances.)

By 'general *zams*', Ibn Mājid simply intended to define a certain standard: 'such is my figure in *zams* of three hours *by normal navigation*; it is up to the reader to adapt it as required.'

He therefore comes closer to al-Mahrī's 'definite *zam*' especially as he distinguishes further between 'heavy' and 'light' *zams*, the typical heavy *zam* evidently being used in conditions of dead calm and in the absence of current.

However, his use of these qualifying terms in connection with particular regions, and thus according to their specific meteorology, is more unexpected. The following extract comes from the 'particularity of particularities' (or 'nature of proportions') *ḍarība al-ḍarā'ib*, in which Mājid associates these distances with variations in the altitudes of stars (being supposedly measurable by astronomical observation of distance from the meridian, which would result in the finding of a component of longitude!): 'the estimated distance of the first rhumb is heavy . . . one does not count it from Hadmati to your Muluk (from 2° 35′ to 1° 50′ North, in the Maldives) as one counts it from Bab (el Mandeb) to Zuqur, nor from Muruti to Brawa (eastern Somalia) . . .'.

There are great differences between the cited distances, the 'lightest' being Somalia where the northeast monsoon blows coolly and regularly, with a good running current; this monsoon marks the longest period of the year for navigating in these waters, whereas all sailing ships get underway at the very start of the southwest monsoon, when the breeze is light, in order to avoid its violence later on.

The discrepancies due to the relativity of the unit of measurement were compounded by the variability of the routes described; thus Ibn Mājid declares: 'from such a point in Somalia to Aden there are 20 *zams*, sometimes less in clear easterly monsoon weather . . .'. This passage shows that distances were not necessarily calculated between the norm of the departure point and the arrival point; this is not a problem for the long routes, but probably explains the sometimes surprising speeds on short routes.

Ḍarība (not dated), like *Dhahabiyya* or *Ḥāwiya*, deals in a similar fashion with these observed distances in variable *zams* (unacceptable, as we have seen, because they were assimilated to the observation of longitude). Yet if the *Ḥāwiya* represents Ibn Mājid's early armoury, he speaks of his great age from the very start of the *Ḍarība*; we must therefore conclude that either Mājid was a victim of *perseverare diabolicum* (which is highly unlikely on his part), or that he did not understand the correlation with longitude.

If the distance–time ratio constitutes a relative element, it is nevertheless probable that the theoretical *zam*, based on an eighth of a finger, possessed a value to which we could accord a figure of about 12 nautical miles.

Al-Mahrī, for his part, did establish the 'mathematical value of the *zam*' by standardizing it on the finger: 'the astronomers are well aware that the revolution of the Pole Star (taken by mariners as a standard of 4 fingers) is 6° 6/7′ [which is correct for 1505], thus 1 finger = 1° 5/7′ and 1° = 1 *zam* less a third . . .'; this gives a value of 12.82 nautical miles for the *zam* – an allowable figure.

Tirfas *(and deviations)*

A *tirfa* is the distance covered in each rhumb before the meridian value of a celestial body varies by one finger.

Here again we find ourselves confronting the notion, which is now foreign to us, of a relative unit of measurement that seemed natural to a milieu before abstraction, where individuals were used to relying solely on the observation of concrete data.

Tirfas were classified by their obliquity in relation to the meridian, that is, according to the rhumb of the course: the less angled rhumbs (1–5) were called the *ruḥuwayāt* or *raḥawiyyāt*, the others being the *shaqaqāt*. Mājid cites them notably in connection with routes close to the west or east (thus in situations where the estimated course distance is doubtful), specifying that 'for the *raḥawiyyāt*, dead reckoning estimation is preferable . . . especially if it tallies with observation, while for the *shaqaqāt*, the altitudes alone are preferable'. This was quite logical in view of the complete ineffectiveness of meridian observation where westerly or easterly discrepancies were concerned.

We should also mention the *manākib* ('deviations', 'obliques', or 'intercardinal rhumb lines' in the common European sense of the term), which represented the course *par excellence* between the meridian and distances east or west: the *masāfāt*. (see Figure 6.3.)

The *tirfas* shown in Table 6.2 bring together elements relating to estimated distances that are to be found widely scattered in the books of both Ibn Mājid and al-Mahrī.

The theoretical views of the two writers concerning these matters should emerge unambiguously from such a table. Yet we are immediately surprised to note the definite value allotted to the east or west *tirfas*, which are in reality infinite. With regard to al-Mahrī, we have already seen that we cannot date his writings precisely and thus judge the progress of his experience. He is often content to report information collected from various navigators without verification. In his commentary on the *Tuḥfa*, he gives the figures from different schools, including that of the mariners of

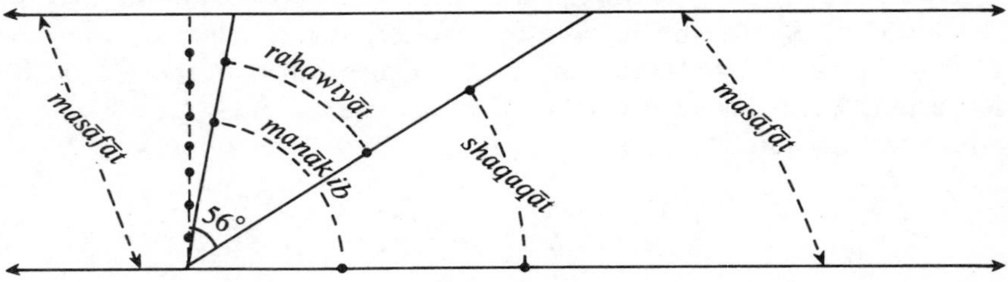

Figure 6.3

Table 6.2 Tirfas (expressed in zams*)*

	Accurate value	*The Ancients*	*Mg*	*Minhāj and Tuḥfa*	*From commentary on Tuḥfa*	*Shūlīān*
Pole	8	8	8	8	8	8
1st rhumb	8,16	10	10	9	9,6	10
2nd	8,65	12	12	10	11,4	12
3rd	9,62	14	14	11.	13,4	14
4th	11,32	16	6 to 16	12	16	16
5th	14,4	18 to 20	18 to 20	20	20	19
6th	20,9	22 to 25	21 to 25	20	35	24
7th	41	30 to 40	30 to 40	35	42	40
between 7 & 8	83	30 to 50	40 to 50	66	72	?
8th	infinite	40 to 60	50 to 64	...	infinite	

Coromandel, listing their approximate figures although basing them on 'the quarter circle neglected by the pilots ... and that is my school ...'.

Previously he had rectified the figures for the first four rhumbs by adding approximate fractions to them and again by using the quarter sine method. In this way, we note that if the values for the first four rhumbs are the least incorrect of the table, aligning it with the data of the Coromandel mariners yields figures that are notably incorrect, except for the seventh rhumb (no reading is given for the following rhumb). As it would be an undoubted exaggeration to hold the copyists responsible for this accumulation of so-called errors of approximation, it would seem that al-Mahrī's science (which is accurate in other respects) must have given way before a question so much more elementary that he illustrated it by building a rose on the ground on which people could walk in the direction of physically marked out rhumbs.

Charts

The manuscripts mention ephemerides briefly and the mariners did have charts (these are never referred to in the texts and are now completely lost, but the Portuguese saw them). However, the mariners of around 1500 navigated the Indian Ocean without charts or ephemerides; they used an approximate calendar and copious nautical instructions. Their own experience did the rest.

In fact the charts of the time would probably have been useless for locating the vessel at sea, their accuracy in regard to the distances between coasts being inferior to the uncertainty of the estimated position corrected by astronomical observation.

The manuscripts of Ibn Mājid and al-Mahrī, which epitomize the nautical instructions used by mariners of the time, provide nautical distances (and also terrestrial distances in the case of Mājid) at the height of each finger (Pole Star, $\beta\gamma$ Ursa Minor $\varepsilon\xi$ Ursa Major). The keying of these distances to the original system of meridians permitted the location of the places, all of which tied in fairly well with the coastal routes (curiously enough, considering the expression of these orientations in true headings and round rhumbs), although there are sometimes differences of detail in a particular area – the 'Berber gulf', for example. The map in Figure 6.4 compares a representation of the coastlines taken from the works of Ibn Mājid and al-Mahrī with the true outlines.

Al-Mahrī orders his distances methodically, and Ibn Mājid occasionally does likewise. The two writers sometimes complement each other, but with some divergencies when dealing with the same region. Co-ordinating the whole was not easy. One of the most revealing examples concerns the height

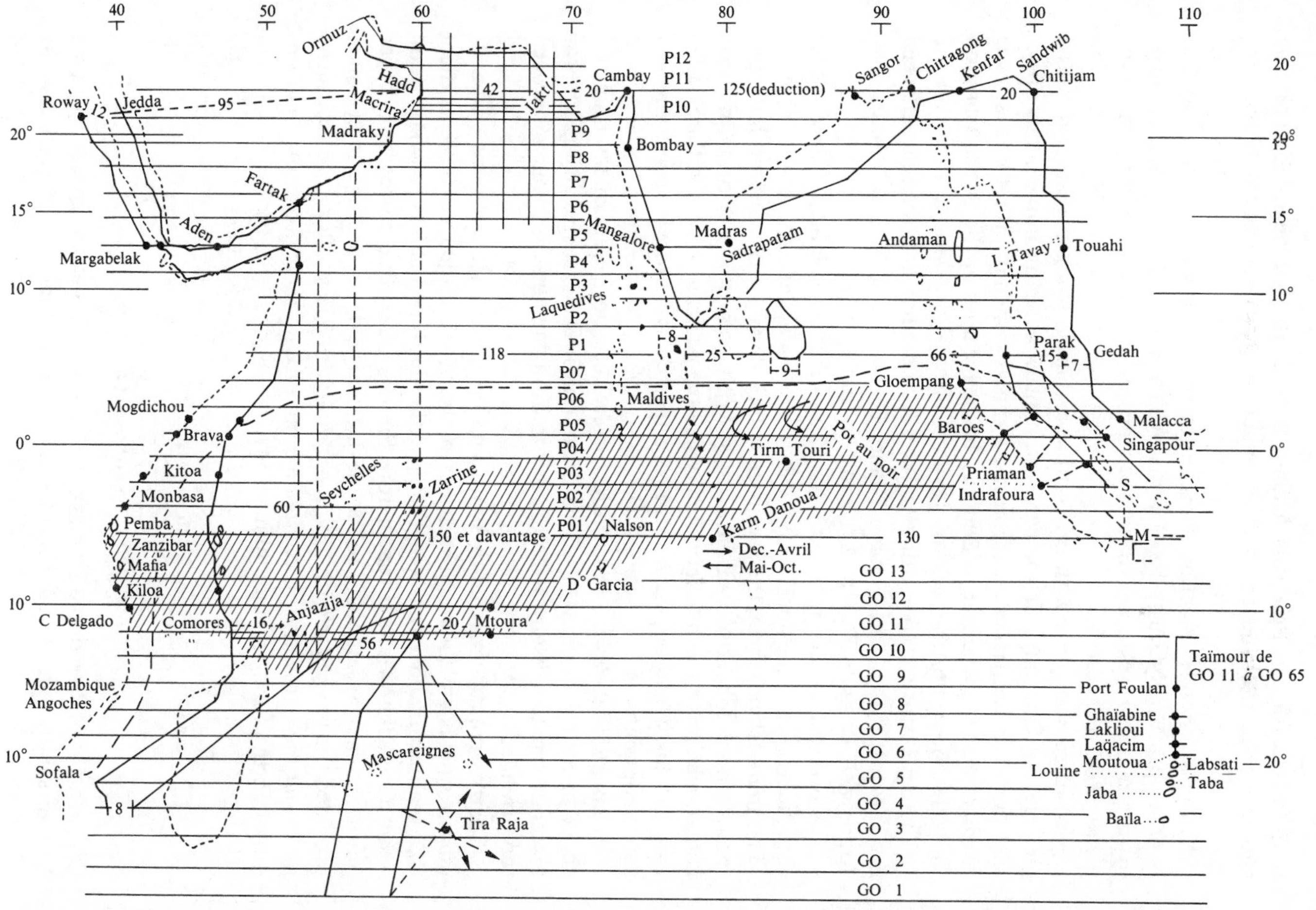

Figure 6.4 P = Polar meridian PO = horizontality of β and γ Ursa Minor GO = horizontality of ε and ξ Ursa Major

of 5 fingers for the Pole star, extended from Bargamlah (modern Margabeleh, near Aseb on the Bab el Mandeb Strait) to Tawahi (Tavoy in Burma).

Errors of latitude reveal on the map the zones that were unknown to the Arabs – principally Australia (Timor), which is indicated by a vertical line in its presumed position (without distances) and whose toponymy is in places relatively recent.

Madagascar appears with two outlines; that showing the western coast only is from Ibn Mājid. In the Far East, confusion begins immediately after Malacca. The western coast of Sumatra shows some important errors. Ibn Mājid and al-Mahrī differ by 2 fingers in the location of the Sunda Islands. Bali still appears to the west of Java.

There is confusion too, although less so, to the north of the line between Sri Lanka (Ceylon) and Nicobar Island, because according to Mājid, 'few Arabs visit Bengal, Siam and the east of India . . .'.

The presence of the mythical isle of Tirm Turi is explained – even more convincingly than the uncertainty over the Seychelles and especially over the Mascarene Islands – by the fact that sailing ships never ventured into the 'pot au noir' ('cauldron of darkness'). With regard to Karm Danwa (or Diwa), which is shifted longitudinally like the African coast and the Sunda Islands, this would appear to testify to the relatively recent migrations of the Indonesians.

In his *Qibla al-Islām*, Ibn Mājid corrects certain beliefs of his time. A check of his orientations confirms the outlines reconstructed in Figure 6.4 (except for places distant from the sea, and for Madagascar, which is much too extended).

Charts were thus marred by serious uncertainties, and we know that the manuscripts are silent about their actual use at sea. Moreover it seems that the Arab geographers knew nothing of the mariners' charts. Nevertheless it must be acknowledged that before the rise of Iberian cartography, the mariners' chart (as opposed to the marine chart), developed by simple folk, provided the navigator with a kind of Identikit picture of regions to which he would have travelled in complete ignorance of their geography had he only had tradition to guide him.

The instruments

The compass (and declination)

Out of sight of directional seamarks, alongside radio-electrical aids to navigation, sailors still use the instrument that Europeans and others know as the 'compass'. The same word, within the meaning of the period, was used by Mājid when writing about the Mediterraneans.

The presence of a magnetic needle, housed in a binnacle, is taken for granted although no one can state with certainty how exactly it was positioned. Two points, however, merit attention:

1 *Ibra* and *Samaka* certainly denote the needle, but the second term is only used twice.
2 We believe we can put forward the hypothesis of a block or housing with an axial support, on the evidence of a passage (admittedly unique) in a commentary on the defects of the compass: '. . . resulting from the heaviness of the rose and the poor quality of its "cupola"'. And unless it were supported on an axis, how could the needle have floated freely without hitting the sides of the container? How come in that case that when discussing the defects of the compass, there is no mention of any container? A seafarer will realize at once that the 'heaviness' makes the compass insufficiently sensitive to right itself rapidly after rolling or yawing when changing course.

In the daytime, it was possible to use a strip of cloth to indicate the relative wind, so as to steer in relation to that and help keep the vessel on course.

In the hypothesis of a needle resting by means of a block on an axis in a box or binnacle, how would bearings be found for the relevant course? There are two basic configurations:

1 The dish of the binnacle *fixed to the ship* is graduated, but in reverse; in Figure 6.5(a), if the ship is heading NW, the NW graduation will be to the right of N and the needle will point there.
2 Conversely, on a graduated rose borne by the needle, and so *fixed to the needle*, a single mark suffices, on the binnacle which need not be circular: a mark that stands for the forward direction (our 'line of trust') and which must be in line, or approximately in line, with a graduation which is the course (Figure 6.5(b)).

Solution (2) is the more convenient, because the helmsman always reads the course in front of him, almost unconsciously, whereas in (1) he has to refer to the point of the needle whose position varies from that of the course, and he will thus find it less easy to keep the ship on that course.

Steering by the stars is a variant of (1). So why should the two methods not have coexisted in this transition period? This would explain why the texts indiscriminately use the terms *al-ḥaqqa* (strictly speaking, the binnacle only), *Bayt al-ibra* (strictly, the location of the pointer) and *al-dāʾira* (strictly, the rose).

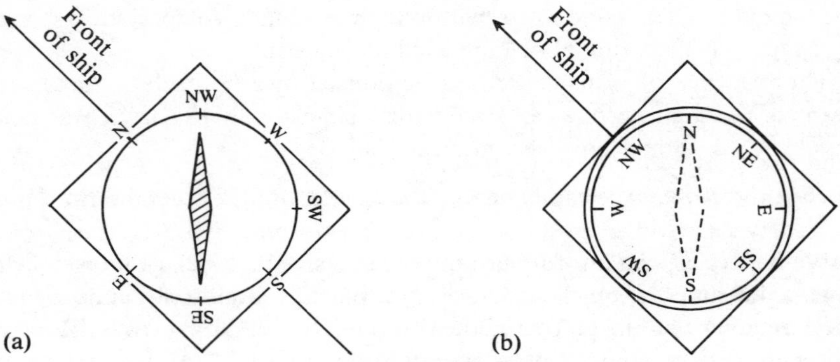

Figure 6.5

Finally, there is the question of how the compass was illuminated. Doubtless naphtha was lit for certain celebrations – for example on the arrival at Grand Nicobar: '. . . light the naphtha and dress the ship' – but was there a properly protected night light for the compass?

Declination

If iron or steel influence the compass, this 'deviation', which varies with the course, combines with magnetic 'declination' (the influence of the earth's magnetic field, independent of the course) to give a 'variation'.

Although Ibn Mājid and al-Mahrī warn of errors between course and compass (drift, etc.), which are prolifically commented on, we search in vain for a formal definition of declination. But in two cases, we may wonder whether the navigators had detected some inexplicable interference. The first case, taken from Ibn Mājid, brings us back to the *samaka* which is definitely the needle, because 'it [the route] is only distorted by . . . or by defects in the housing of the needle whose fish is called the fish of the box (*samaka*)'.

Further on: '. . . the pilot thinks he is plying a (given) route but departs from it because of his lack of knowledge or the bad positioning of a binnacle or [because of] a needle touching the "*farqadeuse*" stone as stated, east or west *farqad*, or *faraqad*, referred to Ursa Minor.'

Al-Mahrī is less vague: '. . . it could be that some roses indicate NNW . . .'.

In fact, since routes advised in such excellent fashion by credible men led dependably to port (excepting individual errors of application), why worry that the needle did not point exactly north? Did many even notice it?

The woods

In about the first half of the sixteenth century, two techniques emerged for measuring the altitude of a celestial body: the measurement of the angle formed by the line of sight of the celestial body with that of the horizon; and the locating of the celestial body on a small board (or boards) graduated in 'fingers', whose lower edge had to be aligned with the horizon.

While we would spare the reader who is unfamiliar with nautical matters from details of the various methods which can be used to sight the horizon and a celestial body simultaneously with one eye, we should nevertheless bear in mind the substantial inherent problem of the constant instability of the ship, and the related unsteadiness of the arm that was holding the measuring instrument: it was necessary to take a quick sighting of objects (points or lines) that were sometimes indistinct. In short, prior to the development of electronic plotting, neither the wood, nor even a sextant, gave precise altitudes, the skill of the operator none the less acting to correct this inaccuracy.

Can we then establish from the texts in our possession the relative frequency with which graduated equipment (quadrant, astrolabe) and woods were used at the time of Ibn Mājid and al-Mahrī? The term 'wood' (*khashabāt* – more rarely *khushb* or *khushub* – plural of *khashaba*) referred to the apparatus for measuring the distance of a star from the horizon. The singular was used frequently in an expression describing the situation in which stars were found at the same altitude: 'fī khashaba wāḥida'. Divergent views among present-day commentators mean that the analysis of the texts concerning the use of woods must be approached with the greatest care.

When Barros talks of unexpected Arabic instruments (such as a quadrant) being used to measure the altitude of the sun, is this the result of a compulsion for 'sensational reporting', or a deception on the part of the informer, who reveals shortly afterward that he himself only uses the woods? The same, though more subtly, is the case in Celebi's *Muḥīt* (a translation with commentary and analysis of some of the books of Mājid and al-Mahrī) written in Turkish in 1553, translated into German by Hammer-Purgstall and from there into English by Princep, who adds to his translation a commentary on the description of the measuring instruments. Celebi discusses in detail the characteristics of the graduations of a wooden apparatus equipped with a graduated wire, which, he explains – according to al-Mahrī – could take the slack.

Al-Mahrī for his part also refers to the simultaneous use of the two techniques:

> the hand altitude [taken with the apparatus], which is the measuring woods, and the division (degrees of arc of a circle) altitude [taken with the apparatus] is not altered by the increase in altitude of the stars, unlike [the case of] measurement by hand ...

The term *khaṭba* for *khashabāt* is rarely used except by al-Mahrī. This quotation appears to allude to apparatus of the astrolabe type, based on the true vertical, and al-Mahrī's statement is obviously logical for measurements which are known to have been made on land.

Al-Mahrī refers subsequently to another apparatus with wire: 'in proportion to the raising of the hand, the wire which is in the measure slackens as the apparatus is brought closer to the eye, and the measurement becomes smaller ...' How was it that the wire slackened, whereas its purpose was to remain taut? Khoury's answer is to attribute to *khayṭ* the meaning of imaginary wire – a theoretical line.

Whatever the case, we should now examine what Ibn Mājid and al-Mahrī have to say concerning the use of the woods, the technique which appears to have been the most widely employed – indeed virtually the only technique – of their era. They do not mention it a great deal, but what exactly do they say about it?

> the [necessary] condition of measurements is that on the four large woods they be low, on the four medium ones they be standard [or normal, usual = without correction]; between the star and the wood a wire [must be left], and between the wood and the water also a wire like the sharp edge of a knife, in the observer's sight; and the condition for the small woods is that they be high ...

> between the observed star and the direction of your face put 7 rhumbs as from the north to *al-taïr* [which would make 8 rhumbs] and the large woods are low in measurement, spread your hand to the maximum extent, and the four small woods are high, contract your hand to the maximum, for the four medium ones the measurement is normal, this being in order to dilate the 'section' of the horizon and the reduce its upper part ...

> the best measurement is made with average woods neither too big nor too small ...

Thus it is possible to conclude that there was a set of three small boards increasing by series of 4 fingers, each board constituting a rigid unit, as in Figure 6.6 for example (since we do not know its true configuration). In place of a step-ladder gradation, the fingers could have been shaded in alternate dark and light bands (whence *khaṭba*). All divisions are imaginable

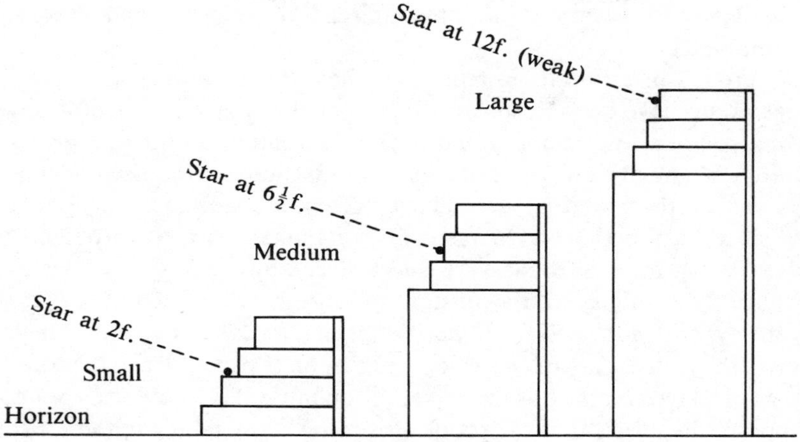

Figure 6.6

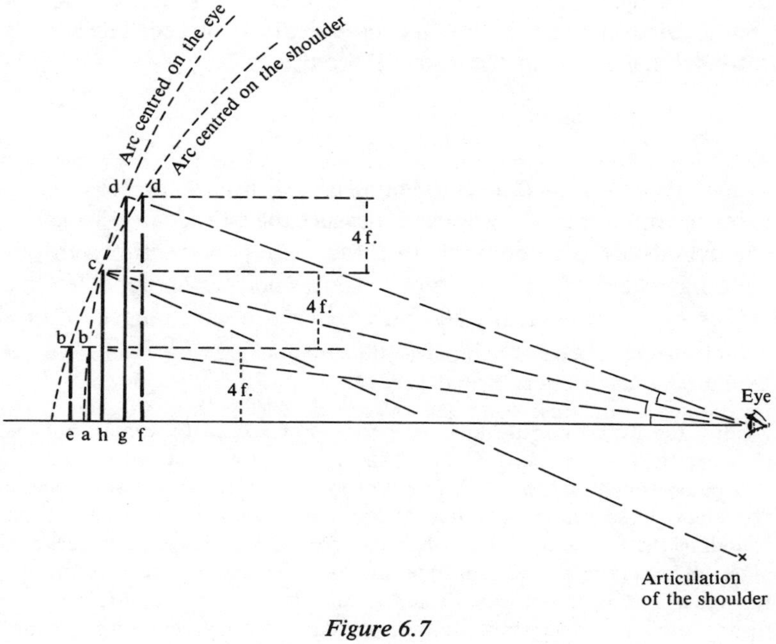

Figure 6.7

since it was necessary to be able to read the fingers, and even some sub-multiples.

Figure 6.7 offers an interpretation of how the apparatus was used. The ideal would have been to spread three jointed elements, each of 4 fingers, over a sector whose radius would have been centred on the eye, giving the chords *ab'* (small woods), *hc* (medium woods) and *gd'* (large woods). Since this was not the case (that would have made it a quadrant or an astrolabe), and since each of the woods was held by its upper part, the problem to be solved was as follows: to measure, successively, by means of the boards, 4*f*, 8*f* and 12*f* of angles with a constant distance between them of 4*f*, as seen by the eye: i.e. *aob'*, *aoc'* and *aod'*. Keeping a constant tension, the hand travels through the arc of circle bcd, centred on the shoulder, but in this way the 4*f* displayed on the boards (equal, by definition) are obviously not right. Suppose, therefore, that the problem is resolved by taking c, the fourth of the medium woods (thus 8*f*) as the point of departure, at normal hand tension. Then, at 4*f* vertically from *c*, let two parallel lines be drawn to the horizon. The fourth of the small woods (4*f*) and the fourth of the large woods (12*f*) intersect the arc centred on the eye, at *b'* and *d'* respectively, where the top of the boards should consequently be placed. The hand is thus 'spread' from *f* to *g* and 'contracted' from *e* to *a*.

Other instruments

We have already seen that Ibn Mājid and al-Mahrī mention the use of instruments other than the woods to measure the altitude of celestial bodies.

The hypothesis of a tool with an actual wire is not entirely unfounded. It could have been of a similar type to the *kamāl*, which appeared in about the 1540s – the wire evidently being used to measure the tangent of the angle of elevation and therefore the elevation also.

As Tibbets has already noted (1971):

> Neither Ibn Mājid nor al-Mahrī ever speaks of *kamal*, or *kamāl*, but many are nevertheless convinced of its use in that era. Amongst other reasons for this conviction, we can see Mājid's propensity for superlatives, including '*kamalān*' = excellently, a source of misunderstanding; thus when Mājid is rounding the Laccadive Islands (*fāl*, or *fālāt*), and writes that, because of seasonal imperatives at certain times of year, they should not be rounded too far out to sea, he comments 'do not let the Pole Star fall and (if necessary) turn northwards, certainly do not deviate (southwards) by 3 *kamalān* [meaning 'strong'] . . .

'Strong' (*kamalān*) is ambiguous. Mājid uses it first in his rhyming works; the 'strong' or 'weak' values are expressed differently from the normal usage of the term.

As for the astrolabe in the strict sense, some maintain that it was used by the Arab mariners, on the evidence of the only altitude in figures (round degrees) said to have been 'taken by astrolabe'. Mājid quotes some co-ordinates in degrees, but he took them from geography books. Al-Mahrī gives some altitudes measured with the 'division instrument'. Compared with the accumulation of thousands of altitudes in fingers measured with the 'woods', however, it is evident that the standard measuring instrument was not the astrolabe.

The quadrant (another circle or part-circle split into equal divisions) is also one of the instruments to which the texts may be referring.

The calendar

In seas that are subject to marked seasonal regimes, navigation is obviously completely dependent on the seasons. But how can a particular first day in the solar year be accurately determined, since the stars move in precession relative to the sun?

The question of the calendar was a source of such problems for the human race that an acceptable solution was not found until the Gregorian reform at the end of the sixteenth century. So where did that put the mariners of the Indian Ocean a century earlier?

According to the computations given in the nautical manuscripts, the first day of *nawrūz* (or *nayrūz*, or *nīrūz*) was determined by the appearance at dawn of the mansion of 'Diadem' (Libra) at 15° declination. This first day of *nīrūz* was around 20 November of the modern calendar.

The problems involved in drawing up an invariant calendar start here, because *nīrūz* had 365 whole days, and leap years were unknown. The first day of *nīrūz* thus moved forward by nearly three months in four centuries (the great Arab astronomers wrote in around the tenth century). Compared with this difference, the discrepancy due to precession becomes negligible. Nevertheless, this over-short *nīrūz* was used in Ibn Mājid's time and still is used in the Indian Ocean (although it operates differently from one region to another and is no longer based on the Diadem).

The second problem was that the appearance of a star varies according to latitude and to its declination, a phenomenon of which Mājid was aware. Now the astronomers 'of the great books', as he records, marked out in a regular mathematical fashion each heliacal rising and setting, without taking account of the declinations, as if they had been operating at the equator, whereas they had actually been observing at more than 25° north. In the nautical manuscripts their statements concerning the lunar mansions are copied almost day for day.

Towards the end of the fifteenth century, at 15° north, α Libra did indeed appear at around the present 20 November, and a mariner such as Ibn Mājid, who was constantly scrutinizing the star-filled skies, could very well have observed it. As this coincided, at least to within about ten days, with the assertions put forward in the tenth century, he would have been tempted to relativize the phenomenon 'it is sometimes said that the date of the voyages goes back by one degree per year . . .'. Al-Mahrī, however, saw it completely differently: 'it changes by a quarter of a day per year . . .'. Another proof of the difference in their characters!

How did the mariners of old cope with the dimly understood irregularities of this calendar based on a star? Bearing in mind, on the one hand, their technical heritage (falling rapidly into disuse among modern sailors), and on the other hand, the lively practice of holding meetings for captains in the form of 'seminars', on board their vessels or at the ship-brokers', which would have provided an opportunity for exchanges of information of all sorts, it is possible to envisage the idea of a consensus, in around 1450, which laid down that a vessel should cast off from particular regions toward other particular regions at certain dates of the *nīrūz*, which were nearly always close to within ten days, and very occasionally five days. Some ages later, as a result of repeated experience on given lines, discussion at meetings and the authority of certain celebrated pilots, corrections of five to ten days were gradually applied to the preceding consensus, which was finally revised overall, whence the slide towards modern deviations.

This consensus modelled the calendar for voyages on the calendar of the monsoons (the very term 'monsoon', which is of Arabic origin, implies the idea of seasonal periods; thus the dates of voyages were known as *mawāsim*).

The dividing into periods of the characteristic winds was, of course, expressed in *nīrūz*. The listing of the periods for voyages that was grafted on to this basic division was, on the other hand, quite complex. The outline which follows takes into account numerous micro-climates, which could cause the scheme to be reversed or even cancel the 'closure of the seas'. Moreover, a text sometimes mentions a wind that is inconsistent with the place and the season, but the interpretation of such passages depends, amongst other things, on the local meaning of the terms used.

The closure of the seas, *ghalaq al-baḥr*, is the season when navigation stops, spent as far as was possible at home in the port where the vessel was fitted out. From the beginning of June to mid-August the south-west monsoon rages. On modern seasonal charts, one of the curves indicating the force of the winds east of Socotra in July has an elongated shape marking out the area swept by the strongest winds (known as the 'haricot', or 'bean', by French sailors), which should be avoided by low-powered ships heading west. The season of the south-west monsoon and the wind itself (as well as its numerous derivatives) constitute the *kaws* (although its equivalent *dabūr* or *dabbūr* is more often applied to the wind itself).

With the end of the closure, in August–September, the Great Season (*al-mawsim al-kabīr*) begins, where good weather occurs just about everywhere. It includes the easily managed end phase of the south-westerly winds (*damanī* or *dimanī*), the entire north-east monsoon (*aziyab* or *saba*) from October to April, and finally the equally manageable start of the south-west monsoon, from the end of April to the end of May, still called the start (or head) of *kaws* or the end of the (Great) Season (*awwal*, or *ra'S al-kaws*, or *ākhir al-mawsim* (*al-kabīr*)). The end of *kaws* (*ākhir al-kaws*) marks the end of the manageable start of this wind: the extreme end of the season.

Nautical instructions

Whereas in modern times, nautical instructions refer to an essential document in the navigator's library that contains all the information needed at sea which is not directly connected with charts and measurements, the writings of Ibn Mājid and al-Mahrī, being comprehensive collections of information and advice to the mariners in their particular seas, comprised, together with the instruments described above and personal experience, the only useful aid to navigation.

The following section is thus an account of the essential substance of the nautical instructions used by the Arab mariners of the sixteenth century in the Indian Ocean, focusing on the most important problems that they faced at sea.

THE TECHNIQUES OF PLOTTING A POSITION AT SEA USING DEAD RECKONING AND ASTRONOMICAL OBSERVATION

A bearing, or more accurately an estimation of the ship's position at sea, depended on the estimated distance covered, verified as soon as possible by the measurement of the altitude of known and observable celestial bodies, all with reference to the nautical instructions and the experience of the navigation officer.

Thus what mattered to the navigator were estimates of the course and the true speed and the celestial altitudes. Now, as we have seen, distances were evaluated in *zams*. That is why the most important passages of Ibn Mājid and al-Mahrī's manuscripts as far as the mariner was concerned were those regarding the accuracy of the course and the altitude of the stars.

We should remind ourselves that until the commercialization of the chronometer, reliable in all climates and for prolonged periods – that is to say, until about 150 years ago – sailors could only observe latitude. Of course procedures using triangulation would have permitted them to work out approximately the longitude of an important port, but not that of a ship.

As this discussion concerns Arabic navigation, which was principally carried out between coasts that were, broadly speaking, oriented toward the north, a merely approximate knowledge of longitude did not have to be a serious drawback, and we shall refer to it only occasionally. However, it can be appreciated that the co-ordination between observed latitude and longitude estimated by distance was already in itself a technical feat.

The accuracy of the course

How accurately did the mariners keep to their course on long voyages? The answer depends on a number of practical contingencies.

The finest subdivision of the rose in rhumbs (which is the same in the Indian Ocean today) was at best every 2° (in good weather, the most sophisticated modern ships can hold a course to $\frac{1}{2}$° maximum). Ibn Mājid appears to report navigation over long distances to $\frac{1}{4}$ rhumb, that is, an accuracy of slightly less than 3°. He lists the types of route: coastal, direct on the open sea, and what he calls, in this instance, a 'deduced' route (by comparison with another presumed to be correct). He shows himself critical of the estimated distances – the *tirfas* – accepted by the 'ancients':

> a vessel goes south-eastward from Muscat and Hadd [until] there are 4 *zams* between it and the reef to the north of the Laccadives [Figure 6.8] . . . the route of a second vessel wishing to reach this reef is [set] at 4/7 rhumb between SE and SE1/4E [in reality from SE1/4E towards SE: these approximations are customary with Ibn Mājid], and it reaches the reef after a rapid course of 7 *tirfas* and it would have navigated 28/7 *of* 4 *zams* (more than the first) . . . therefore the *tirfas* are false . . . because for one route as for two, the two distances are equal at 117 *zams* . . . .

Figure 6.8 The current reconstructions include a single scale in latitude and in 'longitudinal distances'.

Routes and orientations of Mājid ---------
of Sulaymān al-Mahrī ....
Crossover points of their paths ⊗

The 'true' contours are marked with solid lines and the 'real' places are marked with thick points

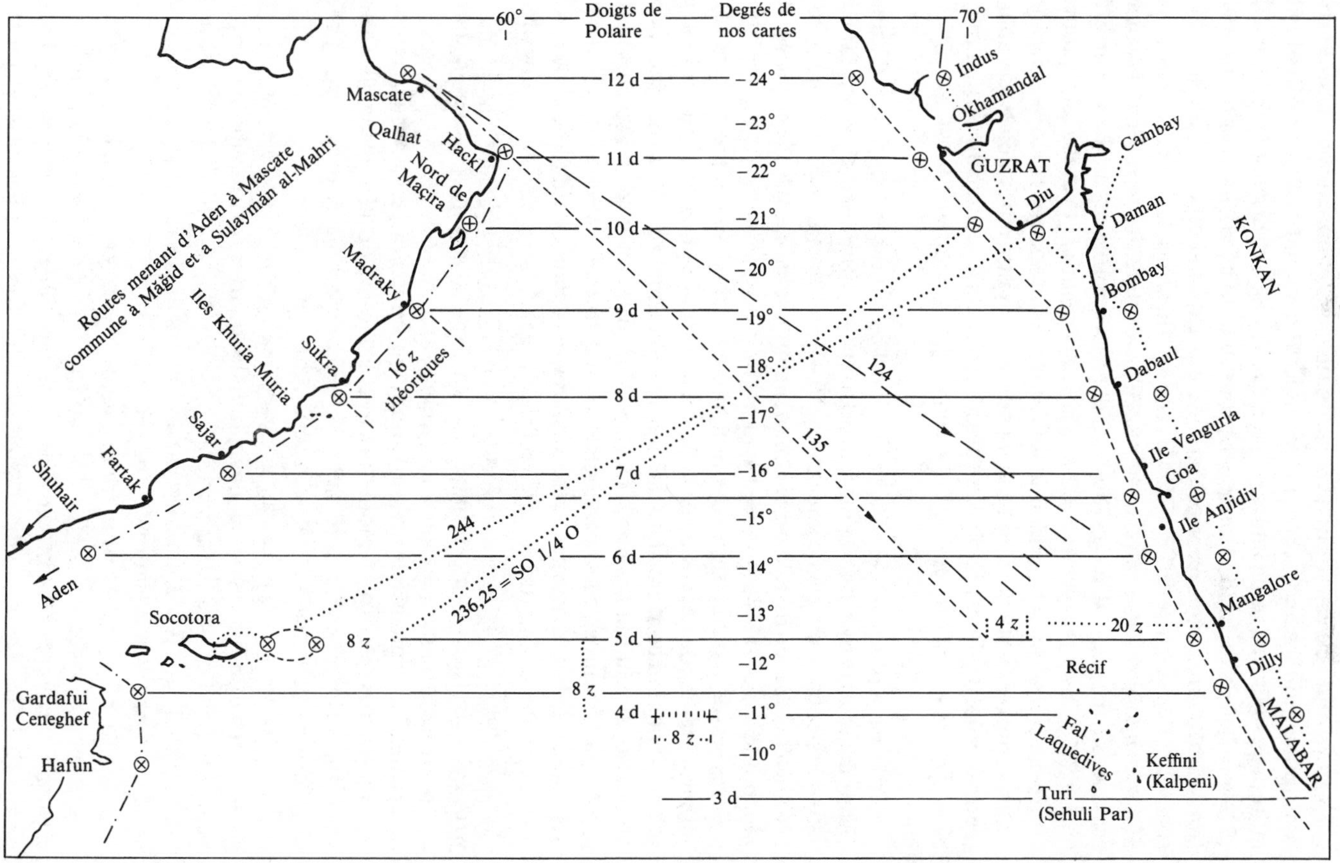
60°
Doigts de Polaire
Degrés de nos cartes
70°
12 d
11 d
10 d
9 d
8 d
7 d
6 d
5 d
4 d
3 d
–24°
–23°
–22°
–21°
–20°
–19°
–18°
–17°
–16°
–15°
–14°
–13°
–12°
–11°
–10°
Mascate
Qalhat
Hackl
Nord de Maçira
Madraky
Routes menant d'Aden à Mascate commune à Māğid et a Sulaymān al-Mahri
Iles Khuria Muria
Sukra
16 z théoriques
Sajar
Fartak
Shuhair
Aden
Socotora
8 z
Gardafui Ceneghef
Hafun
244
236,25 = SO 1/4 O
124
135
8 z
Indus
Okhamandal
GUZRAT
Cambay
Diu
Daman
KONKAN
Bombay
Dabaul
Ile Vengurla
Goa
Ile Anjidiv
Mangalore
Dilly
MALABAR
4 z
20 z
Récif
Fal
Laquedives
Keffini (Kalpeni)
Turi (Sehuli Par)

Here is our explanation of this somewhat elliptical but important passage. A southeasterly course does indeed lead to 4*z* of the reef (whose latitude is supposed to be known, taken as 5*f*, just as Muscat is at 12*f*, since 12 – 5 = 7). For the second vessel, the route would be at 6/8 rhumb, without Ibn Mājid's customary approximations; but let us accept 5/7 instead. The two vessels have travelled 7 *tirfas*; if the southeasterly *tirfa* is 16*z* and the SE1/4E *tirfa* is 18*z*, the difference is 2*z*. To calculate the additional distance covered in the second case, Ibn Mājid proportions this difference at 2/7 (7/7 – 5/7 = 2/7), this gives: 2 × 7 × 2/7 = 28/7, thus 4*z*. (Yet the text quoted above says '28/7 of 4*z*', which is why it seems necessary to correct it: the original probably stated *aʿnī arbaʿa* and the copyist transcribed *ʿan arbaʿa*, then to make it tally with 28/7, he presumed that the proportion could only be 4/7 – and not 5/7, or better 6/8 – which, when multiplied by the 7 *tirfas*, does indeed give 28/7.)

Verification of the extra distance travelled by the second ship is easy using 2 rhumbs, either by subtracting from the SE1/4E journey 5/7 of the difference between the two journeys (18 – 16 = 2, and 2 × 7 × 5/7 = 10; 126 – 10 = 116); or by adding to the first ship's journey 2/7 of 14, giving 4, which added to 112 does indeed make 116.

That leaves the question of the figure of 117 given in the text as the common value for both journeys. The figure for the first is unquestionably 112 + 4, which is confirmed by the preceding calculation. So is this the result of another slip in the copying, from 16 to 17? Whatever the reason may be, Ibn Mājid's demonstration is correct to within one *zam*.

Finally we should mention that no pilot would have dared to head directly for this immense and formidable reef that breaks the surface but is invisible for unfathomable depths, and against which a Portuguese vessel had been wrecked with the loss of all hands on her second voyage. Ibn Mājid does not even feel the need to issue a warning.

Al-Mahrī also mentions fifths of rhumb in similar circumstances, but the two navigators refer to no more than four examples of this type in all. Consequently it is difficult to use these arguments as a basis for asserting the operational reality of fine subdivisions on ocean routes.

On the other hand, where an enclosed sea is concerned, we can cite an example from Mājid involving navigation by quarters of rhumb: in the Red Sea, at the end of one of the various routes plied down from Jedda toward Siban (Jabal Tir), a 245 m peak that dominates its region and is surrounded by steeply plunging inshore depths (Figure 6.9).

In the Red Sea, banks of more or less dense reefs extend a long way out to sea; on the Arabian side, they rise up from great depths, and on the 'foreign' side they are often preceded by soundable depths. When sailing back up the Red Sea, however, the mariners looked for an easier passage on the

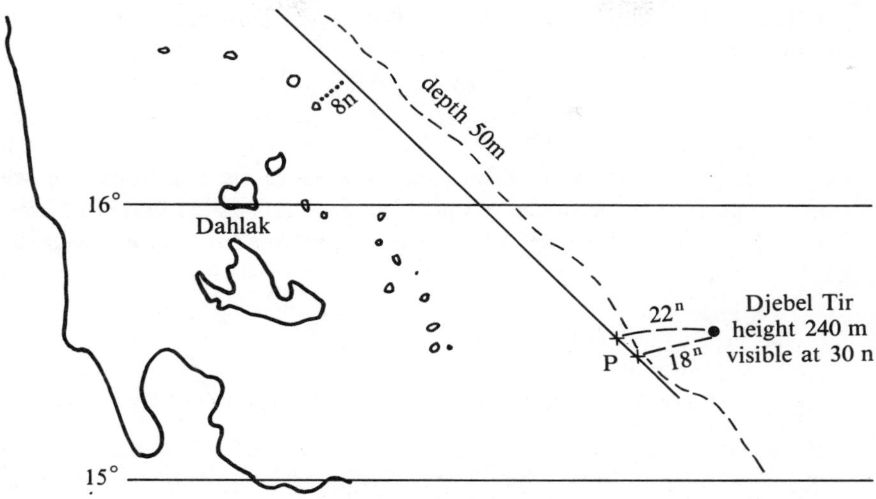

Figure 6.9

Arabian side, because in the evening the reefs there were easier to detect thanks to the setting sun, even though it was at a low angle; moreover, the following winds on the way back were subject to frequent inversions, whereas on the way down, the wind is less irregular (is this the reason why Mājid increases the number of celestial altitudes for the voyage back and gives only a few for the descent?).

Some of these routes ended west of Siban. But after sailing more than 300 nautical miles from Jedda, or after approximately 17° (around $7\frac{1}{2}f$ of the Pole Star), caution is essential: where are we longitudinally? (Figure 6.9 shows the soundable depths around the Dahlak, on which, in places, scarcely more than some low rocks covered with sand and rare tufts of undergrowth can be distinguished.) Knowing that the route is SE1/4S, Ibn Mājid advises that if the sounding line indicates a shifting towards the west, to hold between 35 m and 24 m depth, as necessary: 'by heeling towards the SE by 1/4, 1/3 or 1/2 rhumb'. The manoeuvre described ensured a safe distance from land at shallow depths.

Finally, avoiding the dangers of the Arabian coast by locating their position with soundings when out of sight of seamarks on Huatib and Hajouat, they then strove not to miss the extraordinary seamark of Siban before confronting further dangers to the south.

In conclusion, the example of the route to the redoubtable Fal reef (Laccadives) that was purely depicted in the mind, and the example of the immediate contingencies to be finely negotiated in the Red Sea, support the hypothesis of a compass arrangement allowing the real sustained use of 1/4 rhumb.

The altitudes of stars

In a system of navigation by dead reckoning where the ship's position was generally verified by reference to the altitudes of stars cited in the 'nautical instructions', these altitudes constituted the 'purple passages' of the Arabic nautical manuscripts.

Preliminary remarks

Four points need to be underlined.

1 If the ecliptical co-ordinates of the stars are said to be fixed, and are approximately so, their equatorial co-ordinates – the only ones valid for the observation of latitude – are, on the contrary, unstable. But the movements of the latter are sufficiently slow (of the order of 15′ in 40 years) to have gone unnoticed by the mariners of the time.
2 The Arab mariners used only the stars, precisely because of their fixity; the simple identification of the stars meant that long experience (helped, if notebooks were lost, by the prodigious memory of simple people in permanent contact with nature) was all that these authentic long-haul voyagers needed.
3 Despite the demands of science the ephemerides used by modern sailors are still based on a geocentric universe (their computations being considerably simplified). Consequently we are easily able to reconstruct the procedures used by the sailors of old.
4 When considering the measurement of celestial altitudes in the mid-sixteenth century, we always need to bear in mind the relative imprecision of the instruments, the instability of the platform and the absence of corrections (refraction etc.).

To enter some little way into the minds of these mariners as they navigated the ocean ('such stars are at such an altitude, so I am at such a place'), it is necessary to think back to the great empiricism inherent in the rudimentary methods available at the time (even today the local pilot who takes charge of a ship in delicate areas is still referred to by the Spanish as 'el practico').

Paired altitudes

The woods, which were, as we have seen, the only instrument in regular use, could not measure beyond 12f, nor could they go much below 3*f* (the mariners had detected abnormal effects, due to refraction, at very low altitudes; according to Ibn Mājid: 'it is no good if a star is low over the water . . .'). Consequently the range of meridians was very restricted. However, the mariners had noticed (Figure 6.10) that at a given reference height at latitude L defined by the meridian of a given star *a*, two stars *b* and *c* appear at a particular moment at the same height *h*. In the case of Figure 6.11, the most usual configuration, *c* was setting and *b* rising. But they could equally well be rising at the same time, *b* and *d*, or setting, *c* and *e*; moreover, their declinations could also be related at *b* and *c* or *d* and *e*, or *b* and *d*, *b* and *e*, etc. They were said to be 'on a single wood or in equality (*iʿtidāl*)', but other expressions were used that were either synonymous or bore diverse nuances according to the various situations that were encountered.

Compared with the meridians, whose theoretical efficiency is 100 per cent, the efficiency of celestial pairings can be anywhere from 0 to 100 per cent, because it depends upon the declinations and azimuths of the paired stars. If the empiricism of Ibn Mājid the 'instinctive' seems to have led him to a real degree of clairvoyance, because he often accompanies his pairings with proportions that appear to imply a degree of accuracy, al-Mahrī missed this evidence – surprisingly for him – confining himself to the

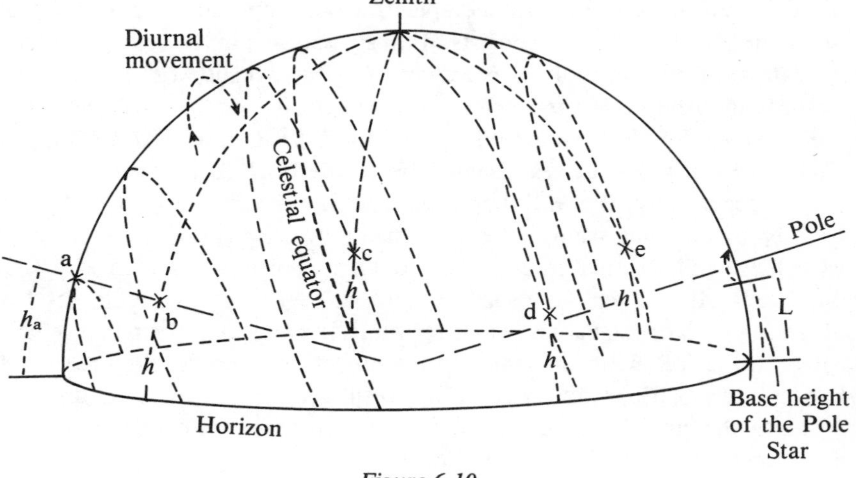

Figure 6.10

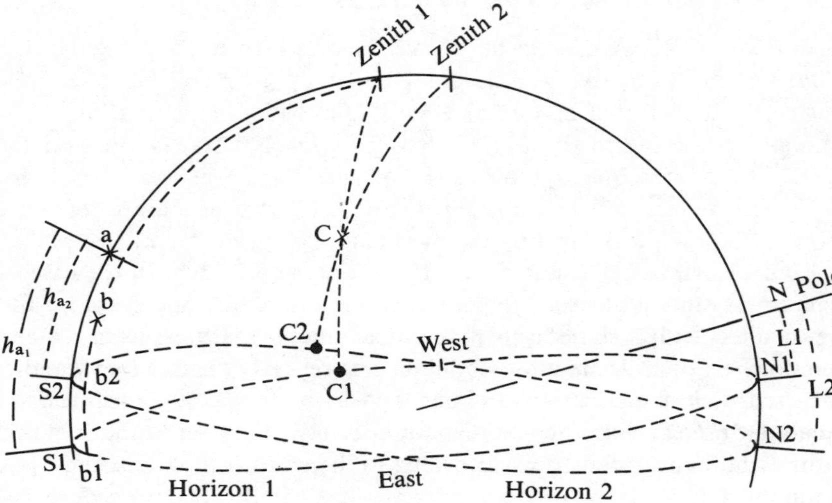

Figure 6.11

comment that 'the measuring of altitude is much more satisfactory if the observed star is meridian at the moment of observation; the reason for this accuracy is that there is then a finishing [perfection, interruption], neither increase nor decrease ... conversely to coastal measurements, which are uncertain because of the speed of their movement ...'

In fact, the momentary relative stability of a star at its highest point (even in equatorial regions) permits a surer observation than the upward velocity of stars that are a large distance apart on the meridian. The list of stars recommended by al-Mahrī consists of nine positions; in the four meridians, he claims the paternity of the meridian of α Paon (valid at the time of the westerly monsoon covering the three-month closure of the seas). The pairs are those of $\beta\gamma$ Ursa Minor, $\varepsilon\xi$ Ursa Major, $\alpha\beta$ Centauri and Canopus-Achernar; but the altitudes disagree from one work to another.

In his usual way, Ibn Mājid does not propose a coherent list of pairings. Only by picking out examples as and when they arise in the manuscripts do we achieve an inventory comprising about sixty pairings, some of which are duplicates. It is then necessary to undertake the considerable task of obtaining each of the values for each couple and checking them mathematically. In the following we shall confine ourselves to giving the broad outlines of the pairings and stating the results of the checks, in order to evaluate the usefulness, or 'profitability', of these venerable navigators after having described their techniques.

The inclusion in the *Ḥāwiya* (written in his youth by Ibn Mājid, if it is entirely his) of many pairings – besides the classic Ursa Minor and Major, we find $\alpha\beta$ Centauri, Vega-Sirius, Achernar α Phoenix and Achernar-Vega – suggests that it may have been Ibn Mājid himself who inaugurated the procedure. Then he applied himself doggedly to developing it, but in a fragmented and sometimes esoteric fashion. Ultimately, there are few pairings that cover a wide range of latitudes and are accompanied by considerations relating to the appropriate periods for voyages and observations.

Schematically, three pairing profiles are recognizable:

1 One of the stars is close to the meridian, the other is distant from it, its speed of ascension standing for local time for the given moment, since the slowness of its companion implies a certain delay. This is the type of the 'staff' or support, of captains ... *'asā*, or *ukkāẓ al-rabbābīn*: Achernar not far from its meridian and Sirius away from azimuth. The latitudinal band goes from 25° 36′ to 19° north, then descends in 'immobilization' from one of the stars (the supporting procedure is described later). The passages concerning the pairing do not present any great difficulty, provided the various names of the stars and the geographical locations are known, and one is familiar with the style of Ibn Mājid. Elsewhere, he gives the equality of Capricorn and Canopus, with the astrolabe in support, (in whole degrees!).

2 The two stars are at some point of declination; cases where one is lower than 45° are, however, very rare, the other remaining indifferent. The model of this type is the Great Solitary '*fard al-kabīr*' ξ Ursa Major and α Aries. The analysis of this pair raises a great many difficulties. The results are excellent for the easterly monsoon between 19° and 14° 30′, and for the westerly monsoon between 18° and 24° north. Elsewhere the approximation exceeds 20′ even to reach 1° 30′, which is absurd. Ibn Mājid boasts of the value of this pairing in the entire world, be it even in the sea of Roums. Elsewhere he merely gives it as 'weak' in Zang lands and strong in high latitudes. But why does he say nothing of this pairing in the voyages to the Moluccas (the vessels rounded Ceylon very far out to sea), nor even abreast of Somalia, as we already saw him do with regard to '*bachi*' in those waters?

3 This is primarily a variant of (1) and (2), but sufficiently original to merit separate examination: it is the '*qayyid*-immobilization'. There are situations when twinning occurs at a time when it is impossible to observe: during the day, for example. Mājid solved this by acknowledging inequalities in the pair, so that one of the stars is always observed at a particular height, and, as it were, immobilized along a band of latitude inside which its companion follows a remarkable gradation.

We should also mention Ibn Mājid's idea of '*abdal*-permutation', using pairs for which the differences of right ascension are close to twelve hours – the Great Solitary being one example. As these stars are on nearly opposite meridians, they return once again to equality about twelve hours later. Of course the second equal altitude is different from the first, and the phenomenon does not occur in a single night, except in high latitudes, in winter; and these sailors never observed beyond 25° north or south (Figure 6.12). This particular feature of the permutation has an additional application in aiding the use of immobilization.

Finally, we must ask whether Ibn Mājid grasped the relationship between positional errors and results of the pairings. Although we cannot maintain it with certainty, it is nevertheless the case that in one part of the characteristics of the '*ḍarā'ib*' (in other words, the results), which include his finding of the '*tartīb*-arrangement', for each pair he often came close to the truth: 'when the latitude changes, the altitude of such a pair, or of the companion of the immobilized [star], evolves by so many fractions of a finger per finger of meridian'. But were his margins of latitude too narrow to have enabled him to detect certain inconsistencies in certain pairs?

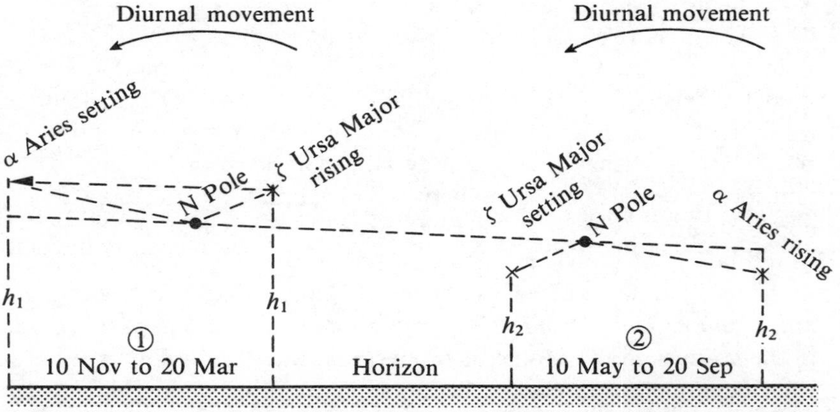

Figure 6.12 In each of the *abdal*-permutation positions the two respective common altitudes or elevations, h_1 and h_2, are not equal

Co-ordination between the measurement of altitude and the reading of the chart

This co-ordination was not always easy as the following two extreme examples show:

1 For the descent southwards (which must already have been situated at the bottom of the chart, since it was called *safil*), the alliance of the references to the Pole Star and to $\beta\gamma$ Ursa Minor is perfect. On the other hand, the link between these and $\varepsilon\xi$ Ursa Major arouses justified controversy. Disagreement between astronomical observations and chart-drawing persisted for a long time, particularly in relation to the south of Madagascar and the Mascarenes (but excepting the Comoros Islands, which were very precise). Is this evidence of a break in Arabic navigation, as in the Moluccas to the east or at Jedda to the north? But the Arab mariners rejoined Sofala in various ways. It is easy to imagine ourselves alongside Mājid on one of the coastal routes, and to experience the full force of a sailor's terror amid currents as violent 'as in Cambay' in the turbulent and perilously shallow waters of the immense Zambezi delta. As for the open-sea route, this headed firstly south-south-east before heading on to the level of Mambone-Chiluan, accurate to within about 20′.

2 In contrast to these random inaccuracies, we have seen precision of Ibn Mājid's altitudes by the Southern Cross in the Red Sea. One of these is of particular interest. First it is a unique example of equivalence to the Pole Star – 7.25*f*: two equal and opposite meridians, giving values of 16° 33′ and 16° 36′ N, which are surprisingly close figures. Furthermore it locates two treacherous reefs in a coral series jutting out from the Farsan (an area that is uncertain on modern charts, so it would be exciting to see one of the bearings defined with the help of a fifteenth-century document – part of the admirable solidarity of sailors of all eras!)

CONCLUSION

These sparse studies and reflections on documents which are themselves singularly lacking in overall coherence, cannot pretend to draw definitive conclusions about Arabic nautical knowledge in the Indian Ocean in around 1500.

As we have mentioned in passing, it still remains to discover, interpret and exploit numerous sources that are dispersed in the archives of the nations which were part of the complex history of navigation in the Indian Ocean.

The preceding pages are only a minor contribution to a far greater collective effort, which will never have the goal of enriching our own navigational science, since we have now entered irreversibly into the domain of electronically assisted navigation. Will our contribution therefore be only a nostalgic farewell from the sailors of the sextant and marine chart to their precursors of the 'woods' and the *tirfa*? A last gesture of complicity between sailors on the gangway, before they hand over to the anonymous servitors of 'central operations'?

To think so would be a grave injustice to those two sailors Ibn Mājid and al-Mahrī (one more of a 'sailor', admittedly, than the other), whom we have come to appreciate in spite of their faults, which indeed make them seem even closer to us. Also, it would overlook the fact that, for all their 'scientific' imperfections, they were the heirs of a prestigious secular tradition of rigorous thought, to which the whole of the present work bears witness.

NOTES

1 Tr. note: a colourful and boastful old sea dog, hero of a French cartoon.

7

The development of Arabic science in Andalusia

JUAN VERNET AND JULIO SAMSÓ

INTRODUCTION

The historical context of this chapter[1] extends from 711 (the date of the first Muslim conquest of the Iberian peninsula) to 1492, the year when Granada was taken by the Catholic kings, who also brought about the demise of the Banū Naṣr, the last independent Muslim dynasty in Spain.

Within this context we shall study the development of the exact sciences and the physical and natural sciences which had Arabic as their language of expression – even though the sources have sometimes been preserved in Latin, Hebrew, Castilian or even Catalan – in a world politically controlled by Islam, excluding a priori medicine but not pharmacology, given its direct relationships with botany. This means, in principle, leaving aside the contributions (very humble, certainly, but extremely interesting from a socio-historical point of view) of the *Mudéjares* (Muslims living in a politically Christian environment) and of the *Moriscos* (Muslims apparently converted to Christianity at the end of the sixteenth and beginning of the seventeenth century); the main reason for this exclusion is the lack of precise studies, even though research in this area had been initiated for medicine.[2] With regard to the geographical context, it should be noted that the term Andalusia, as used here, in no way corresponds to the boundaries of the region known as Andalusia today, but is intended to translate the term *al-Andalus* used by the Arabs to describe Muslim Spain: a political, and often cultural, reality, whose northern border extended to the Pyrenees in the eighth century, but which gradually contracted during the Christian 'reconquest' until it became limited to the Kingdom of Granada from the thirteenth century onwards.

The history of this period, spanning nearly eight centuries, is known in a very uneven way: reasonably well up to the twelfth century and then rather poorly, because periods of decline tend to attract much less attention from historians. In addition, if we consider the development of Arabic science in Andalusia alongside that of eastern science, we note certain interesting differences: first, in Andalusia we find the survival of a modestly important Latin–Visigothic–Mozarabic science (and culture) which dominated until about the middle of the ninth century and survived until at least the eleventh century. An 'easternization' of Andalusian science occurred mainly between 850 and 1031 (the fall of the caliphate of Córdoba): new contributions from eastern science became increasingly rare after the eleventh century,[3] and Andalusian science grew progressively more independent, limiting its cultural interactions generally to North Africa. The eleventh century marked the high point of Andalusian science, whose overall development occurred at least a century after the science of Mashriq. This advance slowed in the fundamentally philosophical twelfth century, and decline began from the thirteenth century at the time of the birth of a scientifically active period in Christian Spain (under Alfonso X). Andalusia hardly benefited at all from the scientific revival that occurred in the Orient from the thirteenth century. Throughout this period, Andalusian scientists cultivated astronomy, botany, medicine and agriculture, in their particular way, often leaving aside mathematics; however, recent research into key figures such as King al-Mu'taman of Saragossa, Ibn Muʿādh al-Jayyānī or Ibn Bājja may require us to change our view in the fairly near future.

THE SURVIVAL OF THE ISIDORIAN CULTURE (711–850)

The Muslim conquerors of Spain were neither men of science nor cultured people. The first waves of invasion involved primarily Berbers whose arabization was very recent;[4] in addition, Hispano-Arab historiographers (notably Ibn al-Qūṭiyya) have shown certain highly placed figures among the Arabs who entered the Iberian peninsula in the eighth century as individuals of a relatively low cultural level. We can, of course, find exceptions: the first Andalusian Umayyad, ʿAbd al-Raḥmān I al-Dākhil (756–88), made attempts to acclimatize oriental plants in the gardens of his Rūṣāfa palace – named after the palace founded by his grandfather Hishām at Damascus – and similar experiments were conducted by his courtiers; we can therefore see here in embryo the botanical gardens established in Spain from the eleventh century (Samsó 1982). But these are very exceptional cases: Muslim tradition attributes to one of the *tābiʿūn*, Ḥanash al-Ṣanʿānī, an aptitude for divination as well as for the determination of the azimuth of the *qibla* for the great mosques of Córdoba and Saragossa, whereas everybody, at least from the tenth century onwards, was aware of the incorrect orientation of the mosque at Córdoba.[5] The problem was obviously too sophisticated for the knowledge of the time in Andalusia, and historical sources related to the conquest contain references to the practice of divination – whether astrological or not (the specific technique is rarely identified) – as much in Muslim circles as in Christian ones (Marin 1986: 509–35; Samsó 1985c). However, there is a certain amount of data which enables us to defend the theory of a surviving Latin–Visigothic astronomical and astrological tradition in Muslim Andalusia: the *Dhikr bilād al-Andalus*, written by an anonymous Maghreb author in the second half of the fourteenth or at the beginning of the fifteenth century, attributes to King Sisebut (612–21) some writings in verse on questions relating to astronomy, astrology and medicine; we know nothing about the medical writings of Sisebut but he is beyond doubt the author of *Epistula metrica ad Isidorum de libro rotarum*, in which he gives an accurate and rational explanation of the eclipses of the sun and moon. Likewise, the famous Hispano-Arab historian al-Rāzī refers to Isidore of Seville's reputation as an astrologer, which can be explained by the astronomical section of his *Etymologies* as well as by his book *De natura rerum* (Samsó 1985b). The encyclopedic work of Isidore is, in fact, more interesting than it may at first appear: it contains, for example, reminiscences of the Babylonian goal years (*années-limites*, *Ziel-jahre*), which are fundamental to astronomical almanacs like that of al-Zarqāllu (Samsó 1979a).

The clearest evidence of the survival of a Latin–Visigothic tradition in the field of astrology is found in an Alfonsine work, the *Libro de las Cruzes*. This book is the Castilian translation of an Arabic astrological text, numerous passages of which have recently been discovered (Vernet 1979e; Muñoz 1981), including thirty-nine lines of verse of an *urjūza* of ʿAbd al-Wāḥid b. Isḥāq al-Ḍabbī, court astrologer to Emir Hishām I (788–96), which correspond very well to chapter 57 of the *Libro de las Cruzes*.[6] We therefore have a text which is, as far as we know, the oldest source on Andalusian astrology and which, in addition, was composed in a period for which we do not possess the slightest clear trace of the introduction into Andalusia of oriental astrological texts, of the Indian, Persian or Greek tradition. It should be added that both the Arabic texts that have been preserved and the Castilian Alfonsine translation stress the fact that 'the system of crosses' (*ṭarīqat aḥkām al-ṣulūb*) was the ancient system of astrological prediction used by the *Rūm* (Romans? Christians?) of Andalusia, Ifrīqiyya and the Maghreb before the introduction of more advanced systems from eastern astrologers. We can thus conclude that the *Libro de las Cruzes* represents the last stage in the evolution of a manual of astrology which originated in early Latin and was in use in Spain and North Africa before the Muslim conquest. This type of astrological technique also survived the period of easternization in Andalusia: we have evidence for believing that it was employed by the astrologers of al-Manṣūr b. Abī ʿĀmir (981–1002) (Vernet 1970), that it was revised later – probably in the eleventh century – by a certain ʿUbayd Allāh who is usually identified with ʿUbayd Allāh al-Istijī (an astrologer contemporary with *qāḍī* Ṣāʿid of Toledo) and that it must still have been appreciated in the thirteenth century, because Alfonso X ordered its translation.[7]

One should not be surprised by this probable Latin origin of the 'system of crosses' because it confirms what we know of the Andalusian culture of the time. Eulogius of Córdoba – well known for his inspiration of the 'voluntary martyrs' Christian movement, beginning in 850 – was a lover of Latin books. He had in his library the *codex* R.II 18 (*Ovetense*) from the Escorial which contains part of the *De natura rerum* by Isidore of Seville, some geography texts (derived from *Etymologies* and other sources), a reference to the eclipses of 778 and 779, the catalogue of the Córdoba church library, etc., all of which is accompanied by marginal notes in Arabic which are also found in other Latin manuscripts containing the *Etymologies*. More spectacular still is the celebrated Isidorian T map preserved in a manuscript in the National Library of Madrid whose legends are written in Arabic: it was drawn either by a Muslim who was very familiar with Isidorian tradition, or by a highly Arabianized Christian (Menendez Pidal 1954). If we pass from geography to history the evidence becomes even clearer, but this chapter is not the most appropriate place to enlarge upon this: it is sufficient to mention, as an example post-dating the period which concerns us here, the Arabic translation made at Córdoba of *Historiarum adversos paganos libri septem* by Paulus Orosius.[8]

Returning to the history of the sciences, we shall consider later the Mozarab cultural elements found in the *Calendar of Córdoba*. First, a reading of the chapter on Andalusian physicians in the *Kitāb ṭabaqāt al-aṭibbāʾ wa-l-ḥukamāʾ* of Ibn Juljul al-Andalusī (Vernet 1979d) is extremely instructive because this author indicates that Andalusian medicine was predominantly practised by the Christians until the time of ʿAbd al-Raḥmān III al-Nāṣir (912–61) and that 'in Andalusia, medicine was practised according to one of the books of the Christians that had been translated. Its title was *Aphorism*, a word meaning summary or compilation'. The term *aphorism* does not here imply a reference to the *Aphorisms* of Hippocrates, because if the definition of Isidore of Seville (*Etym.* 4, 10) is to be accepted, this word denotes a literary style in the medical literature. Moreover, of the six physicians mentioned by Ibn Juljul under the emirates of Muḥammad (852–86), al-Mundhir (886–8) and ʿAbd Allāh (888–912), five were Christian, two of these having names as characteristic as Hamdīn

b. Ubba (i.e. Oppas) and Khālid b. Yazīd b. Rumān. Furthermore, one of these five physicians, called Jawād, is the author of *Monk's Medicine*. This situation changed with the caliphate of ʿAbd al-Raḥmān III, but the Latin medical tradition survived in the person of Yaḥyā b. Isḥāq, author of five notebooks of aphorisms, who consulted a monk about a case of otitis from which the caliph was suffering. All of this is confirmed by the physician Saʿīd b. ʿAbd Rabbihi (d. *c.* 953–77) who, in his *Urjūza fī al-ṭibb*, says that 'the highest limits [of medicine] will only be reached by one who knows [the ancient texts] translated into Arabic' (*al-muʿarrabāt*) (Kühne 1980).

A third area in which the survival of a Latin tradition seems quite clear is agronomy. Until recently it was fairly generally accepted that there existed a direct tradition from Columella amongst Andalusian agronomists and it was even postulated that there was an Arabic translation, made in Spain, of his *De re rustica*. This theory was based on quotations by Ibn Ḥajjāj (*c.* 1073) of an author called Yūnyūs, who was identified as being Iunius Moderatus Columella.[9] But it has been shown that the similarities between quotations from Yūnyūs and certain passages of *De re rustica* are more probably due to the identical nature of the subject treated, contradictions also exist, and greater similarities can be found by comparing the quotations from Yūnyūs with an agronomical work by Vindanios Anatolios of Berito preserved in an Arabic translation that derives from an earlier Syriac translation. Moreover, Yūnyūs is a distortion of the name Vindanios (Rodgers 1978).

However, despite the blow to the theory of a tradition from Columella in Andalusia – which, for some scholars, would have constituted the essential difference between Andalusian agronomy and eastern agronomy – even the most critical authors have not abandoned the idea of a survival of Latin agronomy in Muslim Spain, given that Ibn Ḥajjāj asserts that his statements are founded on the tradition of the *Rūm* (Mozarabs) of Andalusia and that Ibn al-ʿAwwām (twelfth century or first half of thirteenth) says that he collected the opinions of non-Muslim authors, without quoting their names but introducing the quotations with phrases such as 'there are agronomists who say . . .', 'others who say . . .'. One of the anonymous sources has been identified in an Arabic manuscript in the Bibliothèque Nationale in Paris. Its author was clearly a Christian because he eagerly defends the evangelical procedure of fertilizing a barren tree by threatening it with an axe. The text is a short treatise of the tenth century, whose author is a Mozarab, steeped in Arabic culture, who quotes the classical authors in eastern Arabic translation.[10]

THE DEVELOPMENT OF EASTERN CULTURE (850–1031)

The picture we have so far drawn is inevitably one-sided. We have emphasized the survival of the Latin–Visigothic culture because this is the most characteristic feature, but we do not claim it to be the only one. Moreover, the chronological milestones of our exposé are simply points of reference: we have given a sufficient number of examples to demonstrate that Latin culture survived beyond 850, alongside Arabic culture. However, at least after the accession to the throne of the first Umayyad (756), the process of the easternization of Andalusian culture began with a period of Syrian influence, followed by a phase of Iraqi influence, which began in the ninth century and was consolidated under Emir ʿAbd al-Raḥmān II (821–52).[11] Travellers who departed for the East, either to study or to accomplish their duty of pilgrimage, returned with the latest inventions. The Great Mosque of Córdoba, founded in 786 by ʿAbd al-Raḥmān I, became a centre of cultural diffusion and medicine, astronomy and mathematics were slowly introduced into the higher education given in the mosques or in private houses (the *madrasa* appeared much later).[12] We know nothing about the development of other scientific institutions, such as hospitals (there surely were some) and observatories (which perhaps did not exist), but the situation is altogether different with regard to libraries (Ribera 1928a,b). There was a constant interest by certain emirs in books for these: ʿAbd al-Raḥmān II, a reader of works of philosophy and medicine, sent ʿAbbās b. Nāṣiḥ to the East to buy books, and the existence of a royal library is attested from the time of the emirate of Muḥammad (852–86). It was considerably developed under al-Ḥakam II (961–76), even if we reject the total of 400,000 volumes which tradition claims for it in that caliph's day (the same number is reported for the great library of Alexandria). In addition, private libraries appeared in large numbers during the tenth and eleventh centuries at Córdoba, Seville, Almeria, Badajoz, Toledo, Saragossa, etc.

The role of promoting this easternization of scientific culture must perhaps be accorded to ʿAbd al-Raḥmān II. Our anonymous Maghreb author of the fourteenth or fifteenth century tells us that it was he who first introduced astronomical tables to Andalusia (*Huwa awwal man adkhala kutub al-zījāt*), as well as books of philosophy, music, medicine and astronomy (Molina 1983: 138). In fact, it was in this period that ʿAbbās b. Firnās (d. 887) (Terés 1960), or ʿAbbās b. Nāṣiḥ (d. after 844) (Terés 1962), introduced a version of the *Sindhind* tables which is usually identified with that of al-Khwārizmī. It is possible that the *al-daftar al-muḥkam* of which Ibn Firnās speaks in a poem is also a *zīj*.[13] At all events, astrology was fashionable at the court of Córdoba and the emir was surrounded by a court of poet-astrologers such as Ibn Firnās, Ibn Nāṣiḥ, Yaḥyā al-Ghāzāl (Vernet 1979e) and Ibn al-Shamir (Terés 1959). The emir's interest in astrology may have originated from the important astronomical events which took place during his reign, including the solar eclipse of 17 September 833, virtually total at Córdoba, which terrified the townspeople and led them to gather at the great mosque for the ritual prayer of the eclipse. There was also a massive shower of shooting stars between 20 April and 18 May 839. From this time onward at least, the astrologer became a prominent figure who frequently enjoyed the confidence of emirs, and later of caliphs, which aroused the jealousy of the pious *fuqahā'* and of certain poets. There is evidence of anti-astrological dispute, which also became anti-astronomical, in the ninth as well as the tenth century (Samsó 1979b).

At the same time, this was a period in which innovations were continually being introduced. To give just a few examples: easternization in the field of medicine may have much to do with the presence in Córdoba of the physician al-Ḥarrānī, who practised at the court of ʿAbd al-Raḥmān II. Ibn Juljul, who refers to this figure, also mentions his grandsons (?) Aḥmad and ʿUmar b. Yūnus al-Ḥarrānī, who studied at Baghdad between 941 and 962 with Thābit b. Sinān b. Thābit b. Qurra, also a Harranian. There is thus a continuity of tradition, and it has been suggested that, on their return to Andalusia, they may have introduced to that country both the works of Thābit b. Qurra and the techniques of talismanic magic which were to blossom in Spain in the eleventh century with the *Ghāyat al-ḥakīm* (*Picatrix*) of Abū Maslama al-Majrīṭī. In the tenth century also Ibn Juljul used Latin and Arabic sources to write his *Ṭabaqāt al-aṭibbā'*, and amongst the latter is the *Kitāb al-ulūf* by Abū Maʿshar. The interest in this type of astrology is also apparent in the introduction of the *Liber Universus* of ʿUmar b. Farrukhān al-Ṭabarī to Córdoba towards the end of the tenth century (Pingree 1977). During this century the *Rasā'il* of Ikhwān al-Ṣafā' and the *Tabula Smaragdina* (Stern 1961) were also introduced, Yaḥyā b. Isḥāq wrote a manual of medicine in which he brought together all the Greek medicine known in his time (Meyerhof 1935, esp. p. 6), and Ibn Juljul provided a list of the sixteen works of Galen which it was necessary for a student of medicine to know.[14]

Andalusian science began to appear productive. From this point of view, the most outstanding figure in the second half of the ninth century is perhaps ʿAbbās b. Firnās (d. 887), who was not only a poet and astrologer but also carried out experiments in flying at the Ruṣāfa of Córdoba – reminiscent of similar attempts made in England in the eleventh century by the monk Eilmer of Malmesbury; he introduced a new technique for cutting rock crystals, and constructed a kind of planetarium in a room of his house as well as an armillary sphere that he presented to ʿAbd al-Raḥmān II, and, finally, a water-clock equipped with moving robots. This *mīqāta* or *minqāna* enabled one to determine the hour for canonical prayers when there was neither sun nor stars to serve as an indicator; it was given to Emir Muḥammad (Vernet 1980a,b).

ʿAbbas b. Firnās is a quite exceptional figure in the ninth century without, however, being a real man of science but rather a courtier endowed with an encyclopedic curiosity and the skill to exploit his knowledge. The true development of Andalusian science occurred during the following century, especially in its latter half, when we find: a popular calendar, the *Calendar of Córdoba*, which contained the first known evidence of the Andalusian *mīqāt*; the development of a native pharmacology; and the school of Maslama of Madrid, the starting point of Hispano-Arabic astronomy.

The Calendar of Córdoba

The *Calendar of Córdoba*[15] was compiled for al-Ḥakam II, before or after his accession to the caliphate (960), by the physician and historian ʿArīb b. Saʿīd[16] and the Mozarab bishop Rabīʿ b. Zayd (Recemund). This work contains a curious mixture of different traditions: Latin and Mozarab (references to the feasts of Christian saints, customary farming practices in Spain); pre-Islamic Arab (meteorological predictions based on the ancient system of the *anwāʾ*); and Greco-Alexandrian (dietetic references which the text ascribes to the school of Hippocrates and of Galen and which correspond closely to the Hippocratic *Diet* (Samsó 1978). But it also contains the new astronomy created by the Arab-Islamic culture on the basis of the Indo-Iranian and the Ptolemaic traditions. Thus the text gives the date when the sun enters the twelve signs of the zodiac according to the *Sindhind* and the *Asḥāb al-mumtaḥan*, and we have been able to confirm that this refers to the *zīj* of al-Khwārizmī and possibly that of Yahyā b. Abī Manṣūr or Ḥabash al-Ḥāsib (Vernet 1979a, esp. pp. 28–30). Furthermore, the *Calendar* gives a whole series of numerical values which demonstrate the existence in tenth-century Andalusia of the *mīqāt* tradition, revealed here for the first time.[17] Thus the text contains:

1 Twenty-three meridian heights of the sun, distributed throughout the year, which correspond to a latitude of 37; 30° (plotted for Córdoba in one of the manuscripts of the *Toledan Tables*) and an obliquity of 23; 50° (a value rounded from the Ptolemaic figure of 23; 51, 20°).
2 The *shadows* corresponding to the preceding meridian heights, calculated for a gnomon $g = 1$, since the gnomon used had the height of a man. These values appear, nevertheless, to be derived from a table calculated

for $g = 12$ or, rather, from two tables of the same type, calculated probably using arithmetical methods, one giving the shadow corresponding to the entry of the sun in the signs of the zodiac and the other to its passage through the middle of each sign.

3 Twenty-four values (two per month) corresponding to the length of the day and of the night throughout the year, computed by means of the same parameters as those above, using a trigonometrical calculation, with results that are generally correct.

4 Twenty-eight values for the duration of twilight: this series is without doubt the most surprising since it seems to be calculated for an arc of depression of the sun of 17°, using an approximate formula similar to that of Brahmagupta:

$$t = \frac{D}{\text{cotan } h + 1}$$

Here, then, is one example of the extensive evidence that exists to demonstrate the influence of the Indo-Iranian tradition of astronomy in Andalusia, which we shall be emphasizing later in this chapter. However, the four series of numerical values that we considered above employ very different methods and pose a problem concerning the source used by the authors of the *Calendar*: given that neither ʿArīb b. Saʿīd nor Rabīʿ b. Zayd were astronomers, they may have used the *mīqāt* tables for a latitude of 37; 30° which could be for Córdoba or another town of the same latitude (Samsó 1983b).

The development of a native pharmacology

Even if an Andalusian pharmacology can be said to have existed before the era of ʿAbd al-Raḥmān III, a fundamental development occurred during his caliphate. The Andalusian physicians had difficulty in identifying the simples (medicinal plants and resulting medicines) referred to by Dioscorides, in his *De materia medica*, which was known through an Arabic translation made in the East by Iṣṭifan b. Basīl. In 948(?) Caliph ʿAbd al-Raḥmān III received from the emperor of Byzantium (Constantine VII?) a magnificent illustrated manuscript of Dioscorides in Greek which could not be understood because there were no Hellenists in Córdoba at that time. At the caliph's request the Byzantine emperor sent the monk Nicolas to Andalusia and with his help, a group of Andalusian physicians undertook a systematic revision of the botanical nomenclature used in the Arabic version of Dioscorides, succeeding in identifying most of the simples (Vernet 1979b; Meyerhof 1935; Dubler and Terés 1952, 1953, 1957). This had important consequences, amongst which was a rapid expansion of pharmacology and Hispano-Arabic botany which began shortly after the completion of the work on Dioscorides, and one of whose first manifestations was the botanical work of Ibn Juljul, to whom we have already referred more than once; he knew the collaborators of monk Nicolas and he made haste to write a book on the plants and remedies identified and a second on the medicines which had not been mentioned by Dioscorides (Garijo 1990, 1992a, b).

This was also the period that witnessed the first manifestations of the maturity of Andalusian medicine. Let us briefly mention the name of ʿArīb b. Saʿīd, who was the author, in about 964, of a treatise on obstetrics and paediatrics which also contained one of the first Andalusian references to medical astrology. Much more important is the work of Abū al-Qāsim al-Zahrāwī (born after 936; died around 1013), whose *Taṣrīf* contains one of the most important treatises on surgery of the entire Middle Ages, as well as a treatise on pharmacology in which he uses advanced laboratory techniques that were in use among Egyptian or Iraqi artisans and perfumers who had preserved procedures of Mesopotamian origin. His work on pharmacology is also of theoretical interest because, basing himself on the theory of the humours, the four therapeutic qualities of Hippocrates (cold, hot, wet, dry) and the Galenic degrees of those qualities, he investigated the problem of the dosage of simples to be used in a compound medicine: he may have known the *De medecinarum compositarum gradibus* of al-Kindī.[18]

The school of Maslama al-Majrīṭī

Maslama had a similar role in the history of Andalusian astronomy to that of Abū al-Qāsim in the history of medicine. Born in Madrid, he studied at Córdoba, where he died in 1007. An astrologer of renown, he foretold the fall of the caliphate as well as certain details of the politics which preceded the *fitna*. However, his prestige stemmed in particular from his adaptation of the tables of al-Khwārizmī, which are consequently often called the *zīj* of al-Khwārizmī–Maslama. We have already mentioned the introduction of the *Sindhind*, probably in the Khwārizmian version, to Andalusia during the emirate of ʿAbd al-Raḥmān II. This text, known in Spain in its condensed form, without demonstrations, was the object of an adaptation by Maslama and his disciple Ibn al-Ṣaffār (d. 1034), which has been preserved in a Latin translation by Adelard of Bath (Suter 1914; Neugebauer 1962a,b). Establishing the precise contribution of the Andalusian astronomers to this *zīj* is not easy, given that al-Khwārizmī's original text appears to be lost and we can only try to reconstruct it by means of the data preserved in Ibn al-Muthannā's commentary (Millás Vendrell 1963; Goldstein 1967a,b) in the *Liber de rationibus tabularum* of Abraham b. Ezra (Millás Vallicrosa 1947), or in similar texts, such as the *Kitāb fī ʿilal al-zījāt* of al-Hāshimī.[19] The presence in this *zīj* of al-Khwārizmī of material corresponding to the Indo-Iranian, Greco-Arabic and Hispanic traditions has been established. One could contend a priori that the Indo-Iranian material is from the original *zīj*, but this is not always true, notably for the tables of mean motion, since the basic parameters are of Indian origin but the disposition of the tables transmitted shows an important formal modification that is traditionally attributed to Maslama. In fact, the original tables used the Persian solar year, and the date of origin was the beginning of the era of Yazdegerd III (16/06/632), whereas the tables that have been preserved use the Muslim lunar year and the beginning of the Hijra (midday on 16/07/622) as the date of origin. The intervention of Maslama in the tables of eclipses has also been pointed out (Pingree 1976: 165), as well as in the tables for computing the latitude of the planets, although in this last case the results that he obtained were not very good (Kennedy *et al.* 1983: 125–35). A similar situation exists with regard to the part of the *zīj* that was influenced by Ptolemy: on the one hand, al-Khwārizmī was a contemporary of Caliph al-Ma'mūn, i.e. he lived at a time when the *Almagest* and the *Handy Tables* were very well known; on the other hand, there is sometimes a more or less well founded impression that the original material may have been reshaped and have been subject to interpolations by Maslama or someone else. The same applies to certain trigonometric tables, such as the sine table, calculated for a radius of 60 p. This table is the result of the division by two of the table of chords in the *Almagest*. This contradicts the evidence of Ibn al-Muthannā, who states that the value of the radius used in the sine table of al-Khwārizmī was 150 p. We can also postulate a contribution from Maslama for all the Hispanic material, for example, the reference to the Hispanic era (38 BC) in the chronological part of the *zīj* or the use of the meridian of Córdoba for certain tables, such as those for the determination of the conjunction and opposition of the moon and the sun – derived from the original table but modified by Maslama – or the tables for the mean motion of the ascendent node of the moon, which contains a supplementary table for the meridian of Córdoba and for the period between 970 and 1174 (Neugebauer 1962a: 61, 63, 95, 108–10). There is a similar example in the tables of the projection of radii (*projectio radii stellarum*), which comprise nearly a fifth of all the numerical tables of the *zīj*: they are calculated for a latitude of 38; 30° (Córdoba) and do not coincide with the original tables of al-Khwārizmī preserved by the Eastern astrologer Ibn Hibinta (Baghdad, *c.* 950). A recent work shows that the work of Maslama improved the calculation methods of al-Khwārizmī because the tables of the astronomer from Córdoba give accurate results and are much easier to use than those of al-Khwārizmī (Kennedy *et al.* 1983: 373–84; Hogendijk 1989).

Ascribing certain modifications to Maslama is sometimes more problematical, and the intervention of later hands must be considered. This is the case with the table for the visibility of the new moon, based on an Indian theory of visibility but calculated for a latitude of around 41; 35°, much farther north than Córdoba, which could correspond to Saragossa, and was thus probably established in the eleventh century, a period when the exact sciences underwent rapid expansion in that town (Kennedy *et al.* 1983: 151–6; King 1987c: 189–92; Hogendijk 1988b).

Maslama's work in connection with astronomical tables was not limited to the *zīj* of al-Khwārizmī. In his *Ṭabaqāt al-Uman*, Ṣāʿid of Toledo tells us that Maslama 'applied himself to the observation of the heavenly bodies and to understanding the book of Ptolemy entitled *Almagest*' and that he was 'the author of a summary of the part of al-Battānī's table concerning the equation of the planets'.[20] These three affirmations must be treated independently.

1 With regard to the observations of heavenly bodies, we need only recall the evidence of al-Zarqāllu, who stated that Maslama observed the star Qalb al-Asad (Regulus) in 979 and that he established its longitude to be 135; 40°. This evidence agrees with the value of the longitude of this star found in the small table of twenty-one stars which accompanies his commentaries on the *Planisphaerium* (Millás Vallicrosa 1943: 310–11; Kunitzsch 1980); the determination of the longitude of this star was used by Maslama to establish a movement of precession of 13; 10° with respect to the catalogue of stars in the *Almagest*, which permitted him to determine the longitude of the rest of his stars.
2 We know nothing of Maslama's work on the *Almagest* (his disciple Ibn al-Samḥ appears to have written a résumé of it) but this work was obviously well known to the school of Maslama whose interests were not confined to the *Sindhind*: Ibn al-Ṣaffār refers to Ptolemy's *Geography* in his work on the use of the astrolabe and a manuscript related to the translations made in the monastery of Ripoll towards the end of the tenth century gives a structuring of the climates of the earth which may derive from the *Almagest* or *Geography*.[21]
3 We do not know either what Maslama took from the *zīj* of al-Battānī, although the edition of Nallino contains half a dozen tables attributed to Maslama but probably false. However, it is clear that the school of Maslama knew the works of al-Battānī well since, in his treatise on the construction of the equatorium, Ibn al-Samḥ used al-Battānī's

parameters for the longitudes of the apogees of the planets, while the values for the eccentricities and the radii of the epicycles could have been derived from either al-Battānī or the *Almagest* (Samsó 1983c, Comes 1991).

Moreover, Maslama produced a version of the *Planisphaerium* of Ptolemy: given the conceivable connection between Maslama and the monk Nicolas, and thus the possibility that Maslama had learnt Greek, it has been suggested that he may have translated the *Planisphaerium*; it is equally possible that he revised an eastern Arabic translation to which he added his commentaries. The Greek original of this work has not been preserved, and the question cannot be resolved without first studying all the available material, i.e. (1) Maslama's version of the *Planisphaerium* in a Latin translation by Hermann of Dalmatia (1143)[22] and in a Hebrew version; (2) an Arabic version (earlier than Maslama?), preserved in manuscript;[23] (3) Maslama's commentaries on the *Planisphaerium* (Vernet and Catala 1979, Kunitzsch and Corch 1994).

The last text contains a series of additions to the work of Ptolemy: three new methods for dividing the ecliptic of the astrolabe (Ptolemy gives two others); three procedures also for dividing the horizon, analogous to those given for the ecliptic, which fill a gap in the *Planisphaerium*; three methods for locating the fixed stars of the *rete*, or star map, on the astrolabe, using ecliptic, equatorial and horizontal co-ordinates. In a second part of the work, Maslama employs only one trigonometric tool: the theorem of Menelaus on which he had previously written several notes that have been preserved in a Latin translation (Björnbo and Suter 1924: 23–4, 39, 79, 83). He deals with the determination of the right ascension of the beginning of each zodiacal sign, using a similar procedure to the one that he had already described for dividing the horizon from the basis of right ascensions; the determination of the declination of a star; the determination of the degree of culmination of a heavenly body in the sky (using certain formulae of al-Battānī); and the determination of the degree of the zodiac that rises or sets with a heavenly body. Finally he gives a table of 'inclinations' of the fixed stars for a latitude of 38; 30° (Córdoba), whereas in an example in the first part of the work he uses a latitude of 39°.

These commentaries of Maslama on the *Planisphaerium* are not in any way a treatise on the construction of the astrolabe but they doubtless influenced the treatises of Andalusian origin on the construction on this instrument, notably the work of Alfonso X (Samsó 1980a,b,c,d) and that which is wrongly attributed to Māshā'allāh (Viladrich 1982; Viladrich and Marti 1981), for it has been demonstrated that this so-called treatise of Māshā'allāh concerning the construction and use of the astrolabe is in reality a compilation of the thirteenth century, made up of extremely heterogeneous elements including some passages that could possibly be identified with the school of Maslama. This school is represented, with regard to the instrument in question, by the commentaries of Maslama which we have been discussing here, as well as by Ibn al-Ṣaffār's treatise on the use of the astrolabe (Millás Vallicrosa 1955) – very popular on account of its brevity and practicality – and the much more verbose work by Ibn al-Samḥ (Viladrich 1986). This last text is interesting for two reasons: first, it contains quotations from an unknown work on the astrolabe by the eastern astronomer Ḥabash al-Ḥāsib (*c.* 835), which constitutes the first evidence of the knowledge of this author in Andalusia; second, this book of Ibn al-Samḥ was the source used by the collaborators of Alfonso X for the writing of a treatise on the use of the spherical astrolabe – in the absence of an Arabic text to translate, they adapted a treatise on the plane astrolabe to the requirements of the spherical astrolabe (Viladrich 1987).

The tenth century also witnessed the emergence of other innovations in the field of astronomical instruments. The oldest extant sundials date from this era (King 1978a; Barceló and Labarta 1988; Carandell 1984a,b; King 1992; Casulleras 1993; Labarta and Barceló 1995), and one of these instruments is explicitly credited to Ibn al-Ṣaffār; but the important defects in the instrument make it difficult to accept this attribution to a competent astronomer and suggest instead that it was made 'in the manner of Ibn al-Ṣaffār' by a less conscientious craftsman. However, there is no such doubt that Ibn al-Samḥ is the author of the first known treatise on the construction of an equatorium (Comes 1991): the instrument designed by this astronomer consisted of eight plates (one for the sun, six for the deferents of the moon and the five planets, and one for the planetary epicycles) that were placed within the mother of an astrolabe.[24] The plates of the planetary deferents contained, in addition to the geometrical diagram, tables of mean motion in longitude and in anomaly of the corresponding planet which recall the *Zīj al-ṣafāʾiḥ* of Abū Jaʿfar al-Khāzin (d. 961–71) (King 1980): the latter *zīj* could be found on the plates of an equatorium-astrolabe and, in that case, this type of instrument would have been of eastern origin. The question remains open until new elements are discovered.

THE GREAT EXPANSION OF ANDALUSIAN SCIENCE (ELEVENTH CENTURY)[25]

During the tenth century Andalusian science reached a productive level and certain Andalusian men of science acquired a reputation even in the East: an obvious example is Abū al-Qāsim al-Zahrāwī, and another is [Maslama] al-Majrīṭī, who was cited by Ibn al-Shāṭir in the prologue of his *Nihāyat al-sūl*, in the fourteenth century, as one of the authors who had criticized Ptolemy (Kennedy *et al.* 1983: 62). The repercussions in the East of Andalusian scientific successes were much more numerous from the eleventh century: the work of Andalusian agronomist Ibn Baṣṣāl became well known in the Yemen, where, in the mid-fourteenth century, the sovereign *rasūlī* al-Malik al-Afḍal used the complete version of the *Kitāb al-qaṣd wa-l-bayān* instead of the shorter version which has reached us (Serjeant 1963, 1977). We could give many more examples of this type but we shall confine ourselves to the influence in the East of the universal astrolabes developed in the eleventh century by ʿAlī b. Khalaf and by al-Zarqāllu: the *ṣafīḥa* of the latter, in two versions (*zarqāliyya*, a very elaborate instrument; and *shakkāziyya*, a more simple instrument), was well known in the Near East where, at the end of the fourteenth and the beginning of the fifteenth century, developments of the simple version of the instrument appeared in the form of quadrants of the *shakkāzī* type, which were employed by the astronomers at the observatory of Istanbul in the sixteenth century.[26]

The standard of Andalusian science increased considerably after the political crisis of 1031, which did not lead to a cultural crisis: three scientific centres of the greatest importance sprang up in Saragossa, Toledo and Seville. The level of easternization of Andalusian culture became more pronounced at this time: a good example is the *Kitāb al-anwāʾ wa-l-azmina wa maʿrifat aʿyān al-kawākib* of ʿAbd Allāh b. Ḥusayn b. ʿĀṣim, known as al-Gharbāl (d. 1012),[27] which is a totally different work from the *Calendar of Córdoba*. In fact, whereas the latter text is a mixture of elements from Arab, Mozarab and Hellenistic cultures, as we have seen, in Ibn ʿĀṣim's book Arabic elements quite clearly predominate and reading it reminds us more of the *Kitāb al-anwāʾ* of Ibn Qutayba than of any other similar text. In addition, this is the period when the survival of the Mozarab culture – the revision of the *Libro de las Cruzes* and the use of Latin sources by the agronomist Ibn Ḥajjāj – became completely residual, and Andalusian students considered that they could acquire an adequate scientific training without needing to travel to the East. The development of local schools is attested by Ṣāʿid of Toledo, whose *Kitāb ṭabaqāt al-umam* supplies sufficient information to enable the reconstruction of the 'genealogical tree' of the schools of Maslama and Abū al-Qāsim al-Zahrāwī, which were enormously important in the development of astronomy, medicine and of Andalusian agronomy in the eleventh century. Moreover, independence with regard to the East is clearly evident if we compare the statistics for journeys undertaken by the Muslims of the Valley of the Ebro:[28] in the tenth century 25 per cent of Muslim travellers from this region departed for the East, whereas in the eleventh century the proportion fell to 11 per cent.

Nevertheless journeys to the East continued, including significantly the case recorded by Ṣāʿid of Toledo of his patron ʿAbd al-Raḥmān b. ʿĪsā Muḥammad (d. 1080), who lived in Cairo, where he met Ibn al-Haytham.

One of the most remarkable characteristics of the eleventh century in Andalusia is the development of mathematics, due especially to the work of three key figures: King Yūsuf al-Mu'taman (1081–5) of the *ṭā'ifa* of Saragossa; the mathematician Ibn Sayyid, master of the great philosopher Ibn Bājja, who wrote his works in Valencia between 1087 and 1096; and the *faqīh* and astronomer Ibn Muʿādh (d. 1093). Until quite recently, all that was known of the first of these three mathematicians was the title of his mathematical work, *al-Istikmāl*, and certain indirect references to its contents (Djebbar 1993); the situation changed with the discovery of four fragments of the work (Hogendijk 1991, 1995), which showed that the *Kitāb al-Istikmāl* is a great mathematical encyclopedia, which bears witness to the knowledge of the best literature and contains original contributions. We only know the work of the second mathematician, Ibn Sayyid, through indirect references.

But without doubt the best known of the eleventh-century mathematicians named above is the third: Ibn Muʿādh al-Jayyānī. His *Maqāla fī sharḥ al-nisba* (Plooij 1950), is a text of great interest and is an important link in the chain of Arabic commentaries on the notion of *ratio* exposed by Euclid in Book V of his *Elements.*[29] In addition, Ibn Muʿādh's *Kitāb majhūlāt qisī al-kura* (Villuendas 1979), is without doubt the oldest treatise of the medieval West concerning spherical trigonometry and in which that discipline becomes totally independent of astronomy (the work contains no reference to astronomy except in the prologue).

The mathematical revival was accompanied by an identification of astronomical research. In this field we must first stress that the influence of the *Sindhind* remained predominant; in relation to this, Ṣāʿid of Toledo emphasized the work carried out by the school of Maslama and by others, amongst whom he placed himself. A small part of that work has been preserved and studied, for example the Latin translation of the canons written by Ibn Muʿādh for his *Tabulae Jahen*: based on the system of the *Sindhind* and calculated for the co-ordinates of Jaén, the town where the astronomer was born (Hermelink 1964), these tables also contain original data. Ibn Muʿādh, following al-Khwārizmī, places the solar apogee at 75; 55° from the vernal point: the same parameter was used by al-Zarqāllu in his treatise on the equatorium (Comes 1991: 92).

The *Toledan Tables*, begun under the direction of *qāḍī* Ṣāʿid, seem to have been a collective work, participated in by the most important Andalusian astronomer of all time, Abū Isḥāq b. al-Zarqāllu (also called al-Zarqiyāl/Azarquiel by Ṣāʿid), but they have disappointed researchers for the tables of mean motions are the only original work, while the rest is derived from the *zīj* of al-Khwārizmī–Maslama and of al-Battānī; however, certain elements attributed to the latter could also have been derived directly from Ptolemy, whose influence can be noted in the tables of retrograde motion and the tables of the co-ordinates of the stars. Lastly, the tables of computations for the trepidation of the sphere of the fixed stars are also found in the *Liber de motu octave spere*, attributed until very recently to Thābit b. Qurra. It is nevertheless possible that these tables, which are only found in some manuscripts of *Liber de motu*, are independent of this work and derive from the work of the astronomers of Toledo (Mercier 1987; Samsó 1994).

These negative findings lead us to certain considerations: it is well known, for example, that al-Zarqāllu devoted twenty-five years of his life to making solar observations, first at Toledo and later in Córdoba (Millás Vallicrosa 1943: 241). The result of this work was contained in a lost text on solar theory, certain elements of which have been reconstructed with the help of indirect sources (Toomer 1969, 1987; Samsó 1988, 1994): notably, around 1074, al-Zarqāllu determined the position of the solar apogee (85; 49°) and estimated that its own movement was 1° in 279 solar years; he also designed a solar model based on a moving eccentric (analogous to the deferent of Mercury in the Ptolemaic model), which produced a trepidation of the position of the apogee as well as a variation in the solar eccentricity. The same solar model was used much later by Copernicus, who, like al-Zarqāllu, did not take into account the trepidation of the apogee, thereby proving that the model had been adopted because it justified the variation in the values of the solar eccentricity proposed by astronomers since the time of Hipparchus. Obviously, al-Zarqāllu also established the value of the solar eccentricity for his era (1; 58 p approximately). In view of this degree of research, it is difficult to accept that al-Zarqāllu simply copied the table of the solar equation from the *zīj* of al-Battānī in the *Toledan Tables*, whereas the solar tables of his *Almanac* implied an eccentricity which was not that of al-Battānī but rather of the order of the parameter quoted above. This all fits very well with the hypothesis according to which the *Toledan Tables* were begun towards the end of *qāḍī* Ṣāʿid's life (1029–70) and after he had completed his *Ṭabaqāt al-Umam* (1068) in which he does not mention the tables (Richter-Bernburg 1987). Al-Zarqāllu would have introduced elements derived from his own observations or from those of Ṣāʿid's team, but most of his work on solar theory was probably carried out after the compilation of the *Tables*. Al-Zarqāllu may also have undertaken work on planetary astronomy, because his treatise on the construction of the equatorium, which is preserved in an Alfonsine Castilian translation, also gives planetary parameters that do not always coincide with those of the *Toledan Tables*:

thus, although the eccentricities of Jupiter, Mars and the moon are Ptolemaic, those of Saturn (2; 51, 23 p or 2; 48, 48 p), Venus (1; 03, 27 p) and Mercury (2; 51,26 p) appear to be original.[30]

The importance of his work on the movement of the fixed stars, which is preserved in a Hebrew version, should also be noted (Millás Vallicrosa 1943: 245–343; Goldstein 1964a,b; Samsó 1987b, 1994). In this work, after several studies, al-Zarqāllu presents us with a model of trepidation derived from that in the *Liber de motu* – although with new parameters – to which he adds, in a fairly artificial manner, a second independent model for calculating the obliquity of the ecliptic that he finds to oscillate between 23; 53° (about the beginning of the Christian era) and 23; 33° (for AD 954–5). The study of the values of the obliquity, which are implicit in the tables of the *Liber de motu*, offers satisfactory results for the time of Ptolemy as well as for the period of the Caliph al-Ma'mūn, but the function gains rapidly increasing values after AD 887 and consequently leads to unacceptable values for the era of al-Zarqāllu. The latter doubtless tried to correct this anomaly by providing a geometric model as well as tables which, while remaining in agreement with the values of the obliquity established by Ptolemy and the astronomers of al-Ma'mūn, gave reasonable values for his own period (23; 33, 49° for the end of 1074).

Finally, with regard to al-Zarqāllu we must also mention his almanac,[31] which can be used to determine, almost without calculation, the true longitude of the sun and planets by means of Babylonian goal-years. It is the first known work of its kind from the Middle Ages and it had a lasting influence in the Muslim and Christian West. However, apart from the solar tables, which may have resulted from observations by al-Zarqāllu himself, the work is an adaptation of a Greek almanac that can be dated between AD 250 and 350 (the presumed author is referred to in the text as Awmātiyūs), and there may also have been an Arabic version in the tenth century before the version of al-Zarqāllu. It should be pointed out that both the geometrical models and the parameters that can be deduced from the planetary tables seem to originate from Ptolemy.

A third area of rapid Andalusian expansion in the eleventh century is alchemy and technology. Abū Maslama al-Majrīṭī is important with regard to the first of these disciplines; his *Rutbat al-ḥakīm* contains descriptions of experiments by the author which imply a certain intuition of the principle of preservation of matter (Holmyard 1924). With regard to technology, the existence of an Andalusian tradition in the field of mechanics has been known for about ten years thanks to the discovery of the *Kitāb al-asrār fī natā'ij al-afkār* of Aḥmad, or Muḥammad, ibn Khalaf al-Murādī, in a unique manuscript which contains a note in the hand of R. Isḥāq b. Sīd, the chief astronomer of Alfonso X.[32] The development of the agronomical tradition is much better known.[33] A school of agronomists, comprising a number of scholars whose chronology is not certain in all cases but whose overall activities appear to have covered some fifty years (1060–1115), emerged first in Toledo, under the patronage of al-Ma'mūn, and later in Seville, under the reign of the Banū ʿAbbād (Attié 1982). The preserved texts are mostly incomplete: they consist of summaries or anthologies written by North African authors.[34] We should mention the physician Ibn Wāfid (999–1074)[35] and Ibn Baṣṣāl, both of Toledo; Abū al-Khayr (Carabaza 1990) and Ibn Ḥajjāj (Attié 1980; Carabaza 1988) of Seville; and al-Ṭignarī (García Sánchez 1987b, 1988, 1990), who studied in Seville and then lived in several Andalusian and North African towns. We must add to this list the name of Ibn al-ʿAwwām, who lived later (his work must be dated at around the end of the twelfth century) and who summarized the contributions of the whole Andalusian school.[36]

Andalusian agronomy inherited a great mixture of ancient agronomical traditions: on the one hand, Babylonian and Egyptian, through the influence of the *Filāḥa Nabaṭiyya* of Ibn Waḥshiyya (El-Faiz 1990); and on the other hand, Carthaginian, Roman and Hellenistic, whose influence was exercised mainly through the Arabic translation of the Byzantine *Geoponika*. The Andalusian sources quote a considerable number of authors from the different traditions mentioned but, in most cases, these are indirect quotations. They also cite other sources, such as the *Filāḥa rūmiyya* and the *Filāḥa hindiyya*, the former of which at least (attributed to a certain Qusṭūs) seems to be a forgery, made around the middle of the tenth century by ʿAlī b. Muḥammad b. Saʿd (Attié 1972). However, as we have already indicated in the first part of this chapter, from the end of the eighteenth century scholars have laid great emphasis on the direct influence of the Latin agronomical tradition.

Andalusian agronomy seems, then, to have been familiar with the best agronomical literature available to the authors of the eleventh century. In addition, contact was never lost with the experience and tradition of the botanical garden, which began in the eighth century in Córdoba and continued in the eleventh century in Toledo and Seville. A third and very important aspect was the theoretical effort undertaken by the Andalusian agronomists to make agronomy a true science. To achieve this end, the Andalusian authors drew on the support of two more highly developed sciences: botany and pharmacology, on the one hand, and medicine on the other. The first of these two disciplines reached its peak in Andalusia in the

ʿUmdat al-ṭabīb fī maʿrifat al-nabāt li-kull labīb (anonymous but written in the eleventh or twelfth century) (Asín Palacios 1940, 1943; Khaṭṭābī 1990; García Sánchez 1994), where we find an excellent attempt at a taxonomic classification of plants by genus (*jins*), species (*nawʿ*) and variety (*ṣanf*), which greatly surpasses the systems of classification in use amongst botanists since Aristotle and Theophrastus. Even if we find no explicit influence of the anonymous botanist (Abū al-Khayr al-Ishbīlī or al-Ṭignarī?) amongst Andalusian agronomists, it is clear that they were greatly interested in the question of the classification of vegetables: Ibn Baṣṣāl, for example, pointed out that grafting could only take place between plants of the same nature and therefore offered a scheme of classification of plants by families; similar efforts can be found in the work of Ibn al-ʿAwwām.

Medicine, like botany, seems to have been linked to agronomy right from the origins of this discipline in Andalusia: a treatise on agriculture has been attributed to Abū al-Qāsim al-Zahrāwī and, although this attribution has recently been disputed, it is undeniable that Ibn Wāfid and al-Ṭignarī were physicians. It is therefore not surprising that Andalusian agronomists developed a theory that seems closely linked to the humoral theory of Hippocrates and Galen. The four humours of the human body (yellow bile, black bile, phlegm and blood) are replaced by the four elements of Empedocles (earth, water, air and fire), the place of fire being given to fertilizer. Each of these four elements is associated with two qualities which are the same as those of classical tradition (the earth is cold and dry, water is cold and wet, and air is hot and wet), except in the case of fertilizer (hot and wet, unlike fire which is hot and dry). The humoral theory held that the human body was healthy when the four humours were in equilibrium and that illness arose from the imbalance of one humour in relation to the others. The same principle was applied in agriculture, where the system of complementarity between the elements of the remedy and the diseased body was also used. The Andalusian agronomists described in great detail the mixtures appropriate to each problem, justifying them theoretically according to the qualities of the soil. The latter, being cold and dry, could only become fruitful by receiving warmth (from the sun and air and from fertilizer) and moisture (from water). The agronomists developed a detailed classification of soils, and made serious efforts to promote the cultivation through human work alone of soils that had previously been considered unusable. Moreover, in the face of the classical tradition which rejected them, Andalusian agronomists stressed the value of black soils, rich in organic material. We also find realistic classifications of different types of water quality, together with descriptions of techniques for recovering, harnessing and using water (Glick 1970): *qanāt* (Oliver Asín 1959; Goblot 1979), wells and *norias* (*nāʿūra*) (Torres Balbas 1940; Caro Baroja 1954). The texts also stress the importance of ploughing, which enables the earth to be warmed by contact with the air, and the use of crop rotation techniques for the same reason. The latter include leaving the land fallow or systematically rotating crops, but fertilizer is the prime method: again there are attempts to classify the different types of fertilizer and detailed formulae for mixtures to suit the particular soil or crops in question.

Generally speaking, according to Lucie Bolens (1981), Andalusian agronomy achieved a high technical level which was not surpassed until the nineteenth century with the development of chemistry: it is interesting to note that between the end of the eighteenth and the middle of the nineteenth century, the work on agronomy by Ibn al-ʿAwwām was published in a Spanish translation and in a French version, not for learned but for utilitarian purposes, the techniques it describes for the development of agriculture in Spain and Algeria being of particular interest.

THE CENTURY OF PHILOSOPHERS

The eleventh century was without doubt the golden age of Andalusian science, but the century that followed marked the beginning of a slow decline. The attempts at political unification under the Almoravids (1091–1144) and then under the Almohads (1147–1232) were not always accompanied by the patronage of cultural activities, even though the most famous philosophers (Ibn Bājja, Ibn Ṭufayl and Ibn Rushd) were physicians to the Almohad caliphs and carried out research under their patronage. There was a long period under the Almohads when the influence of the *fuqahā*ʾ did not facilitate research in astronomy and moreover led to the birth of non-intellectual sentiments. Men of science frequently found themselves forced to leave: this was the case, notably, of the philosopher and physician Mūsā b. Maymūn (Maimonides), who lived in Egypt from 1166 until his death in 1204. There were others too, such as Abū al-Ṣalt Umayya al-Dānī (*c.* 1067–1134), whose rather unhappy stay in Egypt (1095–1112) led him to write scornful commentaries on the knowledge of Egyptian astronomers and physicians (de Prémare 1964–6). The arrival of the Almohads seems also to have caused the departure for the East of Abū Ḥāmid al-Gharnāṭī (1080–1169), an indefatigable traveller whose cosmographic treatise *al-Muʿrib ʿan baʿḍ ʿajāʾib al-Maghrib* should have read *al-Mashriq* in the title instead of *al-Maghrib*: the text contains a large amount of *mīqāt* materials which, unfortunately, does not relate to Andalusia but to Tabaristan.[37]

Some scientific developments of this period seem to have been a continuation of trends from the preceding century. From the tenth century,

Andalusian botany and pharmacology followed in the footsteps of Dioscorides but there were sometimes innovations: Ibn Buklārish, whose work belongs to the beginning of the century, wrote a treatise of pharmacology, the *Mustaʿīnī*, in which the medical material is set out in synoptic tables in the manner of Ibn Buṭlān and Ibn Jazla. Moreover, like Abū al-Qāsim al-Zahrāwī, he was interested in the problem derived from al-Kindī that was also treated by Ibn Rushd: how to calculate the 'degree' of a medicine composed of several simples having different qualities and 'degrees'.[38] However, in most cases, Andalusian pharmacology was concerned with problems already raised in the previous two centuries: Ibn Bājja, author of a list of addenda to the pharmacology of Ibn Wāfid, which seems to be lost, wrote on the classification of plants (Asín Palacios 1940); Maimonides, in his *Sharḥ asmāʾ al-ʿuqqār*, explored the problem of botanical terminology (Meyerhof 1940), which had been the point of departure for the work at Córdoba on the Arabic translation of Dioscorides, as well as the researches of Ibn Juljul. Other authors, such as al-Ghāfiqī (Meyerhof and Sobhy 1932–40) and Abū al-ʿAbbās al-Nabatī (*c.* 1166–1240) (Dietrich 1971), prepared the major work of synthesis that was completed in the following century by Ibn al-Bayṭār: these authors composed treatises on pharmacology of an encyclopedic nature, in which they sought to bring together Dioscorides, Ibn Juljul and the preceding traditions, while adding their personal contribution which related, of course, to plants existing on the Iberian Peninsula. It was also during this century that the major synthesis of Andalusian agronomy appeared: that of Ibn al-ʿAwwām.

The spirit of observation was thus not entirely absent from Andalusian science of the twelfth century, even in the most speculative minds, such as Ibn Rushd (1126–98), whose interest in the observation of nature has often been noted (Alonso 1940; Cruz Hernandez 1960, 1986), together with a certain originality in the presentation of anatomical elements in his *Kitāb al-Kulliyyāt* (*Colliget* in Latin versions), where he does not hesitate to correct his sources nor to employ arguments based on observation (*bi-l-ḥiss*).[39] In fact, he also seems to have been interested in elementary astronomical observations, such as the observation made at Marrakesh in 1153 of the star Suhayl (Canopus), which is invisible from the Iberian Peninsula: by means of a famous argument from Aristotle, he used this to deduce the sphericity of the earth (Gauthier 1948: 5). The observations of sunspots that are attributed to Ibn Rushd and Ibn Bājja are of greater interest; they were interpreted by these two authors as transits of Mercury and Venus in front of the sun (Sarton 1947; Sayili 1960: 184–5; Goldstein 1985d): this interpretation implies, on the part of these authors, a criticism of the positions of Ptolemy and of Jābir b. Aflaḥ on the problem – much discussed in Andalusia in the twelfth century – of the order of the planetary spheres.

Ptolemy had justified the absence of transits of Mercury and Venus in front of the sun by the fact that these two lower planets did not pass through the line between the eyes and the sun (*Almagest* IX, 1), and this was seriously disputed, with reason, by Jābir and by al-Biṭrūjī (Goldstein 1971: I, 123–5). But Jābir postulated a different order of the planetary spheres, placing Mercury and Venus above the sun: in addition to the absence of transits, his basic argument was that these two planets exhibit no observable parallax and therefore could not be closer to the earth than the sun.[40] Al-Biṭrūjī, in turn, proposed the order moon–Mercury–sun–Venus etc. and rejected the argument of the transits because he considered that Mercury (like Venus) had its own light, which implied that a transit would be invisible.

Andalusian astronomy in the twelfth century was divided between authors like Abū al-Ṣalt of Denia (*c.* 1067–1134), Ibn al-Kammād (*c.* 1100) and Ibn al-Hā'im (*c.* 1205), who followed the tradition of al-Zarqāllu, and those who were critical of Ptolemaic astronomy. The criticisms of Ptolemy were ultimately based on positions which were either Ptolemaic (as in the case of Jābir ibn Aflaḥ) or Aristotelian (Ibn Rushd, al-Biṭrūjī, etc.).

In the area of 'orthodox' astronomy, we shall begin with Abū al-Ṣalt of Denia, who wrote on the astrolabe and the equatorium. His work on the latter instrument is the third text of this type to have been preserved, following those of Ibn al-Samḥ and al-Zarqāllu: it seems to have been a development of al-Zarqāllu's equatorium, but the parameters used in the text are Ptolemaic (Kennedy *et al.* 1983: 481–9; Comes 1991: 139–57, 237–51). Ibn al-Kammād, for his part, is the author of some astronomical tables that have been very recently studied (Chabás and Goldstein 1994) and in which the solar tables at least clearly show the influence of al-Zarqāllu (Vernet 1979b; Toomer 1987). The *Zīj al-Kāmil fī al-Ta'ālīm* of Ibn al-Hā'im of Seville is a long collection of canons, without numerical tables, accompanied by meticulous geometrical demonstrations: the author emerges as a faithful disciple of al-Zarqāllu and provides a large quantity of new information about the work of the school of Toledo in the second half of the eleventh century (Samsó 1994b).

With regard to the criticisms of the *Almagest*, the *Islāḥ al-Majisṭī* of Jābir b. Aflaḥ (still unpublished) is probably a key work in the development of 'orthodox' astronomy in twelfth-century Andalusia (Swerdlow 1987; Hugonnard-Roche 1987). This is a book by a theoretician, who criticizes certain aspects of the *Almagest*, for example the fact that Ptolemy does not prove his bisection of the planetary eccentricity. The work also describes two instruments of observation which may herald the arrival of the *torquetum* (Lorch 1976), and he contributes to the European diffusion of the new trigonometry – already introduced into Andalusia by Ibn Mu'ādh in the preceding century – since he uses the 'rule of four quantities', the theorems of sine and cosine, and the 'theorem of Geber'. The *Islāḥ* was well known in Europe through the Latin translation of Gerard of Cremona and two Hebrew translations, and it was frequently cited from the fourteenth century onward: the trigonometry section is generally considered to have been the source of the *De triangulis* of Regiomontanus. However, the European 'exploitation' of this part of the work seems to go back even further, because around 1280 the astronomers of Alfonso X were using the series of trigonometric theorems set out by Jābir (Ausejo 1984). Also, the *Iṣlāḥ* had been introduced into Egypt in the twelfth century by Joseph ben Yehūdah ben Shamʿūn, a disciple of Maimonides, with whom he studied and revised the original work. The book was well known in Damascus in the thirteenth century, and Quṭb al-Dīn al-Shīrāzī (1236–1311) made a summary of it.

The relatively meagre development of mathematical astronomy – after the brilliance of the eleventh century – was in some way compensated for by the birth of a 'physical' astronomy, which does not seem to have been cultivated previously in Andalusia. This was a century dominated by Aristotelian philosophers, and scholars such as Ibn Rushd, Maimonides, Ibn Bājja and Ibn Ṭufayl dreamt of developing an astronomy which could be reconciled with the physics of Aristotle. For Aristotle there could only be three types of motion (centrifugal, centripetal and circular around a centre which, as far as astronomy was concerned, had to coincide with the earth): this implied the rejection of Ptolemaic astronomy based on eccentrics and epicycles, and the wish to return to a system of homocentric spheres. These ideas were accepted, with variations, by the four thinkers just mentioned but, although there are a certain number of indirect quotations which suggest that Ibn Bājja and Ibn Ṭufayl did devise 'physical' astronomical systems, we do not know the details of them: all we have are statements of principle. Ibn Rushd was well aware of the problem, and his case is particularly curious, because in his paraphrase (*Talkhīṣ*) of Aristotle's *Metaphysics*, written in 1174, he seems to accept the Ptolemaic astronomy that he rejected later (after 1186) in his major commentary (*Tafsīr*) on the same work (Sabra 1978, 1984; Carmody 1952). In the *Tafsīr* Ibn Rushd sets out the principles on which astronomical reform must be based (most of which will be adopted by al-Biṭrūjī) and confesses that, even though in his youth he had hoped to carry out the necessary research personally, he had to relinquish the idea because of his advanced age.

However, these thinkers who rejected Ptolemy because of his incompatibility with Aristotle were aware of the predictive capacity of the astronomy of the *Almagest*. Thus Maimonides, who was convinced that the Ptolemaic universe did not coincide with the real universe, also believed that human beings were incapable of achieving a true knowledge of the laws that govern the structure of the cosmos. That is why he used Ptolemaic astronomy, in a totally competent manner, in his book of the *Sanctification of the New Moon*, where he tackled a particularly difficult problem: to determine in advance the visibility of the new moon (Gandz *et al.* 1956). It seems clear that these philosophers knew Ptolemy: Ibn Bājja was able to calculate an eclipse of the moon (*kāna qad ʿarafa waqt kusūf al-badr bi-ṣināʿat al-taʿdil*),[41] and al-Biṭrūjī praised the precision and accuracy of the *Almagest* from which all the numerical parameters employed in his *Kitāb fī al-hayʾa* were derived.

Al-Biṭrūjī was the only representative of the Aristotelian school of twelfth-century Andalusia who succeeded in formulating an embryonic astronomical system in line with the homocentrism of Eudoxus,[42] in which he incorporated a large number of subsequent contributions from Ptolemy to al-Zarqāllu (Goldstein 1964a). He considered first that, if the origin of all celestial motion is the prime mover, situated in the ninth sphere, it is absurd to think that the prime mover transmits to lower spheres movements in opposite directions: a diurnal movement from east to west and a longitudinal movement from west to east. It is necessary to accept that the motion of the ninth sphere – the fastest, the strongest and the simplest of all the movements – is transmitted to the lower spheres, which become progressively slower the farther away they are from the prime mover. The precession of the sphere of fixed stars and the movements in longitude of the planetary spheres are a sort of 'slowing' or 'brake' (*taqṣīr, incurtatio*) which slows down the diurnal motion. Here al-Biṭrūjī set himself a problem that he was incapable of solving: the problem of the transmission of movement between the ninth sphere and the lower spheres. He tried to explain the phenomenon by means of two metaphors which are none the less interesting as attempts to assimilate terrestrial with celestial dynamics. Duhem was the first to draw attention to one of these metaphors and to note that it constituted a return by al-Biṭrūjī to the ancient theory of *impetus* from neo-Platonic dynamics, created in the sixth century by John Philoponus: just as an archer gives the arrow a 'violent tilt' (*al-mayl al-qasrī*) which continues to propel it when it is flying separately from its driving force, one can conceive of a transmission of movement between the celestial spheres even if they are separated from one another (Duhem 1906–13: II, 191). The second of the metaphors also has a neo-Platonic character and derives from the eastern philosopher Abū al-Barakāt al-Baghdādī (eleventh to twelfth century), whose work may have been introduced into Andalusia by Isaac, son of Abraham b. Ezra, who was his disciple in Baghdad: like Abū al-Barakāt, al-Biṭrūjī considered that the circular movement of the celestial spheres is due to the 'desire' (*shawq* in the terminology of al-Biṭrūjī) that each of these spheres experiences for the sphere immediately above and this desire is analogous to that felt by the four elements to occupy their natural place. However, each part of the lower sphere finds itself, at a particular moment, close to another part of the upper sphere and able to satisfy this desire only partially. For that reason, the lower sphere is set in motion and the circular motion results from the effort made by each of the parts to draw nearer to each of the parts of the upper sphere (Samsó 1980c, 1994).

The astronomical system of al-Biṭrūjī is thus founded on the belief that the sphere of the fixed stars moves most rapidly and the sphere of the moon most slowly. There is nothing entirely original in this concept, since Lucretius attributes similar ideas to Democritus, and Alexander of Aphrodisias to the Pythagoreans. Moreover, Martianus Capella (*De nuptiis* VIII, 853) tells us that the peripatetics believed that the planets do not move in the opposite direction to the motion of the celestial sphere but that this sphere overtakes them because it moves at a speed that the planetary spheres cannot achieve. The same ideas are put forward again by Theon of Alexandria and Ibn Rushd. The movement of the ninth sphere is also transmitted to the sublunar world where, in the sphere of fire, it leads to the appearance of shooting stars and, in the sphere of water, to tides and waves. This theory of al-Biṭrūjī on the origin of tides is quoted in the *Kitāb al-madd wa-l-jazr* attributed to Ibn al-Zayyāt al-Ṭādilī (d. 1230), a work which also includes an in-depth study of the daily, monthly and annual cycles of the tides.[43]

We have thus far stressed the physical bases of the system of al-Biṭrūjī. We cannot elaborate here on the details of his models for the sun, the moon, the fixed stars and the planets. It is sufficient to note generally that these are homocentric models in which the planets move at the end of an axis (an arc of circle of 90°) which, in turn, moves on an epicycle whose centre is on a polar deferent. It thus involves a systematic use of the geometric apparatus of Ptolemy, but with the eccentric deferents and the epicycles placed around the pole of the universe: al-Zarqāllu had employed similar solutions in his geometrical models for explaining the variations in the obliquity of the ecliptic. On the whole, the models of al-Biṭrūjī are sometimes ingenious but they do not achieve the precision of those used in the Ptolemaic tradition. Moreover, no tables were ever calculated with these new models. The purely qualitative system of al-Biṭrūjī is not always absolutely consistent with his own principles; he was greatly admired by scholastic philosophers[44] but he does not seem to have been taken seriously by astronomers.

A last point remains to be emphasized: we have seen that, even though the *Kitāb fī al-hay'a* of al-Biṭrūjī was considerably influenced by Aristotle, the physical principles which underlie it are not always in accord with this classical author, and we have been able to discern the influence of neo-Platonic dynamics. This may be due to the indirect influence of Ibn Bājja, the representative in Andalusia of this 'new' physics against Ibn Rushd, a leading defender of Aristotelian orthodoxy. Ibn Bājja seems to have known of the work of John Philoponus through the refutation of al-Fārābī, and one can also envisage the influence of Abū al-Barakāt al-Baghdādī. The ideas of Ibn Bājja are interesting in several respects: he investigated the movement produced by a magnet and the displacement of a weight on an inclined plane, and he showed remarkable intuition in his concept of the driving force, in which certain analogies have been found with the concept of inertia in Newtonian physics. Even though he does not appear to accept the theory of *impetus* and shows support for Aristotelian ideas with regard to 'violent motions', he goes against Aristotle in defending the possibility of 'natural motion' in a vacuum, since he accepts that a body (e.g. planets and fixed stars) can move in the void with finite velocity and needs a period of time t to cover a certain distance d. When motion takes place in a medium it suffers a retardation (*buṭ'*, *tarditas*) proportional to the density/viscosity (?, *quwwat al-ittiṣāl*) of the medium itself, which implies that it needs an extra time (Δt) to cover the same distance d. This new interpretation of Ibn Bājja's ideas (Lettinck, 1994) discards previous hypotheses (Moody 1952; Grant 1965, 1974) according to which the echoes of our author's ideas would have reached sixteenth-century Italian scholars such as Benedetti and Borro and exerted an indirect inflence on Galileo's Pisan dynamics. In spite of this restriction we must acknowledge that Ibn Bājja's theories reached medieval Europe through the great scholastic philosophers of the thirteenth century and they pushed the development of Dynamics in the correct direction: unlike Aristotle, both Ibn Bājja and al-Biṭrūjī have the obvious merit of conceiving a universal dynamics that can be applied to both the sublunar and supralunar world.

THE DECLINE (THIRTEENTH TO FIFTEENTH CENTURIES)

After the fall of the Almohad empire, Muslim Spain found itself reduced to the Nasrid kingdom of Granada (1232–1492),[45] and the decline that had become apparent during the previous period continued even more obviously. Muslim scholars who found themselves in territory conquered by the Christians mostly crossed the frontier either to settle in Granada or to emigrate to North Africa or the East. This was in spite of the policy adopted by Alfonso X (1252–84) to hold on to Muslim men of science after his conquest of Murcia in 1266: if Ibn al-Khaṭīb is to be believed, the king offered considerable compensation to those who converted to Christianity, and this was accepted by figures such as Bernardo el Arabigo, who collaborated in the revision of the Castilian version of the treatise of al-Zarqāllu on the *ṣafīḥa* (*azafea*), made at Burgos in 1278. A much more important physician and mathematician, Muḥammad al-Riqūṭī, refused the royal offer and left for the Granada of Muḥammad II (Samsó 1981). Thus there was no Muslim scientific development in Christian Spain, although we can find occasional exceptional situations: in the second half of the fifteenth century there was a *madrasa* at Saragossa where one could study medicine by reading, in Arabic obviously, the *Urjūza fī al-ṭibb* and the *Qānūn* of Ibn Sīnā (Ribera 1928b). Moreover, despite the limitations, there are documents showing a certain freedom of movement for Muslims, at least in the region of Valencia: some journeyed to Granada or crossed the Strait of Gibraltar to make pilgrimage or to travel for study, and there are also examples of Muslim travellers who arrived in Valencia from Granada or North Africa (Barceló 1984, esp. 102–4). These journeys sometimes had consequences for science: in 1450 a *faqīh* from Paterna introduced a new astronomical instrument to Valencia: the *sexagenarium*, which was used by astronomers in Cairo. This was a device from the equatoria family, with a 'planetary face' (which gave the mean motions of the planets) and a 'trigonometrical face' which contained a sine quadrant permitting the graphical solution of trigonometrical problems to determine planetary equations. The treatise describing the instrument was translated into Catalan, Italian(?) and Latin and is one of the last known cases of scientific transmission through Spain (Thorndike 1951; Poulle 1966).

However, as we have said, the men of science often tended to cross the border. In the thirteenth century, the great pharmacologist Ibn al-Bayṭār left for the Maghreb and Egypt and finally died in Damascus in 1248; the astronomer Muḥyī al-Dīn al-Maghribī was also probably of Andalusian origin but he worked in Syria and, later, at the observatory of Marāgha; a third notable case is that of the mathematician al-Qalaṣādī, who was born at Baza *c.* 1412 and died in Tunisia in 1486. There were also those who stayed in Granada, their only home base in the Peninsula. Certain distinguished monarchs offered them a welcoming reception, notably Muḥammad II (1273–1302), who attracted to his court al-Riqūṭī, to whom we have previously referred, and also the mathematician and astronomer Ibn al-Raqqām (d. 1315), who was of Andalusian origin and settled in Tunisia. The former originated an important school of medicine which gave rise to Muḥammad al-Shafra (d. 1360). Ibn al-Raqqām, in turn, instructed Abū Zakariyyā' b. Hudhayl in mathematics and astronomy, and taught the sultan Naṣr (1309–14) how to calculate almanacs and construct astronomical instruments. Among the illustrious patrons we must also mention

Yūsuf, the brother of Muḥammad II, who was a great lover of books on mathematics and astronomy but was forced to hide his interests from his father Muḥammad I, who disapproved of them.[46]

Nevertheless, the scientific development emerging in Christian Spain during the thirteenth century seems to have been reflected in Nasrid Granada, and there are indications of the beginning of the phenomenon that Garcia Ballester has termed 'reflux of scholasticism' (Garcia Ballester 1976: 21ff.): the introduction into Muslim Spain of a scientific culture developed in Christian Europe in the early Middle Ages using bases that came from the Arab world. This movement, which was later to have important consequences in North Africa, seems to have started here. We can cite, for example, the case of Muḥammad b. al-Ḥājj (d. 1314), who was born in Christian Seville and was praised by Ibn al-Khaṭīb for his knowledge of the language and culture of the *Rūm*. This figure, or his father,[47] a carpenter *mudéjar* of Seville, constructed the great noria at Fes *al-jadīda* for the Marinid sultan Abū Yūsuf (1258–86). This noria attracted the attention of Leo Africanus, who described it, indicating that it only turned twenty-four times a day(?): if this information is correct, it suggests the possibility of a clock set in motion by the noria, like the one built in China in the eleventh century by Su-Sung. On the death of Abū Yūsuf, Ibn al-Ḥājj returned to Granada where he was well received at the court of Muḥammad.

Even more interesting is the case of the surgeon Muḥammad al-Shafra (d. 1360), who was born in Crevillente (Alicante) when the town was already in the hands of the Christians, and who learned surgery 'from a large number of excellent practitioners of this manual art who were Christians', among whom was a certain master Baznad (Bernat?) of Valencia (Renaud 1935).

Within this ambiance, which were the scientific disciplines cultivated by the scholars of Granada? An initial answer to this question can be found in the information given in the *Iḥāṭa* of Ibn al-Khaṭīb (Puig 1983a,b; 1984): from Granada himself, Ibn al-Khaṭīb mentions forty-seven people who showed their interest in the sciences in the Kingdom of Banū Naṣr in the thirteenth and fourteenth centuries. In these forty-seven biographies the most frequent references are to medicine, followed by mathematics and astronomy. This finding corresponds fairly closely with reality, and even leaving aside medicine, the names of Ibn al-Bayṭār (1197–1248) and Ibn Luyūn (1282–1349) can be found in botany and agronomy. The former stands at the summit of Andalusian pharmacology, which had continued to develop since the tenth century: his *Jāmiʿ al-mufradāt* is the most complete treatise of applied botany produced in the Iberian Peninsula in the Middle Ages.[48] It describes 3,000 simples, listed in alphabetical order, and draws information from more than 150 authors, from Dioscorides to al-Ghāfiqī and Abū al-ʿAbbās al-Nabātī. It also includes personal observations by the author but these represent a small percentage of the overall compilation. Ibn al-Bayṭār thus corresponds simultaneously to the peak of this science and the beginning of a decline. The same cannot be said of the second figure, Ibn Luyūn, because the role of Ibn al-Bayṭār in the field of agronomy corresponds with that of Ibn al-ʿAwwām in the preceding century: a major synthesis had already been made, so the next task was to summarize it; and the agricultural *urjūza* of Ibn Luyūn is only an agronomical précis in verse without great interest.[49]

In mathematics there are only two names of note. The first is Ibn Badr, whose dates are uncertain but who seems to have lived in the twelfth or thirteenth century; he is the author of an elementary text of algebra in which he examines the solution of indeterminate equations (Sanchez Perez 1916). Much more important is the work of a writer on many subjects, al-Qalaṣādī (*c.* 1412–86), who is of particular interest because of his texts on arithmetic, algebra and rules of inheritance (*ʿilm al-farāʾiḍ*), which are still as a whole not well known. His *riḥla* with a view to accomplishing his pilgrimage enabled him to study at Tlemcen, Oran and Tunis as well as in the East, which explains the influence in his work of the mathematical treatises of Ibn al-Bannāʾ of Marrakesh (d. 1321) and his use of an algebraic symbolism already employed by various eastern mathematicians and in the Maghreb by the Moroccan Yaʿqūb b. Ayyūb (*c.* 1350) and the Algerian Ibn Qunfūdh (d. 1407).[50]

In the area of astronomy, we would again stress the Andalusian interest in the construction of instruments, and also the fact that contact with the East was not lost, even in this period of decline. Thus Ibn Arqam al-Numayrī (d. 1259) wrote on the linear astrolabe (*al-asṭurlāb al-khaṭṭī*), an instrument invented by the Persian astrolabe maker Sharaf al-Dīn al-Ṭūsī (d. 1213) (Puig 1983b); this same Ibn Arqam is also the author of the first of a series of treatises on the study of horses, a very fashionable discipline in Nasrid Granada.[51] In addition, in 1274, a certain Ḥusayn b. Aḥmad b. Bāṣ (or Mās) al-Islāmī wrote a long treatise on a universal plate, valid 'for all latitudes' (*li-jamīʿ al-ʿurūḍ*), which is identifiable with the tradition of the *azafea* of al-Zarqāllu and, at the same time, with that of the *ṣafīḥa āfāqiyya*, whose plates show the projection of several horizons. Ibn Bās can probably be identified as Ḥasan b. Muḥammad b. Bāṣo (d. 1316), an astronomer who became the chief *muwwāqitūn* at the great mosque of Granada. His son Aḥmad b. Ḥasan was also of the *muwaqqits* at the same mosque, and Ibn al-Khaṭīb praised these two figures for their skill in constructing astronomical instruments, particularly sundials (Renaud 1937; Samsó 1973; Calvo 1990, 1993, 1994). This information is interesting in two respects: on the one hand it constitutes the first clear evidence of the existence of *muwaqqitūn* in Andalusian mosques; on the other hand, the admiration expressed by Ibn al-Khaṭīb for the sundials constructed by Ibn Bāṣo is surprising in view of the poor quality of the instruments of this type known so far (King 1978a). It is very possible that thirteenth- and fourteenth-century Granada saw an important revival of the study of gnomonics and of its application to the construction of sundials: this hypothesis is confirmed by recently completed studies on the *Risāla fī ʿilm al-ẓilāl* of Ibn al-Raqqām (d. 1315), which show the great abilities of this mathematician and astronomer, who applied to the study of sundials analemmic methods not previously known in Andalusia (Carandell 1984a,b; 1988). Ibn al-Raqqām is also the author of astronomical tables (Vernet 1980b) in which he follows the tradition of al-Zarqāllu and Ibn al-Hā'im. These tables are now being studied, and there is every indication that research in depth on this scholar will reveal him as probably the most interesting figure of Nasrid science.

Ibn al-Raqqām is an exception, however. Andalusian science attained its peak in the eleventh century and could still present interesting results in the twelfth, but it did not survive the political decline and the long death throes of the Granadan Nasrids. Al-Qalaṣādī well understood it – as did many other men of science at the end of the thirteenth century – and he left for Ifrīqiyya shortly before the final crisis: his death, in 1486, was followed, six years later, by the end of the entire Andalusian culture.

NOTES

1 Recent overall studies by Vernet (1986, 1993), Samsó (1992, 1994a), Vernet and Samsó (1992).
2 See the work of Garcia Ballester (1976, 1984).
3 The development of these contributions can be traced through translations; see Vernet (1985).
4 Cf. Guichard (1977), who suggests that more Arabs were involved in the first waves of the invasion than is asserted in traditional Spanish historiography, but without changing the fundamental points at issue.
5 See Marin (1981). On the determination of the *qibla* in Andalusia, see King (1978a) and Samsó (1990).
6 See the edited translation of this text in Samsó (1983a).
7 For the techniques employed by the astrologers who followed 'the system of crosses', see Samsó (1979b, 1980d, 1985a) and Poch (1980). See also Castells (1992).
8 From the copious bibliography on this subject we shall limit ourselves to mentioning the recent edition by ʿAbd al-Raḥman Badawī: Ūrūsiyūs, *Tārīkh al-ʿālam*.
9 See Ibn Ḥajjāj for the recent edition of *Kitāb al-muqniʿ fī al-filāḥa*, studied by J. M. Carabaza, 'La edicion jordana de *al-Muqniʿ* de Ibn Ḥaŷŷaŷ. Problemas en torno a su autoría', *Al-Qanṭara* 11 (1990): 71–81.

10 The agronomy text considered to be the work of a Christian author is published in Lopez (1990b).
11 This process has been well described from the point of view of the history of Andalusian culture by Makkī (1961–4).
12 On education in Andalusia, see Ribera (1928b) and Muḥammad ʿAbd al-Ḥamīd ʿĪsā (1982).
13 See Ibn Ḥayyān, *Al-Muqtabas min anbāʾ ahl al-Andalus*, pp. 281–2.
14 Ibn Juljul, *Kitāb ṭabaqāt al-aṭibbāʾ wa al-ḥukamāʾ*, p. 42.
15 See Dozy and Pellat (1961), Martínez Gázquez and Samsó (1981) and Samsó and Martínez Gázquez (1981).
16 On this scholar, see Lopez (1990a).
17 On the Andalusian tradition of *mīqāt*, see also King (1978a), and on the specific problem of the visibility of the new moon, King (1987d).
18 See Hamarneh and Sonnedecker (1963). Concerning al-Kindī's theory of grades and his influence in medieval Europe, see the introduction by M. R. McVaugh to his edition of Arnald of Villanova's *De gradibus*.
19 al-Hāshimī, *Kitāb fī ʿilal al-zījāt*.
20 Ṣāʿid al-Andalusī, *Kitāb Ṭabaqāt al-Umam*, pp. 129–30.
21 See Marti and Viladrich (1981). We have recently had the opportunity of reading the Istanbul Carullah 1279 manuscript, containing the *Kitāb al-Hayʾa* of Qāsim b. Muṭarrif (*c.* 950) which gives a list of the distances and magnitudes of the planets that seems to derive indirectly from Ptolemy's *Planetary Hypotheses*.
22 Ptolemy, *Planisphaerium*.
23 See *Dictionary of Scientific Biography*, 'Ptolemy'.
24 See Comes (1991: 27–68); Poulle (1980a: I, 193–200); and Samsó (1983c), certain errors in which have been indicated by J. L. Mancha in *De Astronomia Regis Alphonsi*, Barcelona, 1987, pp. 117–23.
25 This part of the chapter is an updated summary of Vernet and Samsó (1981: 135–63). Cf. the more recent work of Richter-Bernburg (1987).
26 See Samsó and Catalá (1971–5), King (1974, 1987c, 1988). On the two *ṣafīḥa* of al-Zarqāllu, see Puig (1985, 1986, 1988).
27 The only manuscript has been published in facsimile by the Institute für Geschichte der Arabisch-Islamischen Wissenschaften of the University J. W. Goethe, Frankfurt, 1985. It is also possible that the true author of this *Kitāb al-anwāʾ* was a certain Muḥammad b. Aḥmad b. Sulaymān al-Tujībī and that Ibn ʿĀṣim was the author of a summary of the latter's book. See the partial edition, translation and commentary in Forcada (1993).
28 See the study by J. Vernet and M. Grau in the *Boletín de la Real Academia de Buenas Letras de Barcelona* 23 (1950): 261; 27 (1957–8): 257–8.
29 See *Dictionary of Scientific Biography*, 'Euclid'.
30 See Hartner (1974). It should also be pointed out that the deferent of Mercury in the equatorium of al-Zarqāllu is not a circle but an elipse: see Hartner (1978), Comes (1991: 114ff.) and Samsó and Mielgo (1994).
31 See Millás Vallicrosa (1943: 72–237), Boutelle (1967) and the very important account by Swerdlow in *Mathematical Reviews*, 41 (5149) (1971): 4.
32 See a summary of the matter, together with an up-to-date bibliography, in Vernet (1987) as well as in Vernet and Samsó (1992).

33 Bolens (1981). See the bibliography in that book and in Vernet and Samsó (1981, 1994). In the following we provide only a bibliographical update.

34 A good account of the question of the manuscript sources and their probable authors can be found in García Sánchez (1987a).

35 The attribution of a work on agronomy to this author has been much debated. This work is ascribed in two manuscripts to a certain Abū al-Qāsim b. ʿAbbās al-Nahrāwī, who is probably the celebrated physician and surgeon of the second half of the tenth century, Abū al-Qāsim Khalaf b. ʿAbbās al-Zahrāwī. Recent studies (Forcada, 1995) confirm the interest in agronomy towards the end of the tenth century.

36 See Banqueri, *Libro de Agricultura.*

37 See *Dictionary of Scientific Biography*, I, pp. 29–30, ʿAbū Ḥāmid'. See the edition and Spanish translation of the Muʿrib by Bejarano (1991).

38 On this author see Renaud (1930). See also the more recent work of M. Levey in *Studia Islamica*, 6 (1969): 98–104 and in *Journal for the History of Medicine*, 26 (1971): 413–21. M. Levey and S. S. Souryal have published an English translation of the prologue of *Mustaʿīnī*, containing all the theoretical part of the work, in *Janus*, 55 (1968): 134–66; A. Labarta has published an edited and annotated translation of the same prologue in *Estudios sobre Historia de la Ciencia Arabe*, Barcelona (1980): 181–316; on the sources of Ibn Buklārish, see A. Labarta in *Actas del IV Coloquio Hispano-Tunecino*, Madrid (1983): 163–74.

39 See Rodríguez Molero (1950). The theses of Rodríguez Molero have been discussed by Esteban Torres (1974). See also Ibn Rushd for the critical edition of *Kitāb al-Kulliyyāt.*

40 See Lorch (1975); see also *Dictionary of Scientific Biography*, ʿJābir ibn Aflaḥ', VII, pp. 37–9.

41 See al-Maqqarī, *Nafḥ al-ṭīb*, VII, p. 25.

42 See Kennedy in *Speculum*, 29 (1954): 248.

43 Edited and translated into Spanish by Martinez (1981); see also Martinez (1971).

44 See, for example, Cortabarria Beitia (1982) and Avi-Yonah (1985). See also al-Biṭrūjī for the Latin version of *De motibus celorum.*

45 See Arié (1973: 428–38) for a short survey of the sciences and medicine. See also Calvo (1992).

46 The most important general source for this period is the *Iḥāṭa* of Ibn al-Khaṭīb, the scientific data of which have been examined and analysed by Puig (1983a,b, 1984).

47 The text of Ibn al-Khaṭīb is not entirely clear. For the two interpretations cf. Colin (1933) and Puig (1983a,b).

48 See the French translation by L. Leclerc in *Notices et Extraits des Manuscrits de la Bibliothèque Nationale*, vols 23, 25 and 26 (Paris, 1877–83).

49 See the version edited and translated into Spanish by Eguaras (1975).

50 See Renaud (1944). On Qalaṣādī, see *Dictionary of Scientific Biography*, XI, pp. 229–30, and Souissi (1973).

51 See Colin (1934) and, for a more recent bibliography, Arié (1973) and Viguera (1977).

8

The heritage of Arabic science in Hebrew

BERNARD R. GOLDSTEIN

The medieval Hebrew scientific tradition that reflects the Greek heritage transmitted through Arabic sources began with a period of translations in the twelfth century, and was followed by further study and elaboration based on them. Though the main centres of activity were Spain and southern France, virtually all Jewish communities displayed some interest in the scientific disciplines. Indeed, poets, mystics, legal scholars, as well as philosophers, devoted considerable attention to scientific subjects (Goldstein 1979, 1985a).

Most of these Hebrew texts remain in manuscript form scattered in libraries all over the world, but a sufficient number are available to permit us to describe the character of this tradition. It is also worth noting that many Arabic texts were copied in Hebrew characters, a common practice among Arabic-speaking Jews, and, in some cases, this is their only surviving form. In contrast to literary texts, there are a large number of documents preserved in the Cairo Geniza, most of which were written for a particular occasion and discarded shortly thereafter. The Geniza was originally located in a room in the Cairo synagogue where documents were deposited for subsequent ritual burial, but in fact no such disposition took place, and over 200,000 documents ranging in date from the tenth to the nineteenth century were still there when this valuable collection was transferred to European and American libraries around the turn of the twentieth century. Among these documents are scientific texts representing all disciplines studied in the Middle Ages, for the most part in Arabic written in Hebrew characters, but also some in Arabic written in Arabic characters and some in Hebrew.[1]

The subjects most widely studied in the Jewish community were astronomy, mathematics and medicine, although various branches of physics and biology were also represented, as we learn from the compendious bibliographic studies of M. Steinschneider (1893) and E. Renan (1893) undertaken in the nineteenth century. In addition, most of the large European collections of manuscripts have been catalogued, greatly facilitating detailed examination of them. Among recent studies, we may note an article listing over 100 copies of various Hebrew versions of Avicenna's *Canon of Medicine*, the fundamental text for medical studies in the late Middle Ages (Richler 1982). Similarly, numerous copies survive of Euclid's *Elements* and of Ptolemy's *Almagest* translated from Arabic into Hebrew: these were the basic texts for the study of mathematics and astronomy in the Middle Ages.[2] However, in the subsequent discussion we shall limit our attention to astronomy.

Jews began to contribute to astronomy in Arabic early in the Islamic period, e.g. Māshā'allāh (d. *c.* 815) (Sezgin 1978: 127–9). In the twelfth century an interest in science arose among Jews in Christian countries whose literary language was Hebrew and for whom translations from Arabic were required. The first scholar to provide information for them in matters of astronomy and mathematics was Abraham bar Ḥiyya of Barcelona (twelfth century) (Millás Vallicrosa 1952). Generally, he paraphrased Arabic texts rather than translating them. So, for example, his astronomical tables are based on those of al-Battānī (d. 929), and he also relied on al-Battānī in his introduction to them (Millás Vallicrosa 1959). One of these tables is a list of fixed stars with their co-ordinates. To understand the significance of this list, we must go back to the Greek text of Ptolemy's *Almagest* (*c.* AD 140), where 1,025 stars are listed, that was translated into Arabic during the ninth century (Kunitzsch 1974). Al-Battānī excerpted about half this list and corrected the stellar positions in longitude for the precession from Ptolemy's time to his own. (Precession is the rate by which the longitudes of fixed stars increase over time, and this was already noted by Ptolemy; the other co-ordinate, called latitude, does not change). Bar Ḥiyya shortened the list even more, displaying only the stars of first and second magnitude, where magnitude is to be understood as a measure of a star's brightness.

As Ptolemy's star list was translated, copied and recopied, many errors crept in that seem quite puzzling, but a comparison of the surviving manuscripts in Greek, Arabic and Hebrew reveals the various stages in this transmission and leads to a resolution of most of the problems. For example, a star that in Ptolemy's catalogue is of fourth magnitude is listed by Bar Ḥiyya as of first magnitude, an error that goes back to a confusion between Greek alpha (which had the numerical value 1) and Greek delta (which had the numerical value 4) which were virtually indistinguishable in some hands. Bar Ḥiyya gives the Arabic name of each star (written in Hebrew characters) together with a Hebrew translation of it, a practice followed by many of his successors. From an analysis of the data both in Arabic and Hebrew, it is clear that this medieval tradition of fixed star names and positions was literary and not based on new or independent observations (Goldstein 1985b).

Another influential Arabic text that received much attention in Spain was al-Khwārizmī's astronomical tables. In this case the original ninth-century text is lost and one is forced to depend on a twelfth-century Latin version of a revised Spanish–Arabic version that dates from about the year 1000 (Suter 1914; Neugebauer 1962a). However, we also have a tenth-century Arabic commentary on the original version composed in Spain by Ibn al-Muthannā that is extant only in Hebrew and Latin. One of the Hebrew versions was written by Abraham ibn Ezra (Spain, d. 1167) and it is an important source of information on the early development of Islamic astronomy in the late eighth and early ninth centuries (Goldstein 1967a). It appears that the first astronomical tradition to reach the Arabs in the eighth century derived from Indian sources and that Greek astronomy arrived somewhat later. Ibn al-Muthannā's commentary is an attempt (not always successful) to explain a text that reflects Indian sources by the methods of the Greek tradition. In the introduction to his translation Ibn Ezra wrote (Goldstein 1967a: 149):

> a scholar more eminent than the others in the sciences of geometry and astronomy, whose name is Muḥammad b. Muthannā, composed a distinguished book for one of his relatives concerning the rules of planetary motion which apply to the tables of al-Khwārizmī, and he included short proofs and diagrams whose principles are taken from the *Almagest* . . . There is no difference between Ptolemy's rules for planetary motion and those of the Hindu scholar except in a few places. Where it occurs I will mention how the difference arises.

It is clear that Ibn Ezra was aware of this blending of traditions but his ability to sort out the differences between them was limited by his lack of independent access to the appropriate sources.

The leading Jewish philosopher of the twelfth century, Maimonides, wrote a treatise in Hebrew on the Jewish calendar that depends in part on the works of his Muslim predecessors, notably al-Battānī (Gandz *et al.* 1956). In addition, there are many allusions to astronomy and mathematics in his main philosophic work, *The Guide for the Perplexed*, that was translated from Arabic into Hebrew in his lifetime. Maimonides reports criticisms of Ptolemaic astronomy by Ibn Bājja (Spain, twelfth century) and Jābir ibn Aflaḥ (Spain, twelfth century) (Maimonides 1956: 164, 196). Maimonides adds his own criticisms of Ptolemaic astronomy based in part on the discussion of planetary distances by al-Qabīṣī (tenth century) and concludes that[3]

> Man's faculties are too deficient to comprehend even the general proof the heavens contain for the existence of Him who sets them in motion. It is in fact ignorance or a kind of madness to weary our minds with finding out things which are beyond our reach, without the means of approaching them.

In the thirteenth century a great many texts were translated from Arabic into Hebrew, mainly in southern France, for the use of Jewish scholars there who were ignorant of Arabic. The most prolific translator was Moshe ben Tibbon, a member of an illustrious family of translators that had emigrated from Spain to France in the twelfth century (Romano 1977). An example of his work is his Hebrew version of al-Biṭrūjī's *On the Principles of Astronomy*, composed *c.* 1200 and translated in 1259 (Goldstein 1971). Al-Biṭrūjī set himself the task of reconciling the homocentric planetary models of Aristotle with the eccentric and epicyclic models of Ptolemy. His idea was to consider a modified version of the Ptolemaic models on the surface of a sphere, rather than in the plane of the ecliptic, in order to avoid the criticisms raised by a number of Spanish–Muslim philosophers.

The solution offered by al-Biṭrūjī was itself subject to comment and criticism by Yahuda ben Solomon Kohen of Toledo in an encyclopedic work originally written in Arabic and translated into Hebrew by the author in 1247; by Levi ben Gerson (d. 1344) in his astronomical treatise written in Hebrew that forms Part 1 of Book 5 of his *magnum opus* in philosophy, *The Wars of the Lord*; and by Isaac Israeli of Toledo (*fl. c.* 1310) in his astronomical treatise in Hebrew, *The Foundation of the World* (*Yesod Olam*) (Goldstein 1971: vol. 1, pp. 40–4). In effect, this attempt to replace the Ptolemaic models was rejected because al-Biṭrūjī could not account for all the known astronomical phenomena, and because the Ptolemaic models were highly successful in predicting these events. Moshe ben Tibbon's translation is quite literal and devoid of commentary, and it depended on the formation of a technical vocabulary in Hebrew that did not exist before the twelfth century (Sarfatti 1968).

Due in large measure to the efforts of Moshe ben Tibbon, subsequent generations of Jewish scholars whose only literary language was Hebrew could make original scientific contributions relying on the antecedent Greek and Arabic traditions. Nevertheless, translations from Arabic into Hebrew continued in the fourteenth century, and Samuel ben Judah of Marseilles (d. after 1340), for example, produced a Hebrew version of Ibn Muʿādh's *Treatise on Twilight* written in Spanish in the eleventh century and not extant in the original Arabic (Goldstein 1977a). This treatise concerns an attempt to determine the height of the atmosphere by means of a measurement of the solar depression arc at daybreak or nightfall, where this is defined as the arc from the sun (below the horizon) to the horizon on a circle passing through the observer's zenith. By means of a clear geometric argument, Ibn Muʿādh concluded that the atmosphere reaches up to about 50 miles above the surface of the earth, a value cited by Torricelli in 1644. Samuel ben Judah also revised an earlier Hebrew version of *The Improved Version of the* Almagest (*Iṣlāḥ al-Majisṭī*) by Jābir ibn Aflaḥ, and Samuel tells us something about his motivation for working on this text (Berman 1967: 315):

> When I achieved a good understanding at that time of this honored science [astronomy] and all or nearly all of the other sciences, I realized from the words of Averroes in his book on this science that the good found in them was gleaned from the book of Ibn Aflaḥ ...

A comparison of *The Epitome of the* Almagest by Averroes (Spain, twelfth century) with Ibn Aflaḥ's book on astronomy demonstrates that Samuel ben Judah's assertion has considerable merit.

At about the same time, another translator, Kalonymos ben Kalonymos (Arles, d. after 1328), translated the Arabic version of Ptolemy's *Planetary Hypotheses* (Goldstein 1967b). This work is only partially extant in Greek, and Ptolemy's discussion of cosmic distances that played such an important role in medieval theory only survives in the Arabic and Hebrew versions. Ptolemy's theory assumes that the geometric model that serves to predict a planet's position also reflects the relative distances of that planet from the earth. He then constructed a set of nested planetary spheres with no empty spaces between them that fill the universe such that the outermost sphere, that of the fixed stars, lies at a distance of about 20,000 terrestrial radii.

The most original astronomer to write in Hebrew was Levi ben Gerson (1288–1344) who lived in Orange and occasionally visited nearby Avignon (Goldstein 1974, 1985c). He composed a long treatise on astronomy in which he argued that Ptolemy's models ought to produce agreement with his own observations of planetary phenomena and eclipses or be replaced by more suitable models. For the Ptolemaic tradition he relied heavily on al-Battānī, presumably in the Hebrew version of Abraham bar Ḥiyya. In Levi's *Astronomy* we find tables based on new models that fulfilled the requirements of having a sound philosophical basis and of agreeing with his own observations. Levi rejected the epicyclic model that Ptolemy used extensively, but accepted Ptolemy's equant model that received much criticism by a number of Muslim scholars including Ibn al-Haytham (eleventh century) and Naṣīr al-Dīn al-Ṭūsī (thirteenth century) (Pines 1964b; Ibn al-Haytham 1971; Kennedy 1966). There is no indication that Levi was aware of the important astronomical research being carried out by contemporary Muslim scholars in the eastern Islamic world. Levi was also responsible for a modification of the astrolabe, an instrument widely known in the Islamic world that is used for making observations as well as for transforming coordinates (Goldstein 1977b). This modification involved adding a transversal scale on the rim to allow finer angular subdivisions to be displayed. The transversal scale on an arc of a circle was later used by Tycho Brahe (sixteenth century) on his precise observational instruments (Raeder *et al.* 1946: 29–31). Levi was aware of certain defects in Ptolemy's lunar model, also noticed by Ibn al-Shāṭir (Damascus, fourteenth century), but their solutions were entirely different.[4]

Emmanuel Bonfils of Tarascon (*fl. c.* 1360), who lived a generation after Levi ben Gerson, mentions his debt to Muslim astronomers, notably al-Battānī (Goldstein 1978). His popular tables for the sun and the moon, *The Six Wings*, were even translated from Hebrew into Latin and Byzantine Greek. It is perhaps surprising that he preferred the tables of al-Battānī that depend on Ptolemy's models to those of Levi ben Gerson whose work he also cites.

The impact of science from the eastern Islamic world in the late Middle Ages was also felt. For example, Shelomo ben Eliyahu of Saloniki (*fl. c.* 1380) translated a text called the *Persian Tables* from Byzantine Greek into Hebrew whose ultimate sources lie in the Islamic world (Goldstein 1979: 36). Another Hebrew text (Vatican, MS 381) contains tables that are identical with those in an anonymous Arabic text known from a number of copies (e.g. Paris, Bibliothèque Nationale, MS Ar. 2428).[5] This text uses the year 600 of the Persian era (that corresponds to the year AD 1231) as its radix or starting point, and so it presumably dates from the thirteenth century in the eastern Islamic world. The history of this text in Arabic and Hebrew (and also in Byzantine Greek) awaits further analysis, and for the moment it is not possible to say who the Hebrew translator was, when he lived or where he worked.

A copy from about 1500 (probably written in the vicinity of Venice) of an anonymous Hebrew version of Ulugh Beg's tables also exists among the manuscripts in the Bibliothèque Nationale.[6] This version of a text composed in the mid-fifteenth century is of special interest, for it suggests the possibility that aspects of eastern Islamic astronomy, perhaps even the lunar and planetary models of Ibn al-Shāṭir, may have reached European astronomers via Hebrew intermediaries. So far, the similarities between Ibn al-Shāṭir and Copernicus have been noted, but no route of transmission has been established (Rosińska 1974). The tables of Ulugh Beg are also mentioned in a supplement to a Hebrew prayer book published in Venice in 1520 (Goldstein 1974: 75). A nineteenth-century Arabic copy of Ibn al-Shāṭir's tables written in Hebrew characters in Aleppo, Syria, has also been identified, another indication of the impact of eastern Islamic science on the Jewish community (Goldstein 1979: 38).

Yemenite Jewish scholars were heavily indebted to Muslim scientists, and a number of copies of Arabic texts in Hebrew characters written in Yemen have been found. Included among them are Jābir ibn Aflaḥ's text on astronomy written in twelfth-century Spain and Kushyār ibn Labbān's astronomical tables written in eleventh-century Iran, i.e. Yemenite Jews had access to scientific traditions from diverse regions of the Islamic world.[7]

A number of Jewish scientists, but not all, accepted astrology as a proper scientific discipline and wrote treatises that were widely cited. Abraham ibn Ezra was perhaps the best known expositor of astrology in Hebrew and he depended in large measure on Arabic sources. He also translated into Hebrew an Arabic astrological treatise, Māshā'allāh's *Book of Eclipses*, that includes a discussion of astrological history, i.e. a theory in which historical periods correspond to the time intervals between planetary conjunctions (Goldstein 1964b). Among the opponents of astrology was Maimonides who wrote a polemical work attacking it as inconsistent with both science and religion (Twersky 1972: 463–73).

An important group of astrological texts consisting of almanacs and horoscopes in Arabic (some in Arabic script, others in Hebrew script) have been found among the documents of the Cairo Geniza. The almanacs, all from the twelfth century, are noteworthy in that they follow the Muslim calendar and refer to other calendars used in the medieval world, but not to the Jewish calendar. This suggests that they originated outside the Jewish community and hence tell us something about Muslim tastes in astrology as well as Jewish interest in the subject (Goldstein and Pingree 1981, 1983). An astronomical text from the Geniza that may have been composed with an astrological purpose in mind can be dated to 1299 (Goldstein and Pingree 1982). On the basis of internal evidence, the anonymous author of this Arabic document written in Hebrew characters depended on the astronomical tables of Ibn Yūnus (Cairo, *fl. c.* 1000) that were also popular among Muslim scholars. This text, though brief, is sufficiently detailed for us to notice numerous errors of different kinds that demonstrate the author's limited understanding of astronomy.

Scientific instruments were widely discussed by medieval Hebrew astronomers, and here again the influence of the Arabic tradition can be discerned. For example, al-Ḥadib (*fl. c.* 1400), of Spanish origin but who migrated to Sicily, wrote a description of an equatorium that he invented. Such instruments were designed to allow astronomers to find planetary positions without recourse to complex calculations using astronomical tables. Indeed, many clever adaptations of the planetary models were invented for this purpose, as we know from texts in Arabic, Latin and now Hebrew (Goldstein 1987). Al-Ḥadib cited unnamed Christian scholars as well as al-Zarqāllu (Spain, eleventh century), Ibn al-Raqqām (Tunisia, thirteenth century) and other Muslim scholars.

In sum, medieval Jewish scholars in many different countries, both in Christian Europe and in the Islamic world, depended on a legacy of Arabic science both in the original Arabic and in Hebrew translation. On the basis of this heritage they contributed to various scientific disciplines over the course of many centuries.

NOTES

1 On the Geniza, see Goitein (1967, vol. I, pp. 1–28).
2 Steinschneider (1893: 506, 523). A more complete list of manuscripts can be found at The Institute for Microfilmed Hebrew MSS, The National Library, Jerusalem.
3 Maimọnides (1956: 197–8). On al-Qabīṣī and Maimonides, see Goldstein (1980: 138).
4 On Ibn al-Shāṭir, see Kennedy and Ghanem (1976).
5 The Arabic version of this text is cited in Sezgin (1974: 324) under the name of Abū al-Wafā' although he is mentioned in the introduction, he is not the author. The Hebrew version has not been previously identified or even noted.
6 MS heb. 1091; cf. Goldstein (1979: 38).
7 Goldstein (1985b); on Kushyār see Sezgin (1974: 246); see also Langermann (1987).

9

The influence of Arabic astronomy in the medieval West

HENRI HUGONNARD-ROCHE

At the beginning of his *Epitome astronomiae Copernicanae*, Kepler lists the following components of astronomy, all of which he considers necessary to the science of celestial phenomena (Kepler 1953: 23). The astronomer's task, he says, consists of five main parts: historical, to do with the recording and classification of observations; optical, to do with the shaping of the hypotheses; physical, dealing with the causes underlying hypotheses; arithmetical, concerned with tables and computation; and mechanical, relating to instruments. The first three areas, adds Kepler, involve mainly theory; the last two are more concerned with practical aspects.

In each of the areas identified by Kepler, the contribution of Arabic science was essential to the birth and subsequent development of astronomy in the Latin West. Prior to this contribution, there was indeed no astronomy of any advanced level in those countries.[1] What was understood by astronomy was scarcely more than a collection of imprecise cosmological ideas concerning the shape and size of the world, and some basic notions about the movements of celestial bodies, principally concerning synodical phenomena, such as heliacal risings and settings. The needs of the Church with regard to the regulation of the calendar had nourished a tradition of chronological calculation following the *De temporum ratione* of Bede (d. 735). But this literature of computation, with which the names of Raban Maur, Dicuil or Garlande are associated, was not based on any mathematical treatment of the phenomena. A single example will suffice: in Bede, the planetary movements are represented by simple eccentrics, and the second planetary anomaly thus remains unexplained. In short, the science of the heavens in the early Middle Ages lacked observations, geometrical analysis of celestial phenomena and reflection on the foundations of hypotheses, in other words, the three areas that Kepler related to astronomical theory. Practical astronomy was no better represented: tables were inexistent and instruments (gnomons, sundials) were very basic.

This chapter obviously cannot detail, or even list, all the changes produced in the Latin West by successive translations of Arabic works, nor can it cite all the translations or all the medieval authors who may have been influenced by them.[2] We shall omit, among other things, Arabic influence on the development of trigonometry in the West, on instruments and on the Latin catalogues of stars,[3] as well as the considerable influence exerted on Latin astrology by treatises such as the *Introductorium maius* or the *De magnis coniunctionibus* of Abū Maʿshar (end of the ninth century).[4] This chapter will focus instead on the problems of astronomical theory proper, in order to reveal some essential aspects of Arabic influence on the growth and development of this theory in the medieval West.

THE ASTROLABE AND THE ASTRONOMY OF THE PRIME MOVER

The first evidence of the penetration of Arabic astronomy in the Latin West relates to the stereographic astrolabe. The properties and advantages of stereographic projection, on which this instrument was based, had already been described by Ptolemy in his *Planisphere*, but this text was not known in the Latin world until the twelfth century, through the translation by Hermann of Dalmatia (1143) of a critical Arabic revision of the text by Maslama al-Majrīṭī (*c*. 1000). However, scholars in the north of the Iberian peninsula, who were in contact with Islam, became familiar with this instrument and the treatises relating to it from the end of the tenth century. At this period, the first technical literature appeared in Latin under the names of Gerbert (the future Pope Sylvester II), Llobet of Barcelona and Hermann the Lame. This literature consists of texts describing applications or construction, or construction followed by applications, which are extracts or revisions of earlier Arabic treatises that have still not been clearly identified.[5] A new series of translations in the twelfth century, such as the translation of the treatise of Ibn al-Ṣaffār (d. AH 426 (AD 1035)) by Plato of Tivoli (*fl*. 1134–45), and various original Latin works, such as those of Adelard of Bath (*c*. 1142–6), Robert of Chester (1147) and Raymond of Marseilles (before 1141), gave the Latin West definitive mastery of the instrument. In addition, the inclusion of the astrolabe in university teaching programmes reinforced the educational role of this instrument until the end of the Middle Ages and ensured the success of the Latin translation by John of

Seville (*fl.* 1135–53) of a work attributed to Māshā'allāh (end of the eighth century).

The astrolabe was not only the educational instrument *par excellence* of the Middle Ages, but also an instrument of calculation, permitting the rapid geometrical solution of the principal problems of spherical astronomy. The astrolabe provided an easy demonstration of the daily and annual motions of the sun and of the combination of their effects, covering right and oblique ascensions, the duration of irregular hours, the heliacal rising of stars and the position of the celestial houses in astrology. Bearing in mind the traditional medieval division of astronomy into two distinct areas – the astronomy of the daily motion of the heavenly vault, on the one hand, i.e. the astronomy of the prime mover, and planetary astronomy, on the other – treatises on the astrolabe obviously dealt only with the first of these. Consequently, they contain few technical data: apart from the positions of stars, these are confined to the obliquity of the ecliptic, the location on the zodiac of the apogee of the sun and the position of the first point of Aries (spring equinox) in the calendar, which is associated with the movement of precession. Raymond of Marseilles's treatise on the astrolabe[6] – the oldest Latin text on the subject that is not a pure adaptation from the Arabic – contains two tables of stars, one drawn from the treatises of Llobet of Barcelona and Hermann the Lame, and the other derived from al-Zarqāllu (d. 1100). Raymond demonstrates a marked enthusiasm for this last author, and also borrows from him the position of the apogee of the sun at 17; 50° of Gemini and the value of the obliquity of the ecliptic estimated as 23; 33, 30°, which he prefers to that of Ptolemy (23; 50°). This example already enables us to identify two notable aspects of Arabic influence on Latin astronomy: the major role played by the work of al-Zarqāllu and the questioning of Ptolemaic parameters in relation to the sun.

THE TOLEDAN TABLES AND PLANETARY ASTRONOMY

By the time the treatise of the astrolabe had reached its definitive form, in the middle of the twelfth century, it was far from being the Latin world's only means of access to technical astronomy. A considerable collection of Arabic texts were translated in the course of that century, which opened up to Latin astronomers a much wider field of study in the form of astronomical tables. This designation covers a huge variety of material, which can be divided schematically into three groups: the first comprises elements relating more or less directly to the astronomy of the prime mover (tables of right and oblique ascensions, of declinations, of the equation of time);

the second comprises the planetary tables and is made up of four parts (chronological tables, tables of mean co-ordinates, tables of equations and tables of latitudes); the third group consists of disparate tables relating to conjunctions of the sun and moon, eclipses, parallaxes, the visibility of the moon and other planets, etc.

Three principal sources served to introduce the Latin astronomers to all these subjects: first, the canons and tables of al-Khwārizmī (*c.* 820), as revised by the Andalusian astronomer Maslama al-Majrīṭī and translated by Adelard of Bath (*c.* 1126); next, the tables of al-Battānī (d. AH 317 (AD 929)), first translated by Robert of Chester in a text that remains unfound, and then in a version by Plato of Tivoli, of which only the canons have been preserved;[7] lastly, the tables of al-Zarqāllu, which form the basis of the collection known as the tables of Toledo from their meridian of reference. Translated by Gerard of Cremona (d. 1187), the Toledan tables achieved widespread diffusion throughout the Latin West.[8]

One of the first Latin authors to use tables of Arabic origin was Raymond of Marseilles. In 1141 he composed a work on the motions of the planets, consisting of tables preceded by canons and an introduction in which he claims to draw on al-Zarqāllu. In fact, his tables are an adaptation of those of al-Zarqāllu to the Christian calendar and the longitude of Marseille. As in his treatise on the astrolabe, Raymond utilizes the value of 23; 33, 30° for the obliquity of the ecliptic, which he took from al-Zarqāllu. Furthermore, he is aware of the proper motion of the apogee of the sun as demonstrated by al-Zarqāllu and he reproduces the Arab astronomer's table for the positions of the apogees of the sun and other planets. Appearing some thirty years before the translations of Ptolemy's *Almagest* and the Toledan tables, by Gerard of Cremona,[9] Raymond's work was the first to introduce to the Latin world, through the perspective of a borrowing from al-Zarqāllu, the Ptolemaic method of calculating planetary positions (Figure 9.1), which consists in finding the algebraic sum of the mean motion, the equation of the centre and the equation of the argument, correcting the equation of the argument by means of proportional parts. From his study of the tables of al-Zarqāllu, Raymond of Marseilles understood, however, the clearly stated notion that astronomical tables demand continual correction. Astronomers throughout the Middle Ages found themselves faced with these corrections and the theoretical problems that they involved, and it was one of the aims of Copernicus finally to establish tables that would be permanently valid.

The adaptation of Arabic tables, and particularly the Toledan tables, continued in various parts of the Christian world throughout the twelfth and thirteenth centuries (Millás Vallicrosa 1943–50: 365–94). Thus one can cite tables for the meridian of Pisa compiled around 1145 by Abraham ibn Ezra, tables for the meridian of London in 1149–50 by Robert of Chester

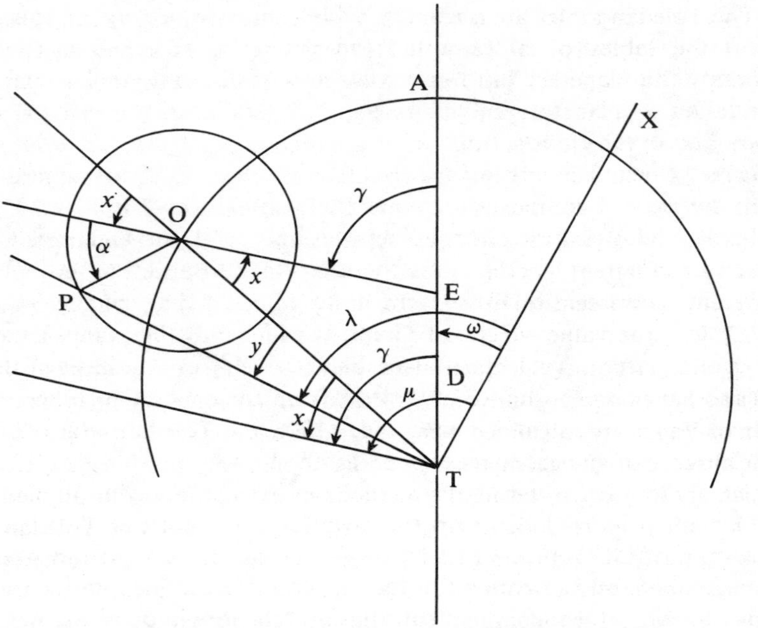

Figure 9.1 Ptolemaic theory of the motion of the planets in longitude (general case: upper planets and Venus). Medieval nomenclature: T, centre of the earth or the world; D, centre of the deferent; E, centre of the equant (TD = DE); O, centre of the epicycle; P, planet; X, origin of the co-ordinates on the ecliptic (first point of Aries); A, apogee on the ecliptic; ω, longitude of the apogee; μ, mean motion; γ, mean centre; α, mean argument; x, equation of the centre; y, equation of the argument; λ, true locus

and in 1178 by Roger of Hereford, and further anonymous tables, for London (1232), Malines, Novara, Cremona, etc. The tables for Toulouse seem to have been particularly well used, notably by Parisian astronomers, because of the proximity of the meridians of Toulouse and Paris. The large number of manuscripts of the Toledan tables dating from the fourteenth and even the fifteenth century testify to their continued use even after the Alphonsine tables had become the preferred source of astronomical reformers in Paris at the beginning of the fourteenth century. The Toledan tables also influenced the almanacs, which were designed not to provide the means of calculating planetary positions but to give the positions themselves. This is the case, for example, in the *Almanach* compiled for Montpellier for the years 1300 and following by Profatius (d. *c.* 1307), who himself records his debt to the Toledan tables.[10]

The Toledan tables are a composite collection, including the parts taken from the tables of al-Zarqāllu alongside extracts from al-Khwārizmī (notably the planetary latitudes), elements from al-Battānī (in particular, the tables of planetary equations), and yet other parts derived from the *Almagest* or the *Handy Tables* of Ptolemy, and from the *De motu octavae spherae*, which was attributed in the Middle Ages to Thābit ibn Qurra.[11] This diversity of composition means that the Toledan tables do not have a coherent underlying astronomical schema and that certain computations are based on different, incompatible parameters. For example, the tables of differences of ascension are calculated for an obliquity of the ecliptic equal to 23; 51°, the value which appears in the *Handy Tables*, whilst the table of right ascension is calculated with the value of 23; 35° used by al-Battānī. As another example, some of the columns which make up the table of equation of Venus are calculated using two different eccentricities for the planet. The absence of any geometrical analysis of planetary motions in the canons, which are limited to stating the methods of calculation to be applied, must have made it more difficult for the early Latin users of the Toledan tables to be critical of them, and they tacitly accepted the new parameters.

The characteristic features of the twelfth- and thirteenth-century Latin tables are therefore identical with those of the Toledan tables, and essentially reflect the modifications applied to Ptolemaic theory by the Arab astronomers of the ninth century. These modifications mainly affected solar parameters, whose definition by Ptolemy had proved very unsatisfactory. Observations made in the East in the ninth century – some 700 years after Ptolemy – had led to different estimations from Ptolemy's[12] for the length of the tropical year, for the speed of precessional movement, for the obliquity of the ecliptic (23; 33° according to the astronomers of al-Ma'mūn and 23; 35° according to al-Battānī, instead of 23; 51, 20° in the *Almagest*), for the eccentricity of the sun (2; 4, 45^{P} for al-Battānī, 2; 29, 30^{P} for Ptolemy) and for the position of the solar apogee (at 65; 30° from the first point of Aries according to Ptolemy, at 82; 17° according to al-Battānī, and 82; 45° according to the *De anno solis* attributed to Thābit ibn Qurra[13]). The discovery of these divergences between the results obtained by Ptolemy and their own findings confronted the Arab astronomers with a delicate problem which echoed on until the time of Copernicus: were these divergences due to errors of observation or to long-term variations in the parameters which would therefore indicate the existence of movements not so far observed? Both interpretations were put forward in the ninth century. The first was supported by al-Battānī, who did not question the kinematic models of Ptolemy, merely adopting a more rapid precessional movement than Ptolemy's (1° in 66 years rather than 1° in 100 years). The second interpretation was represented by the author of *De motu octavae spherae*, who postulated further that the presumed variations of the solar parameters were periodic: to account for this, he imagined a model[14] which produced simultaneously a periodic variation in precession, and thus in the length of the tropical year, and a periodic variation of the obliquity of the ecliptic. Briefly, this model consisted of two ecliptics: a fixed ecliptic inclined at 23; 33° to the equator, which it bisects at two points, the first point of Aries and the first point of Libra; these two points are taken as the centres of two small circles described by the first point of Aries and the first point of Libra of a moving ecliptic (but fixed in relation to the stars), which, in turn, bisects the equator at the equinoctial points. When the moving first point of Aries, which is the origin of the sidereal co-ordinates, completes a revolution on its small circle, the vernal point is drawn into an oscillatory motion on the equator. The parameters of the model were chosen to produce the maximum effect (distance between the first point of moving Aries and the vernal point) equal to ±10; 45°, and the periodicity of the oscillatory movement was 4163.3 Arabic years (4039.2 Christian years). The tables of *De motu* that correspond to this geometrical model were included without amendment in the Toledan tables, thus ensuring until the end of the thirteenth century the unchallenged success of this theory of the oscillatory motion of the equinoxes, known in medieval times as the motion of accession and recession (*accessio* and *recessio* translating the Arabic terms *iqbāl* and *idbār*).[15]

The calculation of planetary motions contained in the Toledan tables is based on three quantities: the mean motion and the two corrections, known as the equation of the centre and the equation of the argument. These two corrections are the translation into computational terms of the irregularities produced by the presence in the Ptolemaic geometrical models of eccentricities and epicycles. They are therefore a function, for each planet, of the eccentricity and of the relation between the radius of the epicycle and the radius of the deferent. It is remarkable that, although the mean co-ordinates given in the Toledan tables (mean motion of the upper planets and mean argument of the lower planets) appear to have been established independently of earlier known tables, the tables of equations are essentially the same as those of al-Battānī and derive from the *Handy Tables* of Ptolemy. The principal exception to the Ptolemaic origin of the tables of planetary equations is the table of equation of the centre of Venus, which is similar to al-Battānī's but completely different from that of the *Handy Tables*. The reason is that, in the table of al-Battānī, the centre of the epicycle of Venus was assumed to coincide with the mean sun and thus the eccentricity of Venus had to be the same as that of the sun. This was the concept – generally accepted by Arab astronomers, according to al-Bīrūnī (d. 1048) (Toomer 1968: 65) – that was duly adopted by the author of the Toledan tables.

With the exception of Venus, therefore, the preservation of equation tables of Ptolemaic origin indicates that the structure of the geometrical planetary models underlying the Toledan tables, and the Latin tables that derived from them, had remained the same since Ptolemy. By contrast, the setting of those models within the reference system of a solar theory associated with the theory of the motion of the fixed stars had involved a complete modification of the Ptolemaic concept. In fact, the Arab astronomers of the ninth century had shown that the position of the solar apogee is variable (in tropical co-ordinates) and had found a value for its movement similar to that for the precessional movement (1° in 66 years). They had therefore assumed that these two movements were identical, i.e. that the solar apogee was fixed, not in relation to the equinox (as Ptolemy had thought) but relative to the sphere of the stars. As a result, the sphere of the stars served as the reference for planetary motions from that time on. Thus, whereas the Ptolemaic tables had been expressed in tropical co-ordinates, the Toledan tables were expressed in sidereal co-ordinates. It was therefore only after having found the true positions of the planets on the sphere of the fixed stars (the eighth sphere in medieval terms), by algebraic summation of the mean motion and the equations, that the positions on the ninth sphere (or sphere of the stationary ecliptic) could be calculated by adding the equation of the motion of accession and recession, to take account of the motion of 'trepidation' of the stars, and consequently of the planetary apogees, in relation to the vernal point. This procedure, inherited from the Toledan tables, was constantly used in Latin astronomy until the end of the thirteenth century.

PLANETARY THEORIES AND THE GEOMETRICAL ANALYSIS OF APPEARANCES

Although the astronomical tables could satisfy the practising astronomer by enabling him to find the position of a celestial body in longitude and latitude at any particular moment, they did not provide any direct information in two of the areas defined by Kepler as constituting the theory of astronomy, i.e. the study of hypotheses and of their causes. These two areas of study developed in the Latin West in the thirteenth century, and once again Arabic influence had a considerable part to play in them. The development of this new field of research was made possible by the appearance of a new type of astronomical text, the *theoricae planetarum*, whose aim was to set forth kinematic models that would represent the celestial motions as faithfully as possible. Instead of the highly technical demonstrations in the

Almagest, Latin astronomers preferred more basic descriptions of the world system according to Ptolemy, as epitomized in two Arabic treatises: the introduction to Ptolemaic astronomy by al-Farghānī, entitled *Differentie scientie astrorum* in the translation of 1137 by John of Seville and *Liber de aggregationibus scientiae stellarum* in the translation by Gerard of Cremona; and second, an analogous treatise composed by Thābit ibn Qurra (d. AH 288 (AD 901)), also translated by Gerard of Cremona and known as *De hiis que indigent antequam legatur Almagesti.*[16] In the same way as these two treatises, the *theoricae planetarum* of the Latin Middle Ages usually restricted themselves to explaining basic astronomical concepts and the general organization of the circles used to represent planetary motions. A notable example of this approach is the most widely known of all the medieval *theoricae*, called the *Theorica planetarum Gerardi*,[17] whose author is unknown but which probably dates from the beginning of the thirteenth century. The geometrical models described in this *Theorica* conform to Ptolemaic constructs, with the exception of those concerning the erroneous determination of planetary stations by the tangents and the theory of planetary latitudes. On the second point, two traditions were known in the Middle Ages: the first was represented by the *Almagest* and followed by al-Battānī and an anonymous translation of the Toledan tables; the other tradition, derived from Indian methods, came into the West via the tables of al-Khwārizmī and the translation of the Toledan tables by Gerard of Cremona. Based on a model of the inclinations of the planes of the various circles representing planetary motions which differed from that of Ptolemy, this second method led naturally to different computational procedures from those in the *Almagest*. These are the procedures discussed in *Theorica Gerardi*, and that work was largely responsible for their dissemination until the beginning of the fourteenth century, at which time the Alfonsine tables restored the primacy of Ptolemaic methods.

A concise example of the medieval *theoricae*, the *Theorica planetarum Gerardi*, gave no indication of the parameters of the geometrical constructions, nor of the periods of revolution of their moving elements. A more elaborate *theorica*, the *Theorica planetarum* of Campanus of Novara (composed between 1261 and 1264), by contrast, combined a detailed theoretical exposé of the Plotemaic kinematics of planetary motions with a description of the appropriate equipment to represent those motions – the first Latin treatise on the equatorium. Included in university programmes during the fourteenth century, the *Theorica* of Campanus aided the widespread diffusion of ideas drawn from the work of al-Farghānī which was, after Ptolemy, its major source. Like al-Farghānī, Campanus augmented his summary of the *Almagest* with information concerning the system of celestial spheres: he completed the description of each planetary model with an evaluation of the dimensions of each part of the model. He was himself the author of astronomical tables for the town of Novara, which were based on the Toledan tables, and he took quite a lot of his parameters from the latter. Thus all the parameters of the planetary apogees were drawn from the Toledan tables, including the solar apogee, which is subject to precessional movement, as for the Arab astronomers. Equally, Campanus adopted the Toledan values for the mean motions of the upper planets and for the mean argument of Mercury, but he used the value from his own Novara tables for the mean argument of Venus. For the distances between station and apogee, he again followed the Toledan tables. Like them also, he adopted Ptolemaic parameters for the eccentricities and the magnitudes of the radii of the epicycles (except in the case of Mars where the difference is probably due to error).

With regard to the dimensions of the world, Campanus derived the basic elements of the comparative dimensions of the spheres of the earth, moon and sun from Ptolemy, and adopted the Ptolemaic principle of the contiguity of the celestial spheres which permits the calculation, step by step, of the relative dimensions of the planetary spheres up to Saturn and, from there, to the fixed stars. However, Campanus based all his estimations in absolute values on the evaluation of the length of a terrestrial degree of latitude ($56\frac{2}{3}$ miles) that he took from al-Farghānī and introduced into the Ptolemaic calculations of basic parameters (the diameters of the earth and sun, the distance from the earth to the sun, etc.). By also using the magnitudes of the planetary bodies provided by al-Farghānī, Campanus was able to calculate the dimensions of all the parts of the world system.

To summarize very broadly, we can say that medieval astronomy in the thirteenth century, as exemplified by the *Theoricae planetarum* of Campanus, was dominated by three major influences: the influence of Ptolemy on the geometrical models and their parameters; the influence of the Toledan tables on the mean co-ordinates of the moving elements in those models; and the influence of al-Farghānī (and through him the influence of Ptolemy's *Planetary Hypotheses*) on the cosmological constitution of the universe. Within this framework, two principal questions remained: the problem of the motion of the sphere of the stars, merely alluded to by Campanus in a reference to both the Ptolemaic movement of 1° in 100 years and the movement of accession and recession (not quantified) attributed to Thābit; and the question of the actual reality of Ptolemy's kinematic models.

THE PROBLEM OF THE FOUNDATION OF THE HYPOTHESES

At the same time as the Latin West discovered, through the *theoricae*, the Ptolemaic hypotheses implicit in the tables and their canons, they learned, through the translations of Michael Scot (d. *c.* 1236), of the commentaries of Averroës (d. 1198) in which those hypotheses were strongly criticized.[18] Aristotelian physics required that the celestial substance undergo no other movement than the uniform rotation of homocentric spheres. It was therefore easy for Averroës to show the contradictions between this physics and the astronomy of eccentrics and epicycles. Simultaneously with the radical criticism by Averroës, the Latin West acquired Michael Scot's 1217 translation of the *De motibus celorum* of al-Biṭrūjī (*c.* 1200), in which the author attempted to reformulate astronomy in accordance with the physics of Aristotle. In principle, the models of al-Biṭrūjī can be seen as a kind of reworking of the homocentric models of Eudoxus – accepted by Aristotle – with the innovation that the inclinations of the axes of the planetary spheres were made variable, the movement of each sphere being governed by that of its pole, which described a small epicycle in the neighbourhood of the pole of the equator.

The discovery of these texts initiated a lengthy medieval debate on the foundation of these hypotheses (Duhem 1913–59: 3, pp. 241–498 *passim*). As early as 1230 echoes of the work of al-Biṭrūjī – albeit still confused – could be found in the writings of William of Auvergne (1180–1249), and a little later in the work of Robert Grosseteste (1175–1253). Albertus Magnus (d. 1280), for his part, was fascinated by a very simplified model of the theory of al-Biṭrūjī, i.e. the attempt to explain all celestial appearances by means of a single driving force that would carry all the celestial bodies in a more or less rapid motion towards the west, which would account for their apparent proper motions towards the east. At the conclusion of his discussion, Albert rejects the criticism of Averroës concerning the eccentrics and epicycles, for the reason that celestial bodies differ from terrestrial bodies in matter and in form. He also rejects the astronomy of homocentric spheres, for 'this astronomy', he says, 'has not been completed by observation of the magnitude of the motions'. He thus gives prominence to the inability of this astronomy to account for appearances quantitatively, a failing that was constantly cited against the hypothesis of al-Biṭrūjī in the Middle Ages and which explains the indifference of astronomers toward it.

The doubts and criticisms concerning Ptolemy raised by the works of Averroës and al-Biṭrūjī, by contrast, prompted a deepening reflection on the status of astronomical theory and led to the appearance of theses which would be studied anew in the sixteenth century as part of the polemic between Ptolemaic and Copernican hypotheses. These theses were clearly articulated by Thomas Aquinas (1225–74), when he stated that the suppositions imagined by the astronomers were not necessarily true even if they seemed to explain appearances, for it may be possible to explain those appearances by some other process not yet conceived. Thomas thus contrasted two ways of explaining a phenomenon: sufficient proof of a principle from which the phenomenon follows, or the demonstration of agreement between the phenomenon and a principle advanced beforehand. Astronomy, according to Thomas, uses the second method, which suffices to explain the most obvious appearances.

In this debate between physics and astronomy – championed at the time of Simplicius by Aristotle and Ptolemy and revived in the guise of the opposition between Ptolemy and al-Biṭrūjī – certain Latin scholastics found the germ of a solution in the work of another Arab author: the treatise on the *Configuration of the World* attributed to Ibn al-Haytham (d. *c.* 1041), of which three anonymous Latin translations survive (one dated 1267).[19] The work is a cosmography without any mathematical treatment in which Ibn al-Haytham returns to the arrangement of solid orbs imagined by Ptolemy in his *Planetary Hypotheses*. Schematically, the sphere of each planet was seen as composed of an orb concentric with the earth into which there is fitted an eccentric orb containing the deferent and the epicycle: the two parts of the concentric orb, which are respectively interior and exterior to the eccentric orb, are of unequal thickness and function, as it were, to 'compensate' the eccentricity and to make the whole of the planetary sphere concentric with the world. Presented by Roger Bacon (d. 1294) in his *Opus tertium* as an *ymaginatio modernorum* created to avoid the difficulties of eccentrics and epicycles, this physical interpretation of Ptolemaic astronomy invalidates the objections of Averroës, according to the author. Conversely, the variations of planetary distances and the non-uniformity of their motions appeared to him to confirm the hypotheses of Ptolemy. This was also the opinion of numerous great medieval scholars, such as Bernard of Verdun, Richard of Middleton and Duns Scotus.

The inability of the system of al-Biṭrūjī to account for simple observations concerning, for example, the eccentricity of the planets – an inability again denounced at the end of the Middle Ages, by Regiomontanus – and, conversely, the ability of the *ymaginatio* inherited from Ibn al-Haytham to respond to the criticisms of Averroës, ensured the triumph of the Ptolemaic hypotheses and their physical interpretation by means of the orbs of Ibn al-Haytham. The most thorough exposition of this interpretation appeared at the end of the Middle Ages in the *Theoricae novae planetarum*, written in 1454 by Georg Peurbach: the description of the celestial orbs contained in this treatise served as an authoritative account of the structure of the heavens until Tycho Brahe (1546–1601) rejected the very existence of the celestial spheres.

THE PROBLEM OF PRECESSION AND THE ABANDONMENT OF THE TOLEDAN TABLES

The second major problem encountered by the medieval astronomers, that concerning the movement of precession, was more difficult to overcome. In his commentary (probably dated 1291) on Gerard of Cremona's translation of the canons of al-Zarqāllu regarding the Toledan tables, the Parisian astronomer John of Sicily[20] enumerated the various hypotheses that he knew relating to precession: the uniform motion estimated by Ptolemy as 1° in 100 years and by al-Battānī as 1° in 66 years; the to-and-fro motion of 1° in 80 years and of 8° amplitude rejected by al-Battānī; and the movement of accession and recession of the *De motu octavae spherae* attributed to Thābit ibn Qurra. He rejected, for his part, the movement of accession and recession and adhered to the Ptolemaic concept of uniform motion, while regarding its exact magnitude as uncertain. In this respect, John of Sicily is representative of the mistrust of Parisian astronomers of the time regarding the theory of *De motu* and, more generally, the Toledan tables.

At the end of the thirteenth century, indeed, the divergence between the positions calculated from these tables or the Latin tables derived from them – notably the tables for Toulouse – and the observed positions of the celestial bodies had become inadmissible. Thus, on the basis of personal observations made to establish his *Almanach*, William of Saint-Cloud[21] estimated the difference between the positions of the moving apogees and those of the fixed apogees on the eighth sphere as 10; 13° for 1290 and 10; 15° for 1292. Noting that this difference was nearly 1° greater than the value which would have resulted from the calculation made according to the law of motion proposed in the *De motu octavae spherae*, he concluded that this law should be rejected, and he accepted that the movement of precession must be considered, at least provisionally, to be uniform at one minute per year (i.e. a value close to that obtained by al-Battānī). Concerning the mean motions of the planets, on the other hand, William supplied empirical corrections to the radices of the Toledan tables, adding or subtracting fixed quantities as follows: +1; 15° for Saturn, −1° for Jupiter, −3° for Mars and +0; 22° for the moon. These same corrections were also proposed by two other Parisian authors, Peter of Saint-Omer and G. Marchionis (Poulle 1980a: 205–9, 260–5) in their treatises concerning equatoria, written in 1294 and 1310 respectively. In addition, Peter of Saint-Omer evaluated the difference between the fixed apogees and the moving apogees at 10; 10°, by reference to the estimations of precessional motion by William of Saint-Cloud, which also seem likely to have inspired Profatius in his treatise on the equatorium written between 1300 and 1306. A collection of texts from the very end of the thirteenth century thus attests to the ending of the comprehensive influence of the Toledan tables: the astronomers of this era no longer considered them sufficient, and they rejected in particular the movement of accession and recession, preferring instead a uniform motion of precession.

The influence of these criticisms was, however, short-lived. At the beginning of the fourteenth century, Latin astronomy replaced the Toledan tables with the Alfonsine tables. The latter were drawn up in Spanish between 1252 and 1272 for Alfonso X of Castile, and only the original canons survive. However, the Latin version, which appeared in Paris around 1320, dominated tabular astronomy from then until the publication of the *De revolutionibus* of Copernicus in 1543. In the first known essay concerning the new astronomy, the *Expositio tabularum Alfonsi regis Castelle*,[22] written in 1321, John of Murs did not refer to planetary parameters, eccentricities and magnitudes of epicycles, but concentrated his study on the values given in the Alfonsine tables for the mean motion of the sun and the movement of the auges of the planets. It was the treatment of the movement of precession, in fact, that most clearly differentiated the Alfonsine tables from the earlier tables. As John of Murs said, they represented an attempt to reconcile the Ptolemaic theory of uniform precessional motion with the Arabic theory of the movement of accession and recession. According to the Alfonsine theory, the motion of the apogees and the stars was made up of two components: a uniform motion in the order of the signs, for which the period was 49,000 years (1° in just over 136 years) and a movement of accession and recession relative to the intersection of the zodiac and the equator, for which the period was 7000 years, with a maximum effect of 9°. The movement of accession and recession of *De motu* was thus conserved as the component that causes the velocity of precessional motion of the apogees and the stars to vary. In addition, this movement of precession was taken into account from the start of operations to compute the planetary positions and not, as in the Toledan tables, at the end when it was necessary to transpose the positions obtained on the sphere of the fixed stars into tropical co-ordinates. More generally, the Alfonsine tables were designed to give the true positions of the planets on the ninth sphere directly, i.e., in tropical co-ordinates.

As far as the planetary equations were concerned,[23] the Alfonsine astronomers made only slight modifications to the Toledan tables, except in the cases of the sun, Venus and Jupiter. The change in the maximal equation of the sun (and consequently in its equation table) arose from the tacit modification (nowhere explained in the canons) of the eccentricity of the sun, which varied from 2; 6^p in the Toledan tables (2; 30^p according to Ptolemy) to 2; 15^p in the Alfonsine tables. The eccentricity (of the deferent) of Venus being traditionally taken as half that of the sun – i.e. 1; 8^p for the Alfonsine astronomers (instead of 1; 15^p for Ptolemy and 1; 3^p in the Toledan tables) – the maximal equation of Venus and the corresponding table of equation were similarly modified. Finally, in the case of Jupiter, the increase in the maximal equation, which changed from 5; 15^p in the Ptolemaic and Toledan tables to 5; 57^p in the Alfonsine tables, corresponded to an increase in the eccentricity from 2; 45^p to 3; 7^p. With regard to the radii of the epicycles, by contrast, parameters derived (by modern computation) from the tabulated values of the equation of the argument show that the Alfonsine tables were based on similar values to those used for the Toledan and Ptolemaic tables.

In short, the new hypotheses did not change the structure of the Ptolemaic planetary models, except as far as the eccentricity of the sun and of Venus and Jupiter were concerned. Once again it was the theory of motion of the sun and the directly linked theory of the movement of the fixed stars that were the essential subject of modification. On this point, the concepts of the *De motu octavae spherae* again played a key role: although they no longer served to describe the actual motion of the equinoxes, they served to describe variations in the velocity of that motion.

THE COPERNICAN REVOLUTION AND ARABIC ASTRONOMY

Once the astronomical tables had been updated by the Alfonsine reforms, the attention of the leading astronomers of the late Middle Ages turned to the analysis of Ptolemy's kinematic models. This was the task, in particular, of the *Theoricae novae planetarum* of Peurbach and the *Epitome in Almagestum Ptolemaei*, started by Peurbach and completed by Regiomontanus (d. 1476). The latter work, which contained a highly detailed analysis of Ptolemy's treatise, was the principal source for Copernicus concerning the results obtained by the Arab astronomers, notably al-Battānī and al-Zarqāllu. In the former work Copernicus could become familiar with the constitution of the solid spheres, as inherited from Ptolemy's *Hypotheses* and Ibn al-Haytham's *Configuration of the World*. There too he could read the description of the movement of accession and recession according to the *De motu octavae spherae* in a chapter on this subject added by Peurbach after his original draft. He could discover there also the representation of the deferent of Mercury as an oval figure, the first mention of which occurs in the treatise on the equatorium of al-Zarqāllu, which had become known in the West through the Spanish translation in the *Libros del Saber* compiled for Alfonso X and which was probably Peurbach's ultimate source.[24]

The question of Arabic influence on Copernican texts[25] focuses on two groups of problems which relate, on the one hand, to the theory of precession and solar theory, and, on the other hand, to planetary theory. As we have seen, the problem of the motion of the sun and stars was the major stumbling block for Latin astronomers throughout the Middle Ages, and it is therefore not surprising that the prime merit ascribed to Copernicus by his disciple Rheticus was to have solved this problem.

The long medieval debate on the solar parameters (eccentricity, position of the apogee and obliquity of the ecliptic) and on the precession or the trepidation of the equinoxes appeared in a new light in the Copernican system, once the earth was seen as responsible not only for the diurnal revolution but also the annual revolution and even, through the motion of its axis, for the westward slide of the equinoxes with respect to the fixed stars and thus the difference in length of the sidereal and the tropical years. Taking into consideration, in his *Commentariolus*, the lengths of the tropical year given by Ptolemy, al-Battānī and the Alfonsine tables, and the corresponding values for precession obtained by the same sources, Copernicus concluded that in all cases the calculation gave a constant sidereal year of 365 days $6\frac{1}{6}$ hours. The model conceived in the *Commentariolus* to account for this result, i.e. the westward movement of the earth's axis accomplishing its revolution in a tropical year, while the great orb carrying the earth turned to the east in a sidereal year, still produced only a uniform precession, because Copernicus, by his own admission, had not at that date discovered the law of precessional motion. It none the less indicated that the sphere of stars is fixed, that the lines of planetary apsides are fixed with respect to it and that it is the motion of the earth's axis which displaces the equinox with respect to the ecliptic. It also demonstrated the return by Copernicus to the concepts of the Arab astronomers, for whom, since the time of Thābit ibn Qurra and al-Battānī, the sidereal year had been constant and the periods of planetary motion had been fixed with respect to the stars.

The analogy does not stop there. When he turned his attention, in the *De revolutionibus*, to a more accurate description of the inequalities in the motions of the earth, Copernicus carried out a historical assessment of the data obtained by his predecessors for the precession, the obliquity of the ecliptic, and the eccentricity and position of the solar apogee, and he took the results of al-Battānī and of al-Zarqāllu for the medieval period.[26] In view of the diversity of values that emerged, Copernicus found himself facing exactly the same problem as the Arab astronomers of the ninth century with their new data for the parameters in question: were the discrepancies in the findings due to error or to variations in the parameters over a long period? In other words, should certain values be rejected, or should they all be integrated in the laws of motion to be determined? On this point, Copernicus was inspired by the example of *De motu octavae spherae*. Like the author of that treatise, Copernicus assumed that the combined observations reflected periodic variations in the relevant motions, and he constructed a model which, like that of the *De motu*, combined a uniform sidereal year and a trepidation of the equinoxes. For Copernicus, however, the trepidation was not a simple one but was composed, as in the Alfonsine tables, of a secular term and a periodic term (having periods of 25,816 and 1717 years of 365 days respectively).

According to Copernicus, however, the variation of the degree of precession was insufficient to explain the variation in the length of the year. It was also necessary to incorporate two long-term inequalities which, according to his assessment, affected the motion of the sun, i.e. a decrease in the eccentricity and a non-uniform motion of the line of aspides. It was in the work of al-Zarqāllu that the Latin astronomers had first discovered the affirmation of the solar apogee's own (but uniform) motion and a clear distinction of the anomalous year confused until then with the tropical year (Ptolemy) or the sidereal year (Thābit, al-Battānī). It was from al-Zarqāllu too – through the intermediary of the *Epitome* of Regiomontanus – that Copernicus adopted[27] the mechanism designed to account for both the variation of the eccentricity (the period of which he assumed to be equal to that of the variation in obliquity of the ecliptic) and the inequality of motion of the line of the apsides: all that was required was to let the centre of the terrestrial orbit, i.e. the mean sun, move on a small circle around a point removed from the real sun by a distance equal to the mean eccentricity in the relevant period (3434 years of 365 days).

It may also be from al-Zarqāllu that Copernicus drew the principle for his model representing the concomitant variations of precession and obliquity of the ecliptic. In fact, al-Zarqāllu had succeeded in making these two variations independent of each other by using, in one case, an epicycle placed around the equinox to make the precession vary (following the method of the *De motu*), and in the other case, a polar epicycle (with its centre on a deferent concentric with the pole of the ecliptic) to make the obliquity of the ecliptic vary.[28] The method of polar epicycles was later generalized by al-Biṭrūjī, who employed it for all planetary motion but with the disastrous consequence that the latitude depended on the longitude (or more accurately, on the argument of the planet). Copernicus, in turn, took up this method of polar epicycles, as part of a complex solution permitted by the fact that these two variations of precession and obliquity could be treated as two perpendicular oscillations of the axis of the terrestrial equator: each of the two variations was then given a small polar circle of appropriate diameter, the earth's axis was made to oscillate back and forth along the diameters of these circles and the two oscillations were combined so as to occur in perpendicular planes and in the relevant periods. The technical procedure used by Copernicus to obtain each of the oscillations is described by Naṣīr al-Dīn al-Ṭūsī (1201–74) in his major treatise *al-Tadhkira fī ʿilm al-hayʾa*, and has consequently become known to modern scholars as the 'Ṭūsī Couple'. This procedure, used by Ṭūsī in planetary theory, thus leads us to the second group of problems relating to Arab influence on Copernican astronomy.

This set of problems is not concerned with the second planetary anomaly, which relates to proving the heliocentric theory, but with the first anomaly, which is explained in Ptolemaic theory by the uniform motion of the eccentric deferent around a point that is not its own centre but the centre of the equant. Such a movement had been strongly criticized as contrary to the principles of physics by Ibn al-Haytham and then by the astronomers associated with the observatory of Marāgha (founded by Hūlāgū in 1259), such as Naṣīr al-Dīn al-Ṭūsī, Mu'ayyad al-Dīn al-ʿUrḍī (d. 1266) and Quṭb al-Dīn al-Shīrāzī (1236–1311), as well as by the Damascene astronomer Ibn al-Shāṭir (1304–75).[29] The method employed by these scientists to avoid the difficulty consisted of breaking down the motion around the centre of the equant into two or more components which were circular motions and which controlled the direction and the distance of the centre of the epicycle in such a way that the centre was as close as possible to the position that it would have occupied in the Ptolemaic model. The Eastern astronomers used two technical procedures to achieve this end: the addition of epicycles to give the Ptolemaic effect of bisection of the eccentricity, and the 'Ṭūsī Couple'. This model permits a rectilinear motion to be produced from circular motions in the following manner (Figure 9.2): if two equal circles rotate around their respective centres D and F so that the circle of centre F revolves in the opposite direction and twice as fast as the circle of centre D, the point H (such that $\widehat{GFH} = -2\widehat{CDF}$) on the circumference of the circle of centre F describes with an oscillatory motion (or motion of libration in the terminology of Copernicus) the diameter AB of a large circle (with centre D and radius double that of the small circles). If this model is in plane, it produces a rectilinear oscillation of H. If it is drawn on a sphere, the diameter AB described by H will be an arc of large circle (provided that the oscillation is weak).

These two technical procedures, the 'Ṭūsī Couple' and the addition of epicycles, were put to work by Copernicus. He used the first, as we have seen, to account at one and the same time for the inequality of the precession and the variation in the obliquity of the ecliptic. For this he used not one, but two, Ṭūsī models, in such a way that the diameters described by

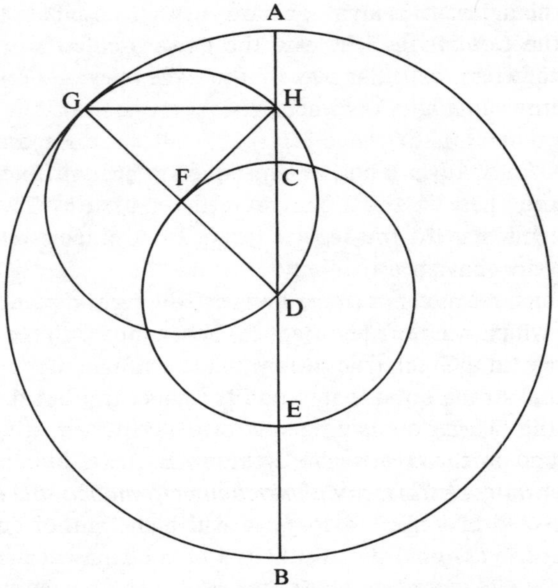

Figure 9.2 Copernicus, *De revolutionibus*, Nuremberg, 1543, fol. 67v

the two resulting oscillations are in perpendicular planes and intersect at the mean North pole to the equator (the radii of the circles and the speeds of rotation being chosen, of course, so that the two oscillatory motions have the necessary amplitude and periodicity). Copernicus also used the Ṭūsī model, as did the author of the *Tadhkira*, to account for the oscillations of the orbital planes in the theory of latitudes.

More striking still is the similar use made by Copernicus and Ibn al-Shāṭir (in his treatise *Nihāyat al-Sūl fī Taṣḥīḥ al-Usūl*) of the other procedure (the addition of epicycles) to represent planetary motion in longitude while avoiding the problems associated with the Ptolemaic equant. Thus, in the *Commentariolus*, all the planetary models are similar, with regard to the first anomaly, to those of Ibn al-Shāṭir in which the combination of a deferent and two epicycles is substituted for the movement of the deferent with respect to the centre of the equant. The only differences between the two authors lie in the values attached to the parameters and, of course, in the fact that the earth was at the centre of the planetary models for Ibn al-Shāṭir while it was the mean sun for Copernicus. A further similarity brings together the models of Copernicus and Ibn al-Shāṭir: both place a 'Ṭūsī Couple' at the tip of the deferent radius of Mercury in such a way as to vary the length of the orbital radius of this planet, by imposing at the centre of the first epicycle an oscillatory motion along a line directed always towards the centre of the deferent. A final similarity is the following: the model of the moon in the *Commentariolus* and the *De revolutionibus* is the same, except for parameters, as the model of Ibn al-Shāṭir.

These numerous analogies suggest that Copernicus was influenced by the Eastern astronomers of the thirteenth and fourteenth centuries. It is true that we do not know of any Latin translation of their works, nor even of any reference to them in the Latin literature of the late Middle Ages. However, it seems that the transmission of certain of these Arabic texts to the Latin West may have been achieved through the intermediary of Byzantine sources which reached Italy in the fifteenth century. Thus a manuscript (Vat. Gr. 211 which was in Italy by 1475) contains a treatise dealing with planetary theory (in a Greek translation, made around 1300 by Chioniades from the original Arabic), that contains Ṭūsī's lunar model and an illustration showing the 'Ṭūsī Couple'. Further evidence of the use of the 'Ṭūsī Couple' is found in the treatise of Giovanni Battista Amico entitled *De motibus corporum coelestium iuxta principia peripatetica sine excentricis et epicyclis*, published in Venice in 1536, in which the author endeavours to revive homocentric astronomy with the aid of models which are all based on the use of Ṭūsī's mechanism.[30]

THE END OF THE INFLUENCE OF ARABIC ASTRONOMY IN THE LATIN WEST

Copernicus marks the end of the long period of influence of Arabic astronomy in the Latin West. He was the last to make constant use of observational results taken from Arab authors, results which helped him to elaborate his estimations of the long-term variations in solar parameters. He was the last, also, to choose the thesis based on the *De motu octavae spherae*, which involved serious use of the collected observations of the past to formulate the laws of motion being sought, rather than using new observations to refute pre-existing theories. Remembering Kepler's three-way division of theoretical astronomy, we note that shortly after Copernicus, the abundant and accurate observations of Tycho Brahe made all reference to the history of ancient observations irrelevant. As for the Ptolemaic geometrical models and their Arabic or Latin variations, Kepler put an end to them. All that remained was the requirement to account physically for the phenomena, which Ibn al-Haytham and the Eastern astronomers of the thirteenth and fourteenth centuries had striven to do: nevertheless, after the refutation of the existence of solid spheres by Tycho Brahe, this requirement was no longer linked by Kepler with an Aristotelian vision of the world but rather with a vision inspired by a Platonic mathematical tradition.

NOTES

1 On the astronomy of the Middle Ages before the arrival of Arabic science in the West, see the synthesis and study by Pedersen (1975).

2 The most recent study of the transmission of Arabic science to the Latin world, with an extensive bibliography, is by Vernet (1985). Despite its age, Haskins (1927) remains useful. See also Carmody (1956).

3 On this last point, see Kunitzsch (1959, 1966).

4 See Lemay (1962). The doctrine of *De magnis coniunctionibus* (translated by John of Seville from *Kitāb al-qirānāt*) which exposes the effects of planetary combinations on the rise and fall of dynasties and earthly kingdoms exerted a persistent influence in the Middle Ages, whose traces can still be found in Rheticus, pp. 47–8, 98–9.

5 The classic study on this subject is in Millás Vallicrosa (1931). See also the work of synthesis entitled 'Las primeras traducciones científicas de origen oriental hasta mediados del siglo XII' in Millás Vallicrosa (1960: 79–115).

6 See the edition of this treatise by Poulle (1964) (with a list of existing editions of Latin treatises on the astrolabe, pp. 870–2). See also Poulle, 'Raymond of Marseilles', in *Dictionary of Scientific Biography*, XI, 1975, pp. 321–3.

7 There is no modern edition of Plato of Tivoli's translation, which was published in Nuremberg in 1537 under the title *De scientiis astrorum*.

8 There is no modern edition of the Toledan tables, but see the detailed analysis by Toomer (1968).

9 An annotated list of the Latin translations attributed to Gerard of Cremona can be found in Lemay, 'Gerard of Cremona', *Dictionary of Scientific Biography*, XV, 1978, pp. 173–92. For the Arabic–Latin tradition of the *Almagest*, see Kunitzsch (1974).

10 The planetary positions calculated from the Toledan tables do in fact coincide well with the values given by Profatius, as demonstrated by Toomer (1973).

11 The Arabic text of this treatise has not been found. The Latin version by Gerard of Cremona appears in Millás Vallicrosa (1943–50: 487–509) (reprinted in Millás Vallicrosa 1960: 191–209) and in Carmody (1960). The attribution of this work, which is definitely not by Thābit, is currently disputed: Millás Vallicrosa has rejected the attribution to al-Zarqāllu, supported by Duhem (1913–59: II, 246f); the attribution to Ibrāhīm b. Sinān, the grandson of Thābit b. Qurra, is supported by Ragep (1993: 400–08). An annotated translation can be found in Neugebauer (1962b).

12 Most of the values that follow are taken from Hartner, 'Al-Battānī', in *Dictionary of Scientific Biography*, I, 1970, pp. 507–16.

13 The Latin version of this treatise has been edited by Carmody (1960), who attributes it to Gerard of Cremona. This attribution is considered doubtful by Morelon, who also thinks that the original Arabic text came from the circle of the Banū Mūsā and not from Thābit: see Thābit ibn Qurra, pp. XLVI–LII.

14 On this model, and on theories of precession generally in the Middle Ages, see Mercier (1976–7), Goldstein (1964a).

15 Analysis of some texts relating to this tradition can be found, for example, in North (1976), vol. 3, pp. 238–70.

16 This translation is published in Carmody (1960). The original Arabic text, with French translation and commentary by Morelon, is in Thābit ibn Qurra.
17 See Gerardus. An English translation by Pedersen is published in Grant (1974: 451–65).
18 The passages of commentary on the treatises of Aristotle in which Averroës criticizes Ptolemaic astronomy are collected in Carmody (1952). On the criticism of Ptolemy by the Arab scholars of Spain, see Sabra (1984).
19 One of these translations, which seems to have been made from a Spanish version (now lost) compiled for Alfonso X, has been published by Millás Vallicrosa (1942: 285–312). On the astronomical concepts of Ibn al-Haytham, see Sabra (1978).
20 See Poulle, 'John of Sicily', in *Dictionary of Scientific Biography*, VII, 1973, pp. 141–2.
21 On this astronomer and the values quoted, see Poulle, 'William of Saint-Cloud', in *Dictionary of Scientific Biography*, XIV, 1976, pp. 389–91, and Poulle (1980a: 68, 209).
22 This important treatise has been published by Poulle (1980b). See also Poulle, 'John of Murs', in *Dictionary of Scientific Biography*, VII, 1973, pp. 128–33.
23 The information that follows has been taken from Poulle (1980a: 26–7, 767–9).
24 Concerning solid spheres and the representation of the deferent of Mercury according to Peurbach (and his Arabic sources), see Hartner (1955).
25 An overall survey of the influence of Arabic astronomy on Copernicus can be found in Swerdlow and Neugebauer, pp. 41–8. For the *Commentariolus*, see also Swerdlow (1973: *passim*.)
26 A good summary of this historical assessment and of the conclusions drawn by Copernicus can be found in Rheticus, pp. 94–8.
27 On the solar theory of al-Zarqāllu and its transmission to the Latin West, see Toomer (1969).
28 See Goldstein (1964a) and the same author's edition of al-Biṭrūjī, *On the Principles of Astronomy*.
29 From the extensive literature on this aspect of Arabic astronomy, we only mention here the studies directly concerned with the comparison of the Arabic and Copernican models: Kennedy (1966), Kennedy and Roberts, Hartner (1971).
30 These two references are taken from Swerdlow and Neugebauer, pp. 47–8. On Amico, see Swerdlow (1972).

Bibliography

Abū ʿAbd Allāh al-Khwārizmī al-Kātib, *Liber mafātīh al-olūm, explicans yocabula technica scientiarum tam arabum quam peregrinorum, auctore Abū Abdallah Mohammed ibn Ahmed ibn Jūsof al-Kātib al-Khowarezmi*, edited and indexed by G. Van Vloten, Lugduni Batavorum, 1895; reprinted, 1968, Leiden.

Albuquerque, L. de (1972) *Quelques Commentaires sur la Navigation Orientale*, Paris, Arquivos do Centro Cultural, Fondation C. Gulbenkian.

Allan, J. W. (1979) *Persian Metal Technology 700–1300 A.D.*, London/Oxford.

Alonso, M. (1940) 'Averroes observador de la naturaleza', *al-Andalus* 5: 215–30.

Arié, Rachel (1973) *L'Espagne Musulmane au temps des Nasrides (1232–1492)*, Paris.

Asín Palacios, Miguel (1940) 'Avempace botánico', *al-Andalus* 5: 255–99.

—— (1943) *Glosario de Voces Romances Registradas por un Botánico Anónimo Hispano-Musulmán (siglos XI–XII)*, Madrid/Granada.

Attié, B. (1972) 'L'origine d'*al-Falāḥa ar-Rūmīya* et du Pseudo-Qusṭūs', *Hespéris-Tamuda* 13: 139–81.

—— (1980) 'Ibn Ḥajjāj était-il polyglotte?', *al-Qantara* 1: 243–61.

—— (1981) 'La bibliographie de *al-Muqniʿ* de Ibn Ḥajjāj', *Hespéris-Tamuda* 19: 47–74.

—— (1982) 'L'ordre chronologique probable des sources directes d'Ibn al-ʿAwwām', *al-Qanṭara* 3: 299–332.

Aubin, G. (1972) *Quelques remarques sur l'étude de l'Océan Indien au XVIe siècle*, Coinbra.

Ausejo, E. (1984) 'Trigonometría y astronomía en el *Tratado del Cuadrante Sennero* (c. 1280)', *Dynamis* 4: 7–22.

d'Avezac, Macaya (1863) 'Coup d'œil historique sur la projection des cartes de géographie', *Bulletin de la Société de Géographie* V (5): 257–361, 438–85.

Avi-Yonah, R. S. (1985) 'Ptolemy vs. al-Bitruji. A study of scientific decision-making in the Middle Ages', *Archives Internationales d'Histoire des Sciences* 35: 124–47.

Bagrow, Leo (1951) *The Vasco Gama's Pilot*, Genoa.

—— (1955) 'A tale from the Bosphorus', *Imago Mundi* 12: 25–9.

Banqueri, J. A. (1988) *Libro de Agricultura*, Madrid, 1802; reprinted with a study by E. García Sánchez and J. E. Hernandez Bermejo, Madrid.

Barani, S. H. (1951) 'Muslim researches in geodesy', *Al-Bīrūnī Commemoration Volume*, Calcutta: Iran Society, pp. 1–52.

Barceló, M. C. (1984) *Minorías Islámicas en el País Valenciano. Historia y Dialecto*, Valencia.

Barceló, C. and Labarta, A. (1988) 'Ocho relojes de sol hispano-musulmanes', *al-Qanṭara* 9: 231–47.

Barmore, F. E. (1985) 'Turkish mosque orientation and the secular variation of the magnetic declination', *Journal of Near Eastern Studies* 44: 81–98.

al-Battānī (1977) *Al Battānī, sive Albatenii Opus Astronomicum* (*al-Zīj al-Ṣābī'*), Arabic edition, Latin translation and commentary by C. A. Nallino, 3 vols, Milan 1899–1907; reprinted in 1 vol., Hildesheim/New York.

Beeston, A. F. L. (1949) 'Idrisi's account of the British Isles', *Bulletin of the School of Oriental and African Studies* 13: 265–80.

Bejarano, I. (1991) *Abū Ḥāmid al-Garnāṭī (m. 565/1169)*, al-Muʿrib ʿan baʿḍ ʿaŷā'ib al-Magrib (*Elogio de algunas maravillas del Magrib*), Madrid.

Bel, A. (1905) 'Trouvailles archéologiques à Tlemcen: Un cadran solaire arabe', *Revue Africaine* 49: 228–31.

Berggren, J. L. (1980) 'A comparison of four analemmas for determining the azimuth of the *qibla*', *Journal for the History of Arabic Science* 4: 60–80.

—— (1981) 'On al-Bīrūnī's method of the *zījes* for the *qibla*', *Proceedings of the XVIth International Congress for the History of Science*, Bucharest, pp. 237–45.

—— (1982) 'Al-Bīrūnī on plane maps of the sphere', *Journal for the History of Arabic Science* 6: 47–96.

—— (1985) 'The origins of al-Bīrūnī's method of the *zījes* in the theory of sundials', *Centaurus* 28: 1–16.

Berggren, J. L. and Goldstein, B. R. (eds) (1987) 'From ancient omens to statistical mechanics: essays on the exact sciences presented to Asger Aaboe', *Acta Historica Scientiarum Naturalium et Medicinalium* 39, Copenhagen.

Berman, L. V. (1967) 'Greek into Hebrew: Samuel ben Judah of Marseilles, fourteenth-century philosopher and translator', in A. Altmann (ed.) *Jewish Medieval and Renaissance Studies*, Cambridge, Mass., pp. 289–320.

al-Bīrūnī, (1954–6) *Al-Qānūn al-Masʿūdī*, 3 vols, Hyderabad.

—— (1958) *Kitāb fī Taḥqīq mā li-l-Hind*, Hyderabad.

—— *Taḥdīd al-Amākin*, critical edition by P. G. Bulgakov, *Majallat al-Makhṭūṭāt al-ʿArabiyya*, Cairo, 1962; English translation by Jamil Ali (1967) *The Determination of the Coordinates of Cities*, Beirut. See also Kennedy (1973).

—— (1985) *Kitāb Maqālīd ʿIlm al-Hay'a. La Trigonométrie Sphérique chez les Arabes de l'Est à la fin du Xᵉ siècle*, edition, translation and commentary by M.-Th. Debarnot, Damascus.

al-Bitrūjī (1952) *De motibus celorum*, critical edition of the Latin translation of Michael Scot, by F. J. Carmody, Berkeley/Los Angeles.

—— (1971) *On the Principles of Astronomy*, an edition of the Arabic and Hebrew versions with translation and analysis by B. R. Goldstein, 2 vols, New Haven/London.

Björnbo, A. and Suter, H. (1924) *Thabits Werk über den Transversalensatz* (*Liber de figura sectore*), Erlangen.

Blachère, R. (1935) *Livre des Catégories des Nations*, vol. XXVIII, Paris.

Bolens, L. (1981) *Agronomes Andalous du Moyen Âge*, Geneva/Paris.

Boilot, D. J. (1955) 'L'œuvre d'al-Bērūnī: essai bibliographique', *Mélanges de l'Institut Dominicain d'Études Orientales* 2: 161–256.

Bonine, M. E. (1990) 'The sacred direction and city structure: A preliminary analysis of the Islamic cities of Morocco', *Muqarnas* 7: 50–72.

Boutelle, M. (1967) 'The Almanac of Azarquiel', *Centaurus* 12: 12–19.

Brice, W. C. (ed.) (1981) *An Historical Atlas of Islam*, Leiden.

Brice, W., Imber, C. and Lorch, R. (1976) 'The Dā'ire-yi Muʿaddal of Seydī ʿAlī Re'īs', *Seminar on Early Islamic Science*, 1, Manchester.

Brieux, A. and Maddison, F. (in collaboration with Ludvig Kalus and Yūsuf Rāghib) *Répertoire des Facteurs d'Astrolabes et de leurs Œuvres*, part 1: 'Islam plus Byzance, Arménie, Géorgie et Inde', 3 vols, Paris, in press.

Bruin, Fr. (1969) 'The Fakhri sextant in Rayy', *Al-Bīrūnī Newsletter* 19, April, Beirut, pp. 1–12.

Calvo, E. (1990) 'La lámina universal de ʿAlī b. Jalaf (s. XI) en la versión alfonsí y su evolución en instrumentos posteriores', *'Ochava Espera' 'Astrofísica', Textos y Estudios sobre las Fuentes Arabes de la Astronomía de Alfonso* X, Barcelona, pp. 221–38.

—— (1992) 'La ciencia en la Granada Nazarí (Ciencias exactas y tecnología', in Vernet and Samsó (1992), pp. 117–26.

—— (1993) *Abū ʿAlī al-Ḥusayn ibn Bāṣo (m. 716/1316), Risālat al-ṣafīḥa al-ŷāmiʿa li-ŷamīʿ al-ʿurūḍ (Tratado sobre la lámina general para todas las latitudes)*, Madrid.

Campanus (1971) *Campanus of Novara and Medieval Planetary Theory: Theorica Planetarum*, ed. with an introduction, English translation and commentary by F. S. Benjamin and G. J. Toomer, Madison.

Carabaza, J. M. (1988) 'Aḥmad b. Muḥammad b. Ḥaŷŷāŷ al-Ishbīlī. Introduccion, estudio y traduccion, con glosario', unpublished PhD thesis, University of Granada.

—— (1990) 'Un agrónomo del siglo XI: Abū-l-Jayr', in E. Garcia Sanchez (ed.) *Ciencias de la Naturaleza en al-Andalus*, vol. I, Granada, pp. 223–40.

Carandell, J. (1984a) 'An analemma for the determination of the azimuth of the *qibla* in the *Risāla fī ʿilm al-ẓilāl* of Ibn al-Raqqām', *Zeitschrift für Geschichte der Arabisch-Islamischen Wissenschaften* 1: 61–72.

—— (1984b) 'Trazado de las curvas de oración en los cuadrantes horizontales en la *Risāla fī ʿilm al-ẓilāl* de Ibn al-Raqqām', *Dynamis* 4: 23–32.

—— (1988) *Risāla fī ʿilm al-ẓilāl de Muḥammad ibn al-Raqqām al-Andalusī*, Barcelona.

—— (1989) 'Dos cuadrantes solares andalusíes de Medina Azara', *al-Qanṭara* 10: 329–42.

Carandell, J., Puig, R., Samsó, J., Vernet, J. and Viladrich, M. (1985) *Instrumentos Astronomicos en la España Medieval. Su Influencia en Europa. Convento de San Francisco, Santa Cruz de la Palma, Junio–Julio 1985*, Madrid.

Cárdenas, A. J. (1974) 'A study and edition of the Royal Scriptorium Manuscript of *El Libro del Saber de astrologia* by Alfonso X, el Sabio', Ph.D. thesis, 3 vols, University of Wisconsin (University Microfilms, Ann Arbor).

—— (1980) 'A new title for the Alfonsine Omnibus on Astronomical Instruments', *La Corónica* VIII (2): 172–8.

Carmody, Francis J. (1952) 'The planetary theory of Ibn Rushd', *Osiris* 10: 556–86.

—— (1956) *Arabic Astronomical and Astrological Sciences in Latin Translation. A Critical Bibliography*, Berkeley/Los Angeles.

—— (1960) *The Astronomical Works of Thābit b. Qurra*, Berkeley/Los Angeles.
Caro Baroja, J. (1954) 'Norias, azudas, aceñas', *Revista de Dialectología y Tradiciones Populares* 10: 29–160.
Carra de Vaux (1892) 'L'Almageste d'Abū-l-Wéfā' Albūzdjānī', *Journal Asiatique* 19: 408–71.
—— (1893) 'Les sphères célestes selon Nasīr Eddin-Attūsī', in Paul Tannery (ed.) *Recherches sur l'Histoire de l'Astronomie Ancienne*, Paris, App. VI, pp. 337–61.
Casanova, P. (1923) 'La montre du Sultan Nour ad-Din (554 de l'Hégire = 1159–1160)', *Syria* 4: 282–99.
Castells, M. (1992) 'Un nuevo dato sobre el *Libro de las Cruzes* en al-*Zīŷ al-muṣṭalaḥ* (obra astronómica egipcia del siglo XIII)', *al-Qanṭara* 13: 367–76.
Casulleras, J. (1993) 'Descripciones de un cuadrante solar atípico en el Occidente Musulmán', *al-Qanṭara* 14: 65–87.
Catala, M. A. (1965) 'Consideraciones sobre la tabla de coordenadas estelares', *al-Andalus* 30: 46–7.
Chabás, J. and Goldstein, B. R. (1994) 'Andalusian Astronomy: *al- Zīj al-Muqtabis* of Ibn al-Kammād', *Archive for History of Exact Sciences* 48: 1–41.
Chumovski, T. A. (1957) *Thalāth Rāḥmanajāt Majhūla li Aḥmad b. Mājid*, Arabic text and Russian translation, Moscow/Leningrad.
Colin, G. S. (1933) 'L'origine des norias de Fès', *Hespéris* 16: 156–7.
—— (1934) 'Un nouveau traité grenadin d'hippologie', *Islamica* 6: 332–7.
Comes, M. (1991) *Los Ecuatorios Andalusíes, Ibn al-Samḥ, al-Zarqālluh y Abū-l-Ṣalt*, Barcelona.
Copernicus, *De revolutionibus*, transl. by Charles Glenn Wallis, Chicago, 1952.
Cortabarria Beitia, A. (1982) 'Deux sources de S. Albert le Grand: al-Biṭrūjī et al-Battānī', *Mélanges de l'Institut Dominicain d'Études Orientales* 15: 31–52.
Cruz Hernandez, M. (1960) 'El pensamiento de Averroes y la posibilidad del nacimiento de la ciencia moderna', *Actas del XII Congresso Internazionale de Filosofia XI*, Florence, pp. 76–7.
—— (1986) *Abū-l-Walīd ibn Rushd: Vida, Obra, Pensamiento, Influencia*, Córdoba.
Dallal, A. (1984) 'Al-Bīrūnī on climates', *Archives Internationales d'Histoire des Sciences* 34: 3–18.
—— (1995a) *An Islamic Response to Greek Astronomy: Kitāb Taʿdīl Hayʾat al-Aflāk of Ṣadr al-Sharīʿa*, edition, translation and commentary, Leiden/New York/Cologne.
—— (1995b) 'Ibn al-Haytham's universal solution for finding the direction of the *qibla* by calculation', *Arabic Sciences and Philosophy* 5: 145–93.
Debarnot, M.-Th. (1978) 'Introduction du triangle polaire par Abū Naṣr b. ʿIrāq', *Journal for the History of Arabic Science* 2: 129–30.
—— (1985) *Al-Bīrūnī: Kitāb Maqālīd ʿIlm al-Hayʾa. La Trigonométrie Sphérique chez les Arabes de l'Est à la fin du Xᵉ siècle*, Damascus.
—— (1987) 'The zīj of Ḥabash al-Ḥāsib: a survey of MS Istanbul Yeni Cami 784/2', in King and Saliba (eds) *From Deferent to Equant*, New York, pp. 35–69.
Deetz, C. H. and Adams, O. S. (1945) *Elements of Map Projection*, 5th edn, Washington, U.S. Coast and Geodetic Survey; reprinted New York, 1969.

Dictionary of Scientific Biography (1970–80), 16 vols, New York.
Dietrich, A. (1971) 'Quelques observations sur la Matière Médicale de Dioscoride parmi les Arabes', *Oriente e Occidente nel Medioevo: Filosofia e Scienze*, Roma, Accademia dei Lincei, pp. 375–90.
Diophante (1984) *Les Arithmétiques*, vols III and IV, edition and translation of Arabic text by R. Rashed, Paris.
Dizer, M. (1977) 'The Dā'irat al-Muʿaddal in the Kandilli Observatory . . .', *Journal for the History of Arabic Science* 1: 257–62.
Djebbar, J. (1993) 'Deux mathématiciens peu connus de l'Espagne du XIe siècle: al-Mu'taman et Ibn Sayyid' in M. Folkerts and J. P. Hogendijk (eds), *Vestigia Mathematica. Studies in Medieval and Early Modern Mathematics in Honour of H.L.L. Busard*, Amsterdam-Atlanta GA, pp. 79–91.
Dodge, Bayard (1970) *The Fihrist of al-Nadīm. A Tenth-Century Survey of Muslim Culture*, Columbia Records of Civilization: Sources and Studies, no. *LXXXIII*, 2 vols, New York and London.
Doncel, M. G. (1982) 'Quadratic interpolations in Ibn Muʿādh', *Archives Internationales d'Histoire des Sciences* 32: 68–77.
Dozy, R. P. A. and de Goeje, M. J. (1866) *Abou-ʿAbdallah Moh. Édrisi, Description de l'Afrique et de l'Espagne*, Leiden; reprinted Amsterdam, 1969.
Dozy, R. P. A. and Pellat, Ch. (1961) *Le Calendrier de Cordoue*, Leiden.
Drecker, Joseph (1927) 'Das Planisphærium des Claudius Ptolemaeus', *Isis* 9: 255–78.
Dubler, C. E. and Terés, E.: *La* Materia Médica *de Dioscórides. Transmisión medieval y renacentista* (vol. I) Barcelona, 1953, (vol. II) Tetuán, 1952 and Barcelona, 1957.
Duhem, P. (1906–13) *Études sur Léonard de Vinci*, 3 vols, Paris.
—— (1913–59) *Le Système du Monde. Histoire des Doctrines Cosmologiques de Platon à Copernic*, 10 vols, Paris.
Eguaras, J. (1975) *Ibn Luyūn: Tratado de Agricultura*, Granada.
Eisler, R. (1949) 'The polar sighting tube', *Archives Internationales d'Histoire des Sciences* 6: 312–32.
El-Faiz, M. (1990) 'Contribution du *Livre de l'Agriculture Nabatéenne* à la formation de l'agronomie andalouse médiévale', *Ciencias de la Naturaleza en al-Andalus* I, Granada: 163–77.
The Encyclopaedia of Islam (1960–) 2nd edn, 8 vols, Leiden.
al-Farghānī: *Kitāb fī-l-ḥarakāt al-samāwiyya wa-jawāmiʿ ʿilm al-nujūm*, Arabic text edited by Golius, Amsterdam, 1669; the Latin version of John of Spain has been edited by F. J. Carmody, *Alfragani Differentie in quibusdam collectis scientie astrorum*, Berkeley, 1943; and that of Gerard of Cremona by R. Campani: Alfragano, *Il libro dell'aggregazione delle stelle*, Citta de Castello, 1910, 'Collezione di Opuscoli Danteschi inediti o rari, 87–90'.
Ferrand, G. (1921) *Instructions Nautiques et Routiers Arabes et Portugais des XVe et XVIe Siècles*, 3 vols, Paris, 1921–8; I and II: Arabic texts, III: *Introduction à l'Astronomie Nautique Arabe*.
—— (1924) *L'Élément Persan dans les Textes Nautiques Arabes*, Paris.

Fischer, Josef (1932) *Claudii Ptolemæi Geographiæ Codex Urbinus Græcus 82*, 3 vols, Leiden.

Forcada, M. (1993) *Ibn ʿĀṣim (m. 403/1013)*, Kitāb al-anwāʾ wa-l-azmina – al-qawl fī-l-šuhūr – (*Tratado sobre los anwāʾ y los tiempos – Capítulo sobre los meses–*), Madrid.

—— (1995) 'Ṣāʿid al-Badgādī y los antecedentes de la agronomía andalusí', *al-Qanṭara* 16: 163–71.

Gandz, S., Obermann, J. and Neugebauer, O. (1956) *The Code of Maimonides. Book Three. Treatise Eight. Sanctification of the New Moon*, New Haven/London.

Garbers, Karl (1936) 'Ein Werk Thābit b. Qurra's über ebene Sonnenuhren', *Quellen und Studien zur Geschichte der Mathematik, Astronomie und Physik* A (4): 1–80. See also Thābit, Treatise 9.

Garcia Ballester, Luis (1976) *Historia Social de la Medicina en la España de los Siglos XIII al XVI*, vol. I: *La Minoría Musulmana y Morisca*, Madrid.

—— (1984) *Los Moriscos y la Medicina. Un Capítulo de la Medicina y la Ciencia Marginadas en la España del Siglo XVI*, Barcelona.

García Sánchez, E. (1987a) 'Problemática en torno a la autoría de algunas obras agrónomicas andalusíes', *Homenaje al Prof. Darío Cabanelas*, Granada, vol. II, pp. 333–41.

—— (1987b) 'El tratado agrícola del granadino al-Ṭignarī', *Quaderni di Studi Arabi* 5–6: 278–91.

—— (1988) 'Al-Ṭignarī y su lugar de origen', *al-Qanṭara* 9: 1–11.

—— (1990) 'Agricultura y legislación islámica: el prólogo del *Kitāb Zuhrat al-Bustān* de al-Ṭignarī', *Ciencias de la Naturaleza en al-Andalus*, Granada, vol. I, pp. 179–93.

—— (1994) 'El Botánico Anónimo sevillano y su relación con la escuela agronómica andalusí', in E. García Sánchez (ed.) *Ciencias de la Naturaleza en al-Andalus. III. Textos y Estudios*, Granada, pp. 193–210.

Garijo, I. (1990) 'El tratado de Ibn Ŷulŷul sobre los medicamentos que no mencionó Dioscórides', in E. García Sánchez (ed.) *Ciencias de la Naturaleza en al-Andalus. Textos y Estudios I*, Granada, pp. 57–70.

—— (1992a): *Ibn Ŷulŷul. Tratado Octavo*, Córdoba.

—— (1992b): *Ibn Ŷulŷul. Tratado sobre los Medicamentos de la Tríaca*, Córdoba.

Gauthier, L. (1909) 'Une réforme du système astronomique de Ptolémée', *Journal Asiatique* 10 sér. 14: 483–510.

—— (1948) *Ibn Rochd (Averroès)*, Paris.

Gerardus (1942) *Theorica Planetarum Gerardi*, ed. by F. J. Carmody, Berkeley.

Glick, T. F. (1970) *Irrigation and Society in Medieval Valencia*, Cambridge, Mass.

Goblot, H. (1979) *Les 'Qanats'. Une technique d'acquisition de l'eau*, Paris/La Haye/New York.

Goitein, S. D. (1967) *A Mediterranean Society*, Berkeley/Los Angeles, vol. I.

Goldstein, B. R. (1964a) 'On the theory of trepidation according to Thābit b. Qurra and al-Zarqāllu and its implications for homocentric planetary theory', *Centaurus* 10: 232–47.

—— (1964b) 'The book of eclipses of Masha'allah', *Physis* 6: 205–13.

—— (1967a) *Ibn al-Muthannā's Commentary on the Astronomical Tables of al-Khwārizmī*, New Haven.

—— (1967b) 'The Arabic version of Ptolemy's *Planetary Hypotheses*', reproduction of the entire Arabic manuscript, which contains the second part of Book I, and a partial English translation, *Transactions of the American Philosophical Society*, N.S. 57/4: 3–55.

—— (1971) *Al-Biṭrūjī: On the Principles of Astronomy*, 2 vols, New Haven.

—— (1974) *The Astronomical Tables of Levi ben Gerson*, New Haven.

—— (1977a) 'Ibn Muʿādh's treatise on twilight and the height of the atmosphere', *Archive for History of Exact Sciences* 17: 97–118.

—— (1977b) 'Levi ben Gerson: on instrumental errors and the transversal scale', *Journal for the History of Astronomy* 8: 102–12.

—— (1978) 'The role of science in the Jewish community in fourteenth century France', *Annals of the New York Academy of Sciences* 314: 39–49.

—— (1979) 'The survival of Arabic astronomy in Hebrew', *Journal for the History of Arabic Science* 3: 31–9.

—— (1980) 'The status of models in ancient and medieval astronomy', *Centaurus* 24.

—— (1985a) 'Scientific traditions in late medieval Jewish communities', in G. Dahan (ed.) *Les Juifs devant l'Histoire: Mélanges en l'Honneur de M. Bernhard Blumenkranz*, Paris, pp. 235–47.

—— (1985b) 'Star lists in Hebrew', *Centaurus* 28: 185–208.

—— (1985c) *The Astronomy of Levi ben Gerson*, New York.

—— (1985d) 'Some medieval reports of Venus and Mercury transits', *Theory and Observation in Ancient and Medieval Astronomy*, London, Variorum Reprints, XV.

—— (1987) 'Descriptions of astronomical instruments in Hebrew', in David A. King and George Saliba (eds) *From Deferent to Equant, A Volume of Studies in Honor of E. S. Kennedy*, Annals of the New York Academy of Sciences 500: 105–41.

Goldstein, B. R. and Pingree, D. (1981) 'More horoscopes from the Cairo Geniza', *Proceedings of the American Philosophical Society* 125: 155–89.

—— (1982) 'Astronomical computations for 1299 from the Cairo Geniza', *Centaurus* 25: 303–18.

—— (1983) 'Additional astrological almanacs from the Cairo Geniza', *Journal of The American Oriental Society* 103: 673–90.

Grafton, A. (1973) 'Michael Maestlin's account of Copernican planetary theory', *Proceedings of the American Philosophical Society* 117: 523–50.

Grant, Edward (1965) 'Aristotle, Philoponus, Avempace and Galileo's Pisan dynamics', *Centaurus* 11: 79–95.

—— (1974) *A Source Book in Medieval Science*, Cambridge, Mass.

Grosset-Grange, H. (1973) 'Analyse des voyages d'Inde à Malacca', *Navigation* 81: 97–109.

—— (1976) 'Une carte nautique arabe', *Acta Geographica* 27: 33–46.

—— (1977) 'Noms d'étoiles, quelques termes particuliers', *Arabica* 1972: 240–5; 1977: 42–6; 1979: 90–8.

—— (1978) 'La côte africaine dans les routiers nautiques arabes', *Azania*, Nairobi, British Institute in Eastern Africa, pp. 1–17.

—— (1979) 'La science nautique arabe', *Jeune Marine*, 1977–9, 16–29 (except 22).

—— (1993) *Glossaire nautique arabe ancien et moderne de l'océan indien*, Paris.

Guichard, P. (1977) *Structures Sociales 'Orientales' et 'Occidentales' dans l'Espagne Musulmane*, Paris.

Haddad, F. I. and Kennedy, E. S. (1971) 'Geographical tables of medieval Islam', *Al-Abhath* 24: 87–102.

Haddad, F. I., Kennedy, E. S. and Pingree, D.: see al-Hāshimī.

Hairetdinova, N. G. (1986) 'On spherical trigonometry in the medieval Near East and in Europe', *Historia Mathematica* 13: 136–46.

Hamarneh, S. Kh. and Sonnedecker, G. (1963) *A Pharmaceutical View of Abulcasis (al-Zahrāwī) in Moorish Spain, with Special Reference to the 'Adhān'*, Leiden.

Hartner, W. (1955) 'The Mercury horoscope of Marcantonio Michiel of Venice: a study in the history of Renaissance astrology and astronomy', *Vistas in Astronomy* I: 84–138; reprinted with additions in W. Hartner (1968) *Oriens-Occidens*, Hildesheim, pp. 440–95.

—— (1971) 'Trepidation and planetary theories. Common features in late Islamic and early Renaissance astronomy', *Accad. Naz. dei Lincei, Fondazione Alessandro Volta, Atti dei Convegni* 13: 606–29.

—— (1974) 'Ptolemy, Azarquiel, Ibn al-Shātir and Copernicus on Mercury. A Study of Parameters', *Archives Internationales d'Histoire des Sciences* 24: 5–25.

—— (1978) *The Principle and Use of the Astrolabe*, reprinted, Paris, Société Internationale de l'Astrolabe (Astrolabica no. 1).

Hartner, W. and Schramm, M. (1963) 'Al-Bīrūnī and the theory of the solar apogee: an example of originality in Arabic science', *Scientific Change*, London, pp. 206–18.

al-Hāshimī, ʿAlī ibn Sulaymān, *The Book of the Reasons behind Astronomical Tables* (*Kitāb fī ʿilal al-zījāt*), reproduction of the manuscript Arabic text, translation and commentary by F. I. Haddad, D. Pingree and E. S. Kennedy, New York, 1981.

Haskins, Charles Homer (1927) *Studies in the History of Mediaeval Science*; reprinted, New York, 1960.

Hawkins, G. S. and King, D. A. (1982) 'On the orientation of the Kaʿba', *Journal of the History of Astronomy* 13: 102–9.

Hermelink, H. (1964) 'Tabulæ Jahen', *Archive for History of Exact Sciences* 2: 108–12.

Hoernerbach, Wilhelm (1938) *Deutschland und sein Nachbarlände nach der grossen Geographie des Idrīsī*, Stuttgart.

Hogendijk, J. P. (1986) 'Discovery of an 11th-century geometrical compilation: the *Istikmāl* of Yūsuf al-Mu'taman ibn Hūd, King of Saragossa', *Historia Mathematica* 13: 43–52.

—— (1989) 'The mathematical structure of two Islamic astronomical tables for "casting the rays"', *Centaurus* 32: 171–202.

—— (1991) 'The geometrical parts of the *Istikmāl* of Yūsuf al-Mu'taman ibn Hūd (11th century). An analytical table of contents', *Archives Internationales d'Histoire des Sciences* 41: 207–81.

—— (1995) 'Al-Mu'taman ibn Hūd, 11th century King of Saragossa and brilliant mathematician', *Historia Mathematica* 22: 1–18.

Holmyard, E. J. (1924) 'Maslama al-Majrīṭī and the Rutbatu'l-Ḥakīm', *Isis* 6: 293–305.

Honigmann, E. (1929) *Die sieben Klimata . . .*, Heidelberg.

Hugonnard-Roche, H. (1987) 'La théorie astronomique selon Jabir ibn Aflah (English abstract)', *History of Oriental Astronomy* (IAU Colloquium 91), Cambridge, pp. 207–8.

Ibn al-Ḥajjāj (1982) *Kitāb al-muqniʿ fī-l-filāḥa*, edited by Ṣalāḥ Jarrār and Jāsir Abū Ṣafya, Amman.

Ibn al-Ḥaytham (1971) *Al-Shukūk ʿalā Baṭlamiyūs*, edited by A. I. Sabra and N. Shehābī, Cairo.

Ibn Ḥayyān (1973) *Al-Muqtabas min anbāʾ ahl al-Andalus*, edited by M. ʿA. Makkī, Beirut.

Ibn Juljul (1955) *Kitāb ṭabaqāt al-aṭibbāʾ wa al-ḥukamāʾ*, edited by Fu'ād Sayyid, Cairo.

Ibn Luyūn (1975) *Tratado de Agricultura*, edition and Spanish translation by J. Eguaras, Granada.

Ibn Mājid (1971) *Kitāb al-Fawāʾid*, edited by I. Khoury, Damascus.

—— (1971) '*al-Ḥāwiya*, introduced and edited by I. Khoury', *Bulletin d'Études Orientales*, Damascus.

Ibn al-Nadīm (1874) *Al-fihrist*, edited by Flügel, Leipzig; English translation by Bayard Dodge, *The Fihrist of al-Nadīm. A Tenth-Century Survey of Muslim Culture*, 2 vols, New York/London, 1970, 'Columbia Records of Civilization, Sources and Studies no. 83'.

Ibn Rushd (1987) *Kitāb al-Kulliyyāt*, critical edition by J. M. Forneas and C. Alvarez Morales, Madrid.

Ibn al-Ṣalāḥ (1975) *Zur Kritik der Koordinatenüberlieferung im Sternkatalog des Almagest*, edition and translation by P. Kunitzsch, Göttingen.

Ibn Yūnus (1804) *Le Livre de la Grande Table Hakémite*, partially edited and translated into French by Caussin, separate edition of the 'Notices et extraits des manuscrits de la Bibliothèque Nationale', Paris, XII.

al-Idrīsī (1970–) *Opus Geographicum*, under the direction of the Istituto Orientali di Napoli, Leiden.

Irani, R. A. K. (1956) 'The *Jadwal al-Taqwīm* of Ḥabash al-Ḥāsib', unpublished Master's dissertation, American University of Beirut.

ʿIsā, Muḥammad ʿAbd al-Ḥamīd (1982) *Tārīkh al-taʿlīm fī al-Andalus*, Cairo.

Janin, L. (1971) 'Le cadran solaire de la Mosquée Umayyade à Damas', *Centaurus* 16: 285–98; reprinted in E. S. Kennedy and I. Ghanem (eds) (1976) *The Life and Work of Ibn al-Shāṭir: An Arab Astronomer of the Fourteenth Century*, Aleppo, pp. 107–21.

—— (1977) 'Quelques aspects récents de la gnomonique tunisienne', *Revue de l'Occident Musulman et de la Méditerranée* 24: 202–21.

Janin, L. and King, D. A. (1977) 'Ibn al-Shāṭir's *Ṣandūq al-Yawāqīt*: an astronomical *Compendium*', *Journal for the History of Arabic Science* 1: 187–256; reprinted in King, D. A. (1987b) *Islamic Astronomical Instruments*, London, Variorum Reprints, XII.

—— (1978) 'Le cadran solaire de la Mosquée d'Ibn Ṭūlūn au Caire', *Journal for the History of Arabic Science*, 2: 331–57; King, D. A. (1987b) *Islamic Astronomical Instruments*, London, Variorum Reprints, XVI.

Jaubert, A. (1836–40) *La Géographie d'Édrisi*, Paris: reprinted Amsterdam, 1975.

Jensen, C. (1971) 'Abū Naṣr's approach to spherical trigonometry as developed in his treatise *The Tables of Minutes*', *Centaurus* 16: 1–19.

Kammerer (1936) *Le Routier de Juan de Castro*, Paris.

Kazemi and McChesney, R. B. (eds) (1988) *Islam and Society: Arabic and Islamic Studies in Honor of Bayly Winder*, New York.

Kennedy, Edward S. (1948) 'Two Persian astronomical treatises by Naṣīr al-Dīn al-Ṭūsī', *Centaurus* 27: 109–20.

—— (1956) 'A survey of Islamic astronomical tables', *Transactions of the American Philosophical Society*, N.S. 46: 123–77.

—— (1958) 'The Sasanian astronomical handbook *Zīj-i Shāh* and the astrological doctrine of transit (*mamarr*)', *Journal of the American Oriental Society* 78: 246–62.

—— (1960) *The Planetary Equatorium of Jamshīd Ghiyāth al-Dīn al-Kāshī (d. 1429). An Edition of the Anonymous Persian Manuscript 75 (446) in the Garrett Collection at Princeton University, Being a Description of Two Computing Instruments: The Plate of Heavens and the Plate of Conjunctions*, Princeton, 'Oriental series, 18'.

—— (1965) 'The crescent visibility table in al-Khwārizmī's zīj', *Centaurus* 11: 73–8.

—— (1966) 'Late medieval planetary theory', *Isis* 57: 365–78.

—— (1968) 'The lunar visibility theory of Yaʿqūb b. Ṭāriq', *Journal of Near Eastern Studies* 27: 126–32.

—— (1973) *A Commentary upon Bīrūnī's Kitāb Taḥdīd al-Amākin: An 11th Century Treatise on Mathematical Geography*, Beirut.

—— (1985) 'Spherical astronomy in Kāshī's Khāqānī Zīj', *Zeitschrift für Geschichte der Arabisch-Islamischen Wissenschaften* 2: 1–46.

—— (1986) 'Geographical latitudes in al-Idrīsī's world map, *Zeitschrift für Geschichte der Arabisch-Islamischen Wissenschaften* 3: 265–8.

Kennedy, E. S. and Debarnot, M.-Th. (1984) 'Two mappings proposed by Bīrūnī, *Zeitschrift für Geschichte der Arabisch-Islamischen Wissenschaften* 1: 145–7.

Kennedy, E. S. and Ghanem, I. (1976) *The Life and Work of Ibn al-Shāṭir*, Aleppo.

Kennedy, E. S. and Id, Y. (1974) 'A letter of al-Bīrūnī: Ḥabash al-Ḥāsib's analemma for the *qibla*', *Historia Mathematica* 1: 3–11; reprinted in E. S. Kennedy *et al.*(1983) *Studies in the Islamic Exact Sciences*, Beirut, pp. 621–9.

Kennedy, E. S. and Kennedy, M. H. (1987) *Geographical Coordinates of Localities from Islamic Sources*, Frankfurt.

Kennedy, E. S. and King, D. A. (1982) 'Indian astronomy in fourteenth century Fez; the versified zīj of al-Qusunṭīnī', *Journal for the History of Arabic Science* 6: 3–45.

Kennedy, E. S. and Regier, M. H. (1985) 'Prime meridians in medieval Islamic astronomy', *Vistas in Astronomy* 28: 29–32.

Kennedy, E. S. and Roberts, V. (1959) 'The planetary theory of Ibn al-Shāṭir', *Isis* 50: 227–35.

Kennedy, E. S. *et al.* (1983) *Studies in the Islamic Exact Sciences*, Beirut.

Kepler, *Gesammelte Werke*, Bd. VII, edited by M. Caspar, Munich, 1953.

Keuning, J. (1955) 'The history of geographical map projections until 1600, *Imago Mundi* 12: 1–24.

Khaṭṭābī, M. ʿA. (ed.) (1990) *ʿUmdat al-ṭabīb fi maʿrifat al-nabāt*, Rabat.

al-Khwārizmī (1926) *Das Kitāb Ṣūrat al-Arḍ des... al-Huwārizmī*, edited by Hans von Mžik, Leipzig.

al-Kindī (1987) *Kitāb fī-l-Ṣināʿat al-ʿuẓmā*, edited by ʿAẓmī Taha al-Sayyid Aḥmad, Cyprus.

King, D. A. (1973a) 'Ibn Yūnus' *Very useful tables* for reckoning time by the sun', *Archive for History of Exact Sciences* 10: 342–94.

—— (1973b) 'Al-Khalīlī's auxiliary tables for solving problems of spherical astronomy', *Journal for the History of Astronomy* 4: 99–100; reprinted in King, D. A. (1986b), XI.

—— (1974) 'An analog computer for solving problems of spherical astronomy: the Shakkāzīya quadrant of Jamāl al-Dīn al-Māridīnī', *Archives Internationales d'Histoire des Sciences* 95: 220–42.

—— (1975) 'Al-Khalīlī's *qibla* table', *Journal of Near Eastern Studies* 34, 81–122; reprinted in King, D. A. (1986b), XIII.

—— (1976) 'Astronomical timekeeping in fourteenth-century Syria', *Proceedings of the First International Symposium for the History of Arabic Science*, Aleppo, II, pp. 75–84 and plates.

—— (1977a) 'Astronomical timekeeping in Ottoman Turkey', in M. Dizer (ed.) (1980) *Proceedings of the International Symposium on the Observatories in Islam, Istanbul 1977*, Istanbul, pp. 245–69.

—— (1977b) 'A fourteenth-century Tunisian sundial for regulating the times of Muslim prayer', in W. Saltzer and Y. Maeyama (eds) *Prismata: Festschrift für Willy Hartner*, Wiesbaden, pp. 187–202; reprinted in King, D. A. (1987b), XVIII.

—— (1978a) 'Three sundials from Islamic Andalusia', *Journal for the History of Arabic Science* 2: 358–92; reprinted in King, D. A. (1987b) XV.

—— (1978b) 'Some medieval values of the *qibla* at Cordova', *Journal for the History of Arabic Science* 2: 370–87; reprinted in King, D. A. (1987b) XV.

—— (1979) 'Mathematical astronomy in medieval Yemen', *Arabian Studies* 5: 61–5.

—— (1980) 'New light on the *zīj al-Ṣafāʾiḥ* of Abū Jaʿfar al-Khāzin', *Centaurus* 23: 105–17.

—— (1983a) 'Astronomical alignments in medieval Islamic religious architecture', *Annals of the New York Academy of Sciences* 385: 303–12.

—— (1983b) 'Al-Bazdawi on the *qibla* in early Islamic Transoxiana', *Journal for the History of Arabic Science* 7: 3–38.

—— (1983c) 'The astronomy of the Mamluks', *Isis* 74: 531–55; reprinted in King, D. A. (1986b), III.

—— (1983d) 'Al-Khwārizmī and new trends in mathematical astronomy in the ninth century', *Occasional Papers on the Near East*, New York University, Hagop Kevorkian Center for Near Eastern Studies, 2.

—— (1984) 'Architecture and astronomy: the ventilators of Cairo and their secrets', *Journal of the American Oriental Society* 104: 97–133.

—— (1985a) 'Osmanische astronomische Handschriften und Instrumente', *Türkische Kunst und Kultur der osmanischen Zeit*, Recklinghausen, II, pp. 373–8; reprinted in King, D. A. (1987b), XIV.

—— (1985b) 'The sacred direction in medieval Islam: a study of the interaction of science and religion in the Middle Ages', *Interdisciplinary Science Reviews* 10: 315–28.

—— (1986a) 'The earliest mathematical methods and tables for finding the direction of Mecca', *Zeitschrift für Geschichte der Arabisch-Islamischen Wissenschaften* 3: 82–146; with corrections, 4 (1987).

—— (1986b) *Islamic Astronomical Astronomy*, London, Variorum Reprints.

—— (1987a) 'A survey of medieval Islamic shadow schemes for simple time-reckoning', *Zeitschrift für Geschichte der Arabisch-Islamischen Wissenschaften* 4.

—— (1987b) *Islamic Mathematical Instruments*, London, Variorum Reprints.

—— (1987c) 'Universal solutions in Islamic astronomy', in J. L. Berggren and B. R. Goldstein (eds) *From Ancient Omens to Statistical Mechanics. Essays on the Exact Sciences Presented to Asger Aaboe*, Copenhagen, pp. 121–32.

—— (1987d) 'Some early Islamic tables for determining lunar crescent visibility', in King and Saliba (eds.) *From Deferent to Equant: a Volume of Studies in the History of Science in the Ancient and Medieval Near East in Honor of E. S. Kennedy*, New York, pp. 185–225.

—— (1988) 'Universal solutions to problems of spherical astronomy from Mamluk Egypt and Syria', in F. Kazemi and R. B. McChesney (eds) *Islam and Society: Arabic and Islamic Studies in Honor of Bayly Winder*, New York/London, pp. 153–84.

—— (1992) 'Los cuadrantes solares andalusíes', in Vernet and Samsó (1992), pp. 89–102.

—— (1993) *Astronomy in the Service of Islam*, Aldershot: Variorum.

King, D. A. and Kennedy, E. S. (1982) 'Indian astronomy in fourteenth-century Fez', *Journal for the History of Arabic Science* 6: 3–45.

King, D. A. and Saliba, G. (eds) (1987) *From Deferent to Equant: a Volume of Studies in the History of Science in the Ancient and Medieval Near East in Honor of E. S. Kennedy*, New York.

Kramers, J. H. (1931–2) 'La question Balkhī-Ibn Ḥawqal et l'Atlas de l'Islam', *Acta Orientalia* 10: 9–30.

Kühne, R. (1980) 'La *Urŷūza fī-l-ṭibb* de Saʿīd ibn ʿAbd Rabbihi', *al-Qanṭara* 1: 279–338.

Kunitzsch, P. (1959) *Arabische Sternnamen in Europa*, Wiesbaden.

—— (1966) *Typen von Sternverzeichnissen in astronomischen Handschriften des zehnten bis vierzehnten Jahrhunderts*, Wiesbaden.

—— (1967) 'Zur Stellung der Nautikertexte innerhalb der Sternnomenklatur der Araber', *Der Islam* 43: 53f; and 56 (1969): 305f.

—— (1974) *Der Almagest. Die Syntaxis Mathematica des Claudius Ptolemäus in arabisch-lateinischer Überlieferung*, Wiesbaden.

—— (1980) 'Two star tables from medieval Spain', *Journal for the History of Astronomy* 11: 192–201.

—— (1981) 'On the authenticity of the treatise on the composition and use of the astrolabe ascribed to Messahalla', *Archives Internationales d'Histoire des Sciences* 31: 42–62.

—— (1986): see Ptolemy.

Kunitzsch, P. and Lorch, R. (1994) *Maslama's Notes on Ptolemy's* Planisphaerium *and Related Texts*, Munich.

Labarta, A. and Barceló, C. (1995) 'Un nuevo fragmento de reloj de sol andalusí', *al-Qanṭara* 16: 147–50.

Lane, E. W. (1908) *The Manners and Customs of the Modern Egyptians*, London.

Langermann, Y. T. (1985) 'The book of bodies and distances of Ḥabash al-Ḥāsib', *Centaurus* 28: 108–28.

—— (1987) *The Jews of Yemen and the Exact Sciences*, Jerusalem (in Hebrew with an English summary).

Langlès (ed.) (1811) *Voyages du Chevalier Chardin en Perse, et Autres Lieux d'Orient*, 10 vols, Paris.

Lemay, R. (1962) *Abu Maʿshar and Latin Aristotelianism in the Twelfth Century*, Beirut.

Lettinck, P. (1994) *Aristotle's* Physics *and its Reception in the Arabic World with an edition of the Unpublished Parts of Ibn Bājja's Commentary on the* Physics, Leiden/New York/Cologne.

Lewis, B. (1943) 'An epistle on manual crafts', *Islamic Culture* XXVII.

López, A. C. (1990a) 'Vida y obra del famoso polígrafo cordobés del s. X 'Arīb ibn Saʿīd', *Ciencias de la Naturaleza en al-Andalus. Textos y Estudios I*, edited by E. García Sánchez, Granada, pp. 317–47.

—— (1990b) *Kitāb fī Tartīb Awqāt al-Girāsa wa-l-Magrūsāt. Un Tratado Agrícola Andalusí Anónimo*, Granada.

Lorch, R. P. (1975) 'The astronomy of Jābir ibn Aflaḥ', *Centaurus* 19: 85–107.

—— (1976) 'The astronomical instruments of Jābir b. Aflaḥ and the *Torquetum*', *Centaurus* 20: 11–34.

—— (1980) 'The *qibla*-table attributed to al-Khāzinī', *Journal for the History of Arabic Science* 4: 259–64.

—— (1982) 'Naṣr b. ʿAbdallāh's instrument for finding the *qibla*', *Journal for the History of Arabic Science* 6: 125–31.

Luckey, P. (1937–8) 'Thābit b. Qurra's Buch über die ebenen Sonnenuhren', *Quellen und Studien zur Geschichte der Mathematik, Astronomie und Physik* (4): 95–148.

Maddison, F. and Turner, A. J. (1976) Unpublished catalogue of an exhibition 'Science and Technology in Islam' held at the Science Museum, London, April–August 1976, in association with the Festival of Islam.

al-Mahrī (1970), *Al-ʿUmda*, edition I. Khoury, Damascus.

—— (1970) *Al-Minhāj*, edition I. Khoury, Damascus.

al-Mahrī (1972), *Al-Risāla – Sharḥ al-Tuḥfa*, edition I. Khoury, introduction in French and bilingual glossary by H. Grosset-Grange, Damascus.
Maimonides (1956) *The Guide for the Perplexed*, translated by M. Friedländer, 2nd edn, New York.
—— (1856–66) *Le Guide des Égarés*, French translation by S. Munk, 3 vols, Paris: reprinted, 1960, Paris.
Makkī, Maḥmūd ʿAlī (1961–4) 'Ensayo sobre las aportaciones orientales en la España Musulmana y su influencia en la formación de la cultura hispano-árabe', *Revista del Instituto Egipcio de Estudios Islámicos* 9–10 (1961–2): 65–231; 11–12 (1963–4): 7–140.
Maqbul, S. (1960) *India and the Neighbouring Territories in the* Kitāb nuzhat al-mushtāq fī-ikhtirāq al-āfāq *of al-Sharīf al-Idrīsī*, Leiden.
—— (1965a) '*Djughrāfiyā*', in *The Encyclopaedia of Islam*, new edn, vol. 2, Leiden, 575–87.
—— (1965b) '*Kharīṭa*', in *The Encyclopaedia of Islam*, new edn, vol. 4, Leiden, pp. 1077–83.
al-Maqqarī (1968) *Nafḥ al-ṭīb*, edited by Iḥsān ʿAbbās, Beirut, vol. VII.
Marín, M. (1981) '*Ṣaḥāba et ṭābiʿūn* dans al-Andalus: histoire et légende', *Studia Islamica* 54: 25–36.
—— (1986) '*ʿIlm al-nujūm* et *ʿIlm al-ḥidthān* en al-Andalus', *Actas del XII Congreso de la U.E.A.I.*, Madrid, pp. 509–35.
Marti, R. and Viladrich, M. (1981) 'Las tablas de climas en los tratados de astrolabio del manuscrito 225 del *scriptorium* de Ripoll', *Llull* 4: 117–22.
Martínez, L. (1971) 'Teorías sobre las mareas según un manuscrito árabe del siglo XII', *Memorias de la Real Academia de Buenas Letras* 13: 135–212.
—— (1981) 'El *Kitāb al-madd wa-l-ŷazr* de Ibn al-Zayyāt', in J. Vernet (ed.) *Textos y Estudios sobre Astronomía Española en el siglo XIII*, Barcelona, pp. 111–73.
Martínez Gázquez, J. and Samsó, J. (1981) 'Una nueva traducción latina del Calendario de Córdoba (siglo XIII)', in J. Vernet (ed.) *Textos y estudios sobre Astronomía Española en el siglo XIII*, Barcelona, pp. 9–78.
al-Masʿūdī, *Murūj al-Dhahab (Les Prairies d'Or)*, edited and translated by Barbier de Meynard and Pavet de Courteille, 1861–77. 9 vols, Paris.
al-Masʿūdī, *Kitāb al-Tanbīh wa-l-Ishrāf*, edited by M. J. de Goeje, Leiden, 1894; reprinted Beirut, 1965; French translation by Carra de Vaux, *Le Livre de l'Avertissement et de la Révision*, Paris, 1897.
Menéndez Pidal, G. (1954) 'Mozárabes y asturianos en la cultura de la Alta Edad Media en relación especial con la historia de los conocimientos geográficos', *Boletín de la Real Academia de la Historia* 134: 137–291.
Mercier, R. (1976–7) 'Studies in the medieval conception of precession', *Archives Internationales d'Histoire des Sciences* 26 (1976): 197–220; 27 (1977): 33–71.
—— (1987) 'Astronomical tables in the twelfth century', in Charles Burnett (ed.), *Adelard of Bath, an English Scientist and Arabist of the Early Twelfth Century*, London, pp. 87–118.
Meyerhof, M. (1935) 'Esquisse d'histoire de la pharmacologie et botanique chez les musulmans d'Espagne', *al-Andalus* 3: 1–41.

—— (1940) 'Un glossaire de matière médicale de Maïmonide', *Mémoires Présentés à l'Institut d'Égypte*, Cairo, vol. 41.

Meyerhof, M. and Sobhy, G. P. (1932–40) *The Abridged Version of* The Book of Simple Drugs *of Aḥmad ibn Muḥammad al-Ghāfiqī by Gregorius Abū 'l-Farag (Barhebraeus)*, Cairo.

Michel, H. (1943) 'L'astrolabe linéaire d'al-Tūsī, *Ciel et Terre*, Brussels, 3–4.

—— (1947) *Traité de l'Astrolabe*, Paris; reprinted 1983.

Michel, H. and Ben-Eli, A. (1965) 'Un cadran solaire remarquable', *Ciel et Terre* 81.

Millás Vallicrosa, José Mª (1931) *Assaig d'Història de les Idees Físiques i Matemàtiques a la Catalunya Medieval*, Barcelona.

—— (1942) *Las Traducciones Orientales en los Manuscritos de la Biblioteca Catedral de Toledo*, Madrid.

—— (1943–50) *Estudios sobre Azarquiel*, Madrid/Granada.

—— (1947) *El Libro de los Fundamentos de las Tablas Astronómicas de R. Abraham ibn ʿEzra*, Madrid/Barcelona.

—— (1952) *La Obra Enciclopédica de R. Abraham Bar Ḥiyya*, Madrid/Barcelona.

—— (1955) 'Los primeros tratados de astrolabio en España', *Revista del Instituto Egipcio de Estudios Islámicos* 3: 55–76.

—— (1959) *Libro del Cálculo de los Movimientos de los Astros de R. Abraham Bar Ḥiyya*, Barcelona.

—— (1960) *Nuevos Estudios Sobre Historia de la Ciencia Española*, Barcelona, Consejo Superior de Investigaciones Cientificas.

Millás Vendrell, E. (1963) *El Comentario de Ibn al-Muthannà a las Tablas Astronómicas de al-Jwārizmī*, study and critical edition of the Latin text of Hugo Sanctallensis, Madrid/Barcelona.

Miller, K. (1926–31) *Mappæ Arabicæ, Arabische Welt- und Länderkarten*, 6 vols, Stuttgart.

—— (1981) *Weltkarte des Arabers Idrisi vom Jahre 1154*, Stuttgart.

Molina, L. (ed.) (1983) *Una Descripción Anónima de al-Andalus*, vol. I, Madrid.

Moody, E. A. (1952) 'Galileo and Avempace. The dynamics of the leaning tower experiment', *Journal for the History of Ideas* 12: 163–93, 375–422.

Morelon, Régis (1981) 'Fragment arabe du premier livre du *Phaseis*', *Journal for the History of Arabic Science* 5: 3–14.

—— (1988) 'Les deux versions du traité de Thābit b. Qurra *Sur le mouvement des deux luminaires*', *Mélanges de l'Institut Dominicain d'Études Orientales* 18: 9–44.

—— (1993): see Ptolemy, *Planetary Hypotheses*.

—— (1994) 'Thābit b. Qurra and Arab astronomy in the 9th century', *Arabic Sciences and Philosophy* 4: 111–39.

Müller, D. H. (1884) *Al-Hamdānī's Geographie der Arabischen Halbinsel*, Leiden.

Muñoz, R. (1981) 'Textos árabes del *Libro de las Cruces* de Alfonso X', in J. Vernet (ed.) *Textos y Estudios sobre Astronomía Española en el siglo XIII*, Barcelona, pp. 175–204.

Mursi, M. (1969) *Thalātha Azhār fī Maʿrifat al-Biḥār*, Cairo.

Mžik, Hans von (1915) 'Ptolemaeus und die Karten der arabischen Geographen', *Mitt. d. K. K. geog. Ges. Wien* 58: 152–75.

—— (1921) 'Idrīsī und Ptolemaus', *Orientalistische Literaturzeitung* 15: 404–5.

—— (1930) *Das Kitāb ʿajāʾib al-aqālīm al-sabʿa des Suhrāb*, Leipzig.
Nadvi, S. (1935) 'Some Indian astrolabe makers', *Islamic Culture* 9(4): 622–6.
—— (1937) 'Indian astrolabe makers', *Islamic Culture* 9 (1): 512–41.
Nafis, Ahmad (1947) *Muslim Contribution to Geography*, Lahore.
Nallino, Carlo Alfonso (1892–3) 'Il valore metrico del grado di meridiano secondo i geografi arabi', *Cosmos di Guido Cora* 11: 20–7, 50–63, 105–21; reprinted in Nallino, C.A. (1944) *Raccolta di Scritti Editi e Inediti*, Rome, pp. 408–57.
—— (1911) *ʿIlm al-falak*, Rome.
—— (1916) 'Un mappamondo arabo disegnato nel 1579 da 'Alī Ibn Aḥmad al-Sharafī di Sfax', *Bolletino della Reale Società Geografica Italiana* V (5): 721–36; reprinted in Nallino, C.A. (1944) *Raccolta di Scritti Editi e Inediti*, Rome, pp. 533–48.
—— (1944) *Raccolta di Scritti Editi e Inediti*, vol. 5, Rome.
Nedkov, Boris (1960) *B'lgariya i c'cednite i zemi prez XII bek spored 'geografiyata' na Idrisi*, Sofia.
Needham, J. and Wang Ling (eds) (1959) *Science and Civilisation in China*, vol. 3: 'Mathematics and the sciences of the heavens and the earth', Cambridge.
Neugebauer, O. (1948) 'Mathematical methods in ancient astronomy', *Bulletin of the American Mathematical Society* 54: 1013–41.
—— (1949) 'The early history of the astrolabe', *Isis* 40: 240–56.
—— (1957) *The Exact Sciences in Antiquity*, 2nd edn, New York; French translation by P. Souffrin, *Les Sciences Exactes dans l'Antiquité*, Arles (1990).
—— (1959) 'The equivalence of eccentric and epicyclic motion according to Apollonius', *Scripta Mathematica* 24: 5–21; reprinted in Neugebauer, O. (ed.) *Astronomy and History: Selected Essays*, New York, pp. 335–51.
—— (1962a) *The Astronomical Tables of al-Khwārizmī*, translation with commentary of the Latin version, Copenhagen.
—— (1962b) 'Thābit ibn Qurra *On the solar year and On the motion of the eighth sphere*', *Proceedings of the American Philosophical Society* 106: 264–99.
—— (1971) 'An Arabic version of Ptolemy's parapegma from the *Phaseis*', *Journal of the American Oriental Society* 91(4): 506.
—— (1975) *A History of Ancient Mathematical Astronomy*, 3 vols, New York.
—— (1983) *Astronomy and History: Selected Essays*, New York.
North, J. (1976) *Richard of Wallingford*, 3 vols., Oxford.
Oliver Asín, J. (1959) *Historia del nombre 'Madrid'*, Madrid.
Pedersen, O. (1974) *A Survey of the Almagest*, Odense.
—— (1975) 'The corpus astronomicum and the traditions of mediaeval Latin astronomy', *Colloquia Copernicana*, Wroclaw, coll. 'Studia Copernicana, XIII', pp. 57–96.
Petersen, V. M. (1969) 'The three lunar models of Ptolemy', *Centaurus* 14: 142–71.
Peurbach, G., *Theoricæ Novæ Planetarum*, Nuremberg, *c.* 1472; reprinted in F. Schmeidler (ed.) (1972) Regiomontanus, *Opera collectanea*, Osnabrück.
Philopon, Jean (1981) *Traité de l'Astrolabe*, edited with commentary by A. Segonds, Paris, coll. 'Astrolabica no. 2'.
Pines, S. (1956) 'La théorie de la rotation de la terre à l'époque d'al-Bīrūnī, *Journal Asiatique* 244: 301–6.

—— (1964a) 'La dynamique d'Ibn Bājja', *L'Aventure de la Science. Mélanges Alexandre Koyré*, vol. I, Paris, pp. 442–68.

—— (1964b) 'Ibn al-Haytham's critique of Ptolemy', *Actes du Xe Congrès International d'Histoire des Sciences*, Paris, pp. 547–50.

Pingree, D. (1968) 'The fragments of the works of Yaʿqūb b. Ṭāriq', *Journal of Near Eastern Studies* 27: 97–125.

—— (1970) 'The fragments of the works of al-Fazārī', *Journal of Near Eastern Studies* 29: 103–23.

—— (1976) 'The Indian and Pseudo-Indian passages in Greek and Latin astronomical and astrological texts', *Viator* 7: 141–95.

—— (1977) 'The *Liber Universus* of ʿUmar Ibn al-Farrukhān al-Ṭabarī, *Journal for the History of Arabic Science* 1: 8–12.

Plooij, E. B. (1950) *Euclid's Conception of Ratio and his Definition of Proportional Magnitudes as Criticized by Arabian Commentators. Including the Text in Facsimile with the Translation of the Commentary on Ratio of Abū ʿAbd Allāh Muḥammad ibn Muʿādh al-Djajjānī*, Rotterdam.

Poch, D. (1980) 'El concepto de *quemazón* en el *Libro de las Cruzes*', *Awrāq* 3: 68–74.

Poulle, E. (1964) 'Le traité d'astrolabe de Raymond de Marseille', *Studi Medievali* 5: 866–904.

—— (1966) 'Théorie des planètes et trigonométrie au XVe siècle d'après un équatoire inédit, le sexagenarium', *Journal des Savants* 129–61.

—— (1980a) *Les Instruments de la Théorie des Planètes selon Ptolémée: Équatoires et Horlogerie Planétaire du XIIIe au XVIe siècle*, 2 vols, Paris, coll. 'Hautes études médiévales et modernes, 42'.

—— (1980b) 'Jean de Murs et les tables alphonsines', *Archives d'Histoire Doctrinale et Littéraire du Moyen Age* 47: 241–71.

—— (1984) *Les Tables Alphonsines avec les Canons de Jean de Saxe*, Paris.

de Prémare, A. L. (1964–6) 'Un Andalou en Égypte à la fin du XVe siècle. Abū-l-Ṣalt de Dénia et son Épître Égyptienne', *Mélanges de l'Institut Dominicain d'Études Orientales* 8: 179–208.

Ptolemy, *Almagest*: Edition of the Greek text by J. L. Heiberg, Leipzig, 1898–1903; French translation by N. Halma, Paris, 1813–16; reprinted, Paris, 1927; English translation in G. J. Toomer (1984) *Ptolemy's Almagest*, New York; edition and German translation of two Arabic versions of the star catalogue, Ptolemäus, Claudius: *Der Sternkatalog des Almagest, Die arabisch-mittelalterliche Tradition, I, Die arabischen Übersetzungen*, edited and translated by P. Kunitsch, Wiesbaden, 1986.

—— *Planetary Hypotheses*: French translation of the first part of Book I by N. Halma *Hypothèses et Époques des Planètes de Cl. Ptolémée*, Paris, 1820. Edition of the Greek text of the first part of Book I and translation from the German of the Arabic of Book II by L. Nix (1907) *Claudii Ptolemæi Opera quæ extant omnia*, vol. II, 'Opera astronomica minora', Leipzig, pp. 68–145. Edition and French translation of the Arabic text for the first treatise of the book by R. Morelon, *Mélanges de l'Institut Dominicain d'Etudes Orientales* 21, 1993: 7–85. Goldstein, B. R. (1967b) 'The Arabic version of Ptolemy's

Planetary Hypotheses', reproduction of the entire Arabic manuscript, which contains the second part of Book I and a partial English translation, *Transactions of the American Philosophical Society*, N.S. 57/4: 3–55.
—— *Phaseis*: French translation of Book II by N. Halma, *Chronologie de Ptolémée ... Apparition des fixes, ou calendrier de Ptolémée*. Paris, 1819, pp. 13–54; edition of the Greek text of Book II by J. L. Heiberg in *Claudii Ptolemæi Opera quæ extant omnia*, vol. II, pp. 1–67; on the contents of Book I see R. Morelon (1981) 'Fragment arabe du premier livre du *Phaseis*', *Journal for the History of Arabic Science* 5: 3–14.
—— *Handy Tables*: commentary of Theon of Alexandria on the manual astronomical tables of Ptolemy translated by N. Halma, I–III, Paris, 1822–5; reprinted Paris, 1990.
—— *Geography: Claudii Ptolemæi Geographia*, edited by C. F. A. Nobbe, 2 vols, Leipzig, 1843–5; reprinted in 1 vol., Hildesheim, 1966. See also Fischer.
—— *Planisphaerium*: J. Drecher (1927) 'Das Planisphærium des Claudius Ptolemæus', *Isis* 9: 255–78. See Kunitzsch and Lorch (1994).
Puig, R. (1983a) 'Dos notas sobre ciencia hispano-árabe a finales del siglo XIII en la *Iḥāṭa* de Ibn al-Jaṭīb', *al-Qanṭara* 4: 433–40.
—— (1983b) 'Ibn Arqam al-Numayrī (m.1259) y la introducción en al-Andalus del astrolabio lineal', in J. Vernet (ed.) *Nuevos Estudios sobre Astronomía Española en el Siglo de Alfonso X*, Barcelona, pp. 101–3.
—— (1984) 'Ciencia y técnica en la *Iḥāṭa* de Ibn al-Jaṭīb. Siglos XIII y XIV', *Dynamis* 4: 65–79.
—— (1985) 'Concerning the *ṣafīḥa shakkāziyya*', *Zeitschrift für Geschichte der Arabisch-Islamischen Wissenschaften* 2: 123–39.
—— (1986) *Al-shakkāziyya – Ibn al-Naqqāsh – Al-Zarqālluh*, Edición, traducción y estudio, Barcelona.
—— (1988) *Los Tratados de Construcción y uso de la Azafea de Azarquiel*, Madrid.
al-Qifṭī, *Tārīkh al-Ḥukamā'*, edited by C. Lippert, Leipzig, 1903.
Raeder, H., Strömgren, E. and Strömgren, B. (1946) *Tycho Brahe's Description of his Instruments*, Copenhagen.
Ragep, F. J. (1993) *Naṣīr al-Dīn al-Ṭūsī's Memoir on Astronomy* (*al-Tadhkira fī 'Ilm al-Hay'a*), with translation and commentary, 2 vols, New York/Berlin/Heidelberg.
Rashed, Roshdi (1984) 'La notion de science occidentale', *Entre Arithmétique et Algèbre. Recherches sur l'Histoire des Mathématiques Arabes*, Paris, pp. 301–19.
—— (1989) 'Problems of the transmission of Greek scientific thought into Arabic: examples from mathematics and optics', *History of Science* 27: 199–209.
—— (1991) 'Al-Samaw'āl, al-Bīrūnī et Brahmagupta: les méthodes d'interpolation', *Arabic Sciences and Philosophy* 1: 101–60.
—— (1993a) *Géométrie et Dioptrique au Xᵉ siècle: Ibn Sahl, al-Qūhī et Ibn al-Haytham*, Paris.
—— (1993b) *Les Mathématiques Infinitésimales du IXᵉ au XIᵉ siècle*. Vol. II: *Ibn al-Haytham*, London.

Regiomontanus, J. (1946) *Epytoma Joannis de Monte Regio in Almagestum Ptolemæi*, Venice; reproduced in F. Schmeidler (ed.) (1972) Regiomontanus, *Opera collectanea*, Osnabrück.

Renan, E. (1893) 'Les écrivains juifs français du XIVe siècle', *Histoire Littéraire de la France*, vol. 31, Paris.

Renaud, H. P. J. (1930) 'Trois études d'histoire de la médecine arabe en Occident. I. Le Mustaʿīnī d'Ibn Beklāreš', *Hespéris* 10: 135–50.

—— (1935) 'Un chirurgien musulman du royaume de Grenade: Muḥammad al-Shafra', *Hespéris* 20: 1–20.

—— (1937) 'Notes critiques d'histoire des sciences chez les musulmans. I. Les Ibn Bāso', *Hespéris* 24: 1–12.

—— (1940) 'Note complémentaire', *Hespéris* 27: 97–8.

—— (1944) 'Sur un passage d'Ibn Khaldūn relatif à l'histoire des mathématiques', *Hespéris* 31: 35–47.

Rheticus, G. J. (1982) *Narratio prima*, critical edition, French translation and commentary by H. Hugonnard-Roche and J. P. Verdet (in collaboration with M. P. Lerner and A. Segonds), Wroclaw, coll. 'Studia Copernicana XX'.

Ribera, J. (1928a) 'Bibliófilos y bibliotecas en la España Musulmana', *Disertaciones y Opúsculos*, vol. I, Madrid, pp. 181–228.

—— (1928b) 'La enseñanza entre los musulmanes españoles', *Disertaciones y Opúsculos*, vol. I, Madrid, pp. 229–359.

Richler, B. (1982) 'Manuscripts of Avicenna's Kanon in Hebrew translation', *Koroth* 8: 145–68.

Richter-Bernburg, Lutz (1982) 'Al-Bīrūnī's *Maqāla fī tasṭīḥ al-ṣuwar wa tabṭīkh al-kuwar: A* translation of the preface with notes and commentary', *Journal for the History of Arabic Science* 6: 113–22.

—— (1987) 'Ṣāʿid, the Toledan tables and Andalusī science', in D. A. King and G. Saliba (eds) *From Deferent to Equant*, New York, pp. 373–401.

Rico and Sinobas, M. (ed.) *Libros del Saber de Astronomía*, 5 vols, Madrid, 1864–7.

Roberts, Victor (1957) 'The solar and lunar theory of Ibn ash-Shāṭir', *Isis* 48: 428–32.

Rodgers, R. H. (1978) '¿Yūniyūs o Columela en la España Medieval?', *al-Andalus* 43: 163–72.

Rodríguez Molero, F. X. (1950) 'Originalidad y estilo de la Anatomía de Averroes', *al-Andalus* 15: 47–63.

Romano, D. (1977) 'La transmission des sciences arabes par les Juifs en Languedoc', in M.-H. Vicaire and B. Blumenkranz (eds) *Juifs et Judaïsme de Languedoc*, Toulouse, pp. 363–86.

Rosenfeld, B. A. (1983) *Muhammad ibn Musa al-Khorezmi*, Moscow.

Rosenthal, F. (1956) 'Al-Kindi and Ptolemy', *Studi Orientalistici in Onore di G. Levi Della Vida*, vol. II, Rome, pp. 436–56.

Rosińska, G. (1974) 'Naṣīr al-Dīn al-Ṭūsī and Ibn al-Shāṭir in Cracow?', *Isis* 65: 239–43.

Ruska, J. (1918) 'Neue Bausteine zur Geschichte der arabischen Geographie', *Geogr. Zeit.* 24: 77–8.

Sabra, A. I. (1978) 'An eleventh-century refutation of Ptolemy's planetary theory', *Science and History. Studies in Honor of Edward Rosen*, Wroclaw, coll. 'Studia Copernicana, 16', pp. 117–31.

—— (1984) 'The Andalusian revolt against Ptolemaic astronomy: Averroes and al-Biṭrūjī', in E. Mendelsohn (ed.) *Transformation and Tradition in the Sciences: Essays in Honor of I. Bernard Cohen*, Cambridge, pp. 133–53.

Ṣā'id al-Andalusī, *Kitāb Ṭabaqāt al-Umam* (*Livre des Catégories des Nations*), French translation and notes by R. Blachère, Paris, 1935.

Saliba, George (1979a) 'The first non-Ptolemaic astronomy at the Maraghah School', *Isis* 70: 571–6.

—— (1979b) 'The original source of Quṭb al-Dīn al-Shīrāzī's planetary model', *Journal for the History of Arabic Science* 3: 3–18.

—— (1980) 'Ibn Sīnā and Abū 'Ubayd al-Jūzjānī. The problem of the Ptolemaic equant', *Journal for the History of Arabic Science* 4: 376–403.

—— (1984) 'Arabic astronomy and Copernicus', *Zeitschrift für Geschichte der Arabisch-Islamischen Wissenschaften* 1: 73–87.

—— (1987a) 'Theory and observation in Islamic astronomy: the work of Ibn al-Shāṭir of Damascus (d. 1375)', *Journal for the History of Astronomy* 18: 35–43.

—— (1987b) 'The height of the atmosphere according to Mu'ayyad al-Dīn al-'Urḍī, Quṭb al-Dīn al-Shīrāzī and Ibn Mu'ādh', in D. A. King and G. Saliba (eds) *From Deferent to Equant*, New York, pp. 445–65.

—— (1994) *A History of Arabic Astronomy. Planetary Theories during the Golden Age of Islam*, New York.

Samsó, Julio (1969) *Estudios sobre Abū Naṣr Manṣūr b. 'Alī b. 'Irāq*, Barcelona.

—— (1973) 'A propos de quelques manuscrits astronomiques des bibliothèques de Tunis: Contribution à une histoire de l'astrolabe dans l'Espagne Musulmane', *Actas del II Coloquio Hispano-Tunecino de Estudios Históricos*, Madrid, pp. 171–90.

—— (1978) 'La tradición clásica en los calendarios agrícolas hispanoárabes y norteafricanos', *Segundo Congreso Internacional de Estudios sobre las Culturas del Mediterráneo Occidental*, Barcelona, pp. 177–86.

—— (1979a) 'Astronomica Isidoriama', *Faventia* 1: 167–74.

—— (1979b) 'The early development of astrology in al-Andalus', *Journal for the History of Arabic Science* 3: 228–43.

—— (1980a) 'Maslama al-Majrīṭī and the Alfonsine book on the construction of the astrolabe', *Journal for the History of Arabic Science* 4: 3–8.

—— (1980b) 'Notas sobre la trigonometría esférica de Ibn Mu'ādh', *Awrāq* 3: 60–8.

—— (1980c) 'Tres notas sobre astronomía hispánica en el siglo XIII', *Estudios sobre Historia de la Ciencia Árabe*, Barcelona, pp. 167–79.

—— (1980d) 'Alfonso X y los orígenes de la astrología hispánica', *Estudios sobre Historia de la Ciencia Arabe*, Barcelona, pp. 81–114.

—— (1981) 'Dos colaboradores científicos musulmanes de Alfonso X', *Llull* 4: 171–9.

—— (1982) 'Ibn Hishām al-Lajmī y el primer jardín botánico en al-Andalus', *Revista del Instituto Egipcio de Estudios Islámicos en Madrid* 21: 135–41.

—— (1983a) 'La primitiva versión árabe del Libro de las Cruces', in J. Vernet (ed.) *Nuevos Estudios sobre Astronomía Española en el siglo de Alfonso X*, Barcelona, pp. 149–61.

—— (1983b) 'Sobre los materiales astronómicos en el *Calendario de Córdoba* y en su versión latina del siglo XIII', in J. Vernet (ed.) *Nuevos Estudios sobre Astronomía Española en el siglo de Alfonso X*, Barcelona, pp. 125–38.

—— (1983c) 'Notas sobre el ecuatorio de Ibn al-Samḥ', in J. Vernet (ed.) *Nuevos Estudios sobre Astronomía Española en el siglo de Alfonso X*, Barcelona, pp. 105–18.

—— (1985a) 'En torno a los métodos de cálculo utilizados por los astrólogos andalusíes a fines del s. VIII y principios del IX: algunas hipótesis de trabajo', *Actas de las II Jornadas de Cultura Arabe e Islámica*, Madrid, pp. 509–22.

—— (1985b) 'Nota sobre la biografía de Sisebuto en un texto árabe anónimo', *Serta Gratulatoria in honorem Juan Régulo*, La Laguna, vol. I: 'Filología', pp. 639–42.

—— (1985c) 'Astrology, pre-Islamic Spain and the conquest of al-Andalus', *Revista del Instituto Egipcio de Estudios Islámicos* 23: 39–54.

—— (1987a) 'Al-Zarqāl, Alfonso X and Peter of Aragon on the solar equation', in D. A. King and G. Saliba (eds) *From Deferent to Equant*, New York, pp. 467–76.

—— (1987b) 'Sobre el modelo de Azarquiel para determinar la oblicuidad de la eclíptica', *Homenaje al Prof. Darío Cabanelas O.F.M. con motivo de su LXX aniversario*, vol. II, Granada, pp. 367–77.

—— (1988) 'Azarquiel e Ibn al-Bannā'', *Relaciones de la Península Ibérica con el Magreb (siglos XIII–XVI)*, Madrid, pp. 361–72.

—— (1990) 'En torno al problema de la determinación del acimut de la alquibla en al-Andalus en los siglos VIII y IX. Estado de la cuestión e hipótesis de trabajo', *Homenaje a Manuel Ocaña Jiménez*, Córdoba, pp. 207–12.

—— (1992) *Los Ciencias de los Antiguos en al-Andalus*, Madrid.

—— (1994a) *Islamic Astronomy and Medieval Spain*, Variorum Reprints, Aldershot.

—— (1994b) 'Le due astronomie dell'Occidente musulmano (1215–1250)', in *Federico II e le scienze*, Palermo, pp. 204–21.

Samsó, J. and Catalá, M. A. (1971–5) 'Un instrumento astronómico de raigambre zarqālī: el cuadrante shakkāzī de Ibn Ṭībugā', *Memorias de la Real Academia de Buenas Letras de Barcelona* 13: 5–31.

Samsó, J. and Martínez Gázquez, J. (1981) 'Algunas observaciones al texto del Calendario de Córdoba', *al-Qanṭara* 2: 319–44.

Samsó, J. and Mielgo, H. 'Ibn al-Zarqālluh on Mercury', *Journal for the History of Astronomy* 25: 289–96.

Sanchez Perez, J. A. (1916) *Compendio de Algebra de Abenbéder*, Madrid.

Sarfatti, G. B. (1968) *Mathematical Terminology in Hebrew Scientific Literature of the Middle Ages*, Jerusalem.

Sarton, G. (1947) 'Early observations of the sun-spots?', *Isis* 37: 69–71.

Saunders, H. N. (1984) *All the Astrolabes*, Oxford.

Savage-Smith, E. (1985) *Islamicate Celestial Globes, Their History, Construction, and Use* (with a chapter on iconography by A. P. A. Belloli), Washington, DC.

Sayili, A. (1960) *The Observatory in Islam and its Place in the General History of the Observatory*, Ankara.

Schoy, Karl (1921) 'Abhandlung des al-Ḥasan ibn al-Ḥasan ibn al-Haitham (Alhazen) über die Bestimmung der Richtung der *Qibla*', *Zeitschrift der Deutschen Morgenländischen Gesellschaft* 75: 242–53.

—— (1922) 'Abhandlung von al-Faḍl b. Ḥātim al-Nayrīzī über die Richtung der *qibla*', *Sitzungsberichte der math.-phys. Klasse der Bayerischen Akademie der Wissenschaften zu München*: 55–68.

—— (1923) 'Gnomonik der Araber', in E. von Bassermann-Jordan, (ed.) *Die Geschichte der Zeitmessung und der Uhren*, vol. IF, Berlin/Leipzig.

—— (1924) 'Sonnenuhren der spätarabischen Astronomie', *Isis* 6: 332–60.

Seco de Lucena, L. (1956) 'El ḥāŷib Riḍwān, la madrasa de Granada y las murallas del Albaicín', *al-Andalus* 21: 285–96.

Sédillot, J.-J. (1834–5) *Traité des Instruments Astronomiques des Arabes composé au treizième siècle par Aboul Hhassan Ali du Maroc* ..., 2 vols, Paris; reprinted, Frankfurt, 1984.

Sédillot, L. A. (1844) 'Mémoire sur les instruments astronomiques des Arabes', *Mémoires de l'Académie Royale des Inscriptions et Belles-Lettres de l'Institut de France* 1: 1–229; reprinted, Frankfurt, 1985.

—— (1853) *Prolégomènes des Tables Astronomiques d'Oloug Beyg*, Paris.

Serjeant, R. B. (1963) *The Portuguese of the South Arabian Coast*, Oxford.

—— (1971) 'Agriculture and horticulture: some cultural interchanges of the medieval Arabs and Europe', *Oriente e Occidente nel Medioevo: Filosofia e Scienze*, Rome, pp. 535–41.

Sévère Sabokht, *Traité de l'Astrolabe*, edited and translated into French by F. Nau, *Journal Asiatique* 3 (13) (1899): 56–101, 238–303.

Sezgin, Fuat (1974) *Geschichte des arabischen Schrifttums, Band V: Mathematik*, Leiden.

—— (1978) *Geschichte des arabischen Schrifttums, Band VI: Astronomie*, Leiden.

Shawkat, Ibrāhīm (1962) 'Kharā'iṭ jughrāfiyya al-ʿArab al-awwal', *Majallat al-Ustadh*, Bagdad, 2.

Shihāb, H. S. (1982) *Fann al-Milāḥa ʿinda-l-ʿArab*, Beyrouth.

—— (1983) *Al-Dalīl al-Baḥrī ʿinda-l-ʿArab*, Kuwait.

—— (1984) *Ṭuruq al-Milāḥa al-Taqlīdiyya fī-l-Khalīj al-ʿArabī*, Kuwait.

Singer, Ch., Holmyard, E. J., Hall, A. R. and Williams, T. I. (eds) (1957) *A History of Technology*, New York/London.

Souissi, M. (1973) 'Un mathématicien tuniso-andalou: al-Qalaṣādī', *Actas del II Coloquio Hispano-Tunecino de Estudios Históricos*, Madrid, pp. 147–69.

Steinschneider, Moritz (1893) *Die Hebräischen Übersetzungen*, Berlin.

Stern, S. M. (1961) 'A letter of the Byzantine Emperor to the Court of the Spanish Umayyad Caliph al-Ḥakam', *al-Andalus* 26: 37–42.

al-Ṣūfī, ʿAbd al-Raḥmān (1953) *Kitāb ṣuwar al-kawākib*, Hyderabad; reprinted, Beirut, 1981; French translation by H. C. F. Schellerup, *Description des Étoiles Fixes composée au milieu du dixième siècle de notre ère par l'Astronome Persan ʿAbd al-Raḥmān al-Ṣūfī*, St Petersburg; reprinted, Frankfurt, 1986.

Suhrāb: see Mžik (1930).

Suter, Heinrich (1900) *Die Mathematiker und Astronomen der Araber und Ihre Werke*, Leipzig, 'Abhand. zu Gesch. der Math. Wiss.', X.

—— (1914) *Die astronomischen Tafeln des Muḥammed ibn Mūsā al-Khwārizmī in der Bearbeitung des Maslama ibn Aḥmed al-Madjrīṭī und der latein. Übersetzung des Athelhard von Bath*, Copenhagen.

Swerdlow, Noël M. (1968) *Ptolemy's Theory of the Distances and Sizes of the Planets: A Study of the Scientific Foundations of Medieval Cosmology*, Ph.D. thesis, Yale University, 1968, University Microfilms International 69-8442.

—— (1972) 'Aristotelian planetary theory in the Renaissance: Giovanni Battista Amico's homocentric spheres', *Journal for the History of Astronomy* 3: 36–48.

—— (1973) 'The derivation and first draft of Copernicus's planetary theory. A translation of the *Commentariolus* with commentary', *Proceedings of the American Philosophical Society* 117: 423–512.

—— (1987) 'Jābir ibn Aflaḥ's interesting method for finding the eccentricities and direction of the apsidal line of a superior planet', in D. A. King and G. Saliba (eds) (1987) *From Deferent to Equant*, New York, pp. 501–12.

Swerdlow, N. M. and Neugebauer, O. (1984) *Mathematical Astronomy in Copernicus's De Revolutionibus*, 2 vols, New York.

Tekeli, S. (1960) '(The) *Equatorial Armilla* of Iz(z) al-Din b. Muhammad al-Wafa'i and (the) *Torquetum*', *Ankara Üniversitesi Dil ve Tarih-Coğrafya Fakültesi Dergisi* 18: 227–59.

Terés, E. (1959) 'Ibn al-Shamir, poeta astrólogo en la corte de ʿAbd al-Raḥmān II', *al-Andalus* 24: 449–63.

—— (1960) 'ʿAbbās b. Firnās', *al-Andalus* 25: 239–49.

—— (1962) 'ʿAbbās ibn Nāṣiḥ, poeta y qāḍī de Algeciras', *Etudes d'Orientalisme dédiées à la Mémoire de Lévi-Provençal*, vol. I, Paris pp. 339–58.

Thābit ibn Qurra, *Œuvres d'Astronomie*, edited and translated by R. Morelon, Paris, 1987.

Thorndike, L. (1951) 'Sexagenarium', *Isis* 42: 130–3.

Tibbets, G. (1971) *Arab Navigation in the Indian Ocean before the Coming of the Portuguese*, London.

Toomer, G. J. (1952) *Revolutions of the Heavenly Spheres*, Chicago.

—— (1968) 'A survey of the Toledan tables', *Osiris* 15: 5–174.

—— (1969) 'The solar theory of az-Zarqāl. A history of errors', *Centaurus* 14: 306–36.

—— (1973) 'Prophatius Judaeus and the Toledan tables', *Isis* 64: 351–5.

—— (1984) *Ptolemy's Almagest*, New York.

—— (1987) 'The solar theory of Az-Zarqāl: an epilogue', in D. A. King and G. Saliba (eds) *From Deferent to Equant*, New York, pp. 513–19.

Torres, Esteban (1974) *Averroes y la Ciencia Médica. La Doctrina Anatomofuncional del* Colliget, Madrid.

Torres Balbas, L. (1940) 'Las norias fluviales en España', *al-Andalus* 5: 195–208.

Turner, A. J. (1985) *The Time Museum. Catalogue of the Collection*, vol. 1: 'Time measuring instruments', part 1: 'Astrolabes. Astrolabe-related instruments', Rockford.

al-Ṭūsī: see Ragep (1993).

Tuulio-Tallgren, O. J. (1936) *Du Nouveau sur Idrīsī*, Helsinki.
Tuulio-Tallgren, O. J. and Tallgren, A. M. (1930) *Idrīsī, la Finlande et les autres Pays Baltiques Orientaux*, Helsinki.
Twersky, I. (1972) *A Maimonides Reader*, New York.
Ünver, A. S. (1975) 'Osmanli Türkler'inde İlim Tarihinde Muvakkithaneler', *Atatürk Konferensları* 5: 217–57.
al-ʿUrḍī, Mu'ayyad al-Dīn, *Kitāb al-Hay'a* (*Tārikh ʿilm al-falak al-ʿarabī*), edited by G. Saliba, Beirut, 1990; coll. 'Silsila Tārīkh al-ʿUlūm ʿInda-l-ʿArab, 2'. English translation and commentary in press.
Ūrūsiyūs, *Tārīkh al-ʿĀlam*, edited by ʿAbd al-Raḥmān Badawī, Beirut (1982).
Vardjavand, P. (1980) 'Rapport sur les résultats des excavations du complexe scientifique de l'observatoire de Marāgha', in M. Dizer (ed.) *International Symposium on the Observatories in Islam, 19–23 September 1977*, Istanbul, pp. 143–63.
Vernet, J. (1956) 'Las tabulæ probatæ', *Homenaje a Millás-Vallicrosa*, Barcelona, vol. II, pp. 501–22.
—— (1970) 'Astrología y política en la Córdoba del siglo X', *Revista del Instituto Egipcio de Estudios Islámicos* 15: 91–100.
—— (1979a) 'La ciencia en el Islam y Occidente', *Estudios sobre Historia de la Ciencia Medieval*: 21–60.
—— (1979b) 'Un tractat d'obstetríacia astrològica', *Estudios sobre Historia de la Ciencia Medieval*: 273–300.
—— (1979c) 'La maldición de Perfecto', *Estudios sobre Historia de la Ciencia Medieval*: 233–4.
—— (1979d) 'Los médicos andaluces en el *Libro de las generaciones de médicos* de Ibn Ŷulŷul', *Estudios sobre Historia de la Ciencia Medieval*: 469–86.
—— (1979e) 'Tradición e innovación en la ciencia medieval', *Estudios sobre Historia de la Ciencia Medieval*: 173–89.
—— (1980a) 'Mármol, obra de Zarquel', *Hommage à Georges Vajda*, Louvain, pp. 151–4.
—— (1980b) 'La supervivencia de la astronomía de Ibn al-Bannā'', *al-Qanṭara* 1: 447–51.
—— (1981) *Textos y Estudios sobre Astronomía Española en el siglo de Alfonso X*, Barcelona.
—— (1985) *Ce que la culture doit aux Arabes d'Espagne*, translated from Spanish by G. M. Gros, Paris; German translation, *Die spanisch-arabische Kultur in Orient und Okzident*, Zürich/Munich, 1984.
—— (1986) *La ciencia en al-Andalus*, Sevilla.
—— (1987) 'Alfonso X y la tecnología árabe', *De Astronomia Alphonsi Regis*, Barcelona, pp. 39–41.
—— (1993) *El Islam en España*, Madrid.
Vernet, J. and Catala, M. A. (1979) 'Las obras matemáticas de Maslama de Madrid', *Estudios sobre Historia de la Ciencia Medieval*: 241–71.
Vernet, J. and Samsó, J. (1981) 'Panorama de la ciencia andalusí en el siglo XI', *Actas de las Jornadas de Cultura Arabe e Islámica (1978)*, Madrid.
—— (1992) *El Legado Científico Andalusí*, Madrid.

—— (1994) 'La Ciencia', in Mª Jesús Viguera Molins (ed.), *Historia de España Menéndez Pidal. Tomo VIII: Les Reinos de Taifas. Al-Andalus en el siglo XI*, Madrid, pp. 565–84.

Viguera, M. J. (1977) *Ibn Hudhayl: Gala de Caballeros, Blasón de Paladines*, Madrid.

Viladrich, M. (1982) 'On the sources of the Alphonsine treatise dealing with the construction of the plane astrolabe', *Journal for the History of Arabic Science* 6: 167–71.

—— (1986) *El* Kitāb al-ʿamal bi-l-asṭurlāb (*Llibre de l'ús de l'astrolabi*) *d'Ibn al-Samḥ*, Barcelona; and 'Dos capítulos de un libro perdido de Ibn al-Samḥ', *al-Qanṭara* 7: 5–11.

—— (1987) 'Una nueva evidencia de materiales árabes en la astronomía alfonsí', *De Astronomia Alphonsi Regis*, Barcelona, pp. 105–16.

Viladrich, M. and Marti, R. (1981) 'En torno a los tratados hispánicos sobre construcción de astrolabio hasta el siglo XIII', in J. Vernet (ed.) *Textos y Estudios sobre Astronomía Española en el siglo XIII*, Barcelona, pp. 79–99.

Villiers, A. (1966) *Sons of Sindbad*, Bath.

Arnald of Villanova (1975) *Aphorismi de gradibus*, edited by M. R. McVaugh, Granada/Barcelona.

Villuendas, M. V. (1979) *La Trigonometría Europea en el Siglo XI. Estudio de la Obra de Ibn Muʿādh: El Kitāb Maŷhūlāt*, Barcelona.

Wieber, R. (1980) 'Überlegungen zur Herstellung eines Seekartogramms anhand der Angaben in den arabischen Nautikertexten', *Journal for the History of Arabic Science* 4 (1): 23–47.

Wiedemann, E. and Frank, J. (1926) 'Die Gebetszeiten im Islam', *Sitzungsberichte der phys.-med. Sozietät zu Erlangen* 58: 1–32; reprinted in E. Wiedemann (1970), *Aufsätze zur arabischen Wissenschaftsgeschichte*, 2 vols, Hildesheim, II, pp. 757–88.

Wiedemann, E. and Juynboll, Th. W. (1927) 'Avicennas Schrift über ein von ihm ersonnenes Beobachtungsinstrument', *Acta Orientalia* 5: 81–167.

Würschmidt, J. (1917) 'Die Zeitrechnung im Osmanischen Reich', *Deutsche optische Wochenschrift*: 88–100.

Youschkevitch, A. P. (1976) *Les Mathématiques Arabes* (*VIIIᵉ-XVᵉ siècles*), Paris.

al-Zarqālluh, *Al-Shakkāziyya – Ibn al-Naqqāsh al-Zarqālluh*, Edition, translation and study by R. Puig, Barcelona, 1986.